Practical Electron Microscopy of Lattice Defects

Practical Electron Microscopy of Lattice Defects

Hiroyasu Saka
Aichi Institute of Technology and Nagoya University, Japan

World Scientific

NEW JERSEY • LONDON • SINGAPORE • BEIJING • SHANGHAI • HONG KONG • TAIPEI • CHENNAI • TOKYO

Published by

World Scientific Publishing Co. Pte. Ltd.
5 Toh Tuck Link, Singapore 596224
USA office: 27 Warren Street, Suite 401-402, Hackensack, NJ 07601
UK office: 57 Shelton Street, Covent Garden, London WC2H 9HE

Library of Congress Control Number: 2021936018

British Library Cataloguing-in-Publication Data
A catalogue record for this book is available from the British Library.

PRACTICAL ELECTRON MICROSCOPY OF LATTICE DEFECTS

ISBN 978-981-123-469-9 (hardcover)
ISBN 978-981-123-470-5 (ebook for institutions)
ISBN 978-981-123-471-2 (ebook for individuals)

For any available supplementary material, please visit
https://www.worldscientific.com/worldscibooks/10.1142/12221#t=suppl

Desk Editors: Aanand Jayaraman/Steven Patt

Typeset by Stallion Press
Email: enquiries@stallionpress.com

Preface

Transmission electron microscopy (TEM) is now well established as a powerful tool for material characterization. Therefore, it is too natural that many textbooks on TEM for material characterization have been published. However, these textbooks are not necessarily suitable for beginners because it is presumed that their readers are familiar with dislocation theory and crystallography. Textbooks on crystallography are very comprehensive and it is neither easy nor necessary for ordinary readers of material science and engineering to understand crystallography in general. Similar thing can be said for dislocation theory.

The main aim of this book is to explain standard knowledge of diffraction contrast to characterize material defects, especially dislocations by transmission electron microscopy to those who have no knowledge of crystallography nor dislocation theory. Therefore, those readers who have knowledge of crystallography and dislocation theory can skip Chapters 1 to 3.

For these reasons, I have tried to explain basis of diffraction contrast as simply as possible, at the cost of strictness. Dynamical theory of diffraction is mentioned only briefly. On the other hand, scanning transmission electron microscopy (STEM), which was developed over the past decades, has gained importance in characterizing material defects. However, most of the textbooks dealing with STEM are devoted to ultra-high-resolution observation of individual atoms (or atom columns) and very few textbooks deal with theory of diffraction contrast in STEM. Furthermore, large area convergent beam electron diffraction (LACBED) is an important technique to determine absolute values of Burgers vectors, the most important issue in dislocation theory, and has been applied to 4H-SiC and GaN successfully. Therefore, LACBED is described in some detail.

It is emphasized that the readers would be able to publish their papers in internationally renown scientific journals with knowledge they obtain from this book.

Hiroyasu Saka
Nagoya 2020

About the Author

Hiroyasu Saka is Professor Emeritus of Nagoya University since 2005. He is currently a visiting Professor at Aichi Institute of Technology since 2011. He was Professor of Quantum Engineering at Nagoya University between 1988 and 2005. He received his Ph.D. degree in Metallurgical Engineering from Nagoya University in 1972. He is an Emeritus Fellow of the Japan Institute of Metals and Materials since 2005. He is also a member of the Japanese Society of Microscopy since 1966.

Acknowledgments

The original Japanese versions of this textbook (結晶電子顕微鏡学; Editions 1 and 2) have been published by Uchida-Rokakuho (内田老鶴圃). I am indebted to Uchida-Rokakuho (内田老鶴圃) for kindly permitting the English version to be published by World Scientific Publishing Co.

I am particularly grateful to Hiroyuki Iwata of Aichi Institute of Technology without whose contribution it would have been impossible to write Chapters 10 and 11. Experiments described in these two chapters were carried out using an electron microscope JEM-2100plus, which was installed in Aichi Institute of Technology through the Ministry of Education, Culture, Sports, Science and Technology of Japan. I am also indebted to D. Cherns and K. Tsuda for the discussion on LACBED.

Finally, I am indebted to H. Fujita, N. Sumida, M. Nemoto, Y. Miura, F. Nagata, K. Kuroda, U. Enokido, K. Sasaki, Y. Kaneko, Moon-Hi HONG, J. van Landuyt, S. Onda and E. Johnson for providing and permitting publishing of TEM micrographs.

Contents

Preface v

About the Author vii

Acknowledgments ix

Chapter 1. The Fundamentals of Crystallography 1

1.1 Crystal Structure and Space Lattice 1
1.2 Miller Indices 8
1.3 Hexagonal Indices 10
1.4 Typical Crystal Structures 13

Chapter 2. Stereographic Projection and the Reciprocal Lattice 23

2.1 Space Projection 23
2.2 Application of Stereographic Projection 29
2.3 Reciprocal Lattice 42

Chapter 3. Crystal Dislocations 49

3.1 Definition of Dislocation 49
3.2 Burgers Vector 52
3.3 Displacement Around a Dislocation 60
3.4 Dissociated Dislocations in FCC 61
3.5 Super-Dislocations in Ordered Alloys and Intermetallic Compounds 70

Chapter 4. Diffraction of Electrons by a Crystal 73

4.1 Plane Wave 73
4.2 Scattering by a Single Atom 76
4.3 Scattering by a Unit Cell: Diffraction 77
4.4 The Ewald Sphere 79
4.5 Structure Factor 82
4.6 Broadening of Reciprocal Lattice Points Due to the Outer Shape of a Crystal 88

Chapter 5. Electron Microscope 93

5.1 Basic Design of the Electron Microscope 93
5.2 Electron Diffraction and Electron Microscope 95
5.3 Electron Diffraction Pattern and its Indexing 97
5.4 Kikuchi Pattern 107
5.5 Analyzing Diffraction Pattern from a Specimen Containing Two Crystals: Analysis of Orientation Relationship 116
5.6 Indexing Laue Zones 125

Chapter 6. TEM Image of a Perfect Crystal 127

6.1 Diffraction Contrast and Bright-Field and Dark-Field Images 127
6.2 Selected Area Diffraction (SAD) 131
6.3 Thickness Contour and Bend Contour 132
6.4 Dynamical Theory of Diffraction 143
6.5 CAT Method to Estimate Sharpness of an Interface in an Artificial Super-Lattice 147
6.6 Supplement 148

Chapter 7. Contrast of Planar Defects and Precipitates 155

7.1 General Theory 155
7.2 Planar Defects 157
7.3 Precipitates 166
7.4 Moire Fringes 170
7.5 Summary of Contrasts from Planar Defects 172
7.6 Supplement 176

Chapter 8. Contrasts of Dislocations 183

8.1 Determining Burgers Vector of Dislocations 183
8.2 Determining the Sign of Burgers Vector 187
8.3 Determining Natures of Dislocation Loop 189
8.4 Miscellaneous Contrasts 196
8.5 Supplement 201

Chapter 9. Dark Field Weak-Beam Method, Stereographic Analysis 207

9.1 Dark Field Weak-Beam Method 207
9.2 Stereographic Observation 223
9.3 Trace Analysis 225
9.4 Effects of Surface 226

Chapter 10. Diffraction Contrast in STEM 231

10.1 General Theory 231
10.2 Diffraction Contrast in STEM 233
10.3 Resolution of Dislocations Observed by Diffraction Contrast in STEM 235
10.4 HAADF 239

Chapter 11. CBED and LACBED 243

11.1 Introduction 243
11.2 Determining Nature of a Dislocation by LACBED 245
11.3 Advantages and Drawbacks of Different Methods to Determine Absolute Values of Burgers Vectors 253

Appendices 257

Bibliography 285

Index 287

Chapter 1

The Fundamentals of Crystallography

1.1 Crystal Structure and Space Lattice

Crystal structure

A crystal consists of atoms or molecules arranged in a pattern that repeats periodically. Figure 1.1 shows a high-resolution transmission electron micrograph of diamond. White spots correspond to carbon atoms. Crystal structure is constructed by displacing a basis, a group of atoms as shown in Fig. 1.2, by constant distances three-dimensionally. Mathematically the translation vector **r**

$$\mathbf{r} = n_1\mathbf{a} + n_2\mathbf{b} + n_3\mathbf{c}, \quad n_1, n_2, n_3 \tag{1.1}$$

is applied to all of n_1, n_2 and n_3. Note that a crystal is three dimensional, so that **a**, **b** and **c** are not on the same plane. For instance, if **a** and **b** are in the plane of paper in Fig. 1.2, **c** is out of paper.

Space lattices

First let us neglect concrete arrangement of atoms in a basis, and pick up an arbitrary point and apply the operation of translation given by Eq. (1.1). Then, different crystal structures such as those illustrated in Figs. 1.3(a)–1.3(c) are all expressed by points as shown in Fig. 1.3(d). This point is referred to as a lattice point. Lattice points

Fig. 1.1: High-resolution electron micrograph of diamond (by courtesy of Professors N. Sumida and H. Fujita).

are arranged in a regularly repeated three-dimensional framework as shown in Fig. 1.4. This is referred to as a lattice space.

Definition 1.1: Crystal structure = basis + space lattice.

Unit cell

The unit of the repetition in lattice space is referred to as a unit cell. It is noted that many different unit cells can be chosen as shown in Fig. 1.5: the choice is not unique. However, survey of all existing crystals has shown that their space lattices belong to one of the 14 unit cells referred to as Bravais lattices (Fig. 1.6 and Table 1.1). Here, $\mathbf{a}$, $\mathbf{b}$ and $\mathbf{c}$ are the fundamental vectors of a unit cell and α, β and γ are angles between $\mathbf{b}$ and $\mathbf{c}$, $\mathbf{c}$ and $\mathbf{a}$, and $\mathbf{a}$ and $\mathbf{b}$, respectively. That is, $\alpha = \mathbf{b}^\wedge\mathbf{c}$, $\beta = \mathbf{c}^\wedge\mathbf{a}$ and $\gamma = \mathbf{a}^\wedge\mathbf{b}$. Values of $\mathrm{a} = |\mathbf{a}|$, $\mathrm{b} = |\mathbf{b}|$, $\mathrm{c} = |\mathbf{c}|$, α, β and γ are referred to as lattice constants or lattice parameters. It is customary that when all of α, β and γ are 90°, they are not written explicitly. Also, when $\mathrm{a} = \mathrm{b} = \mathrm{c}$, both b and c are denoted

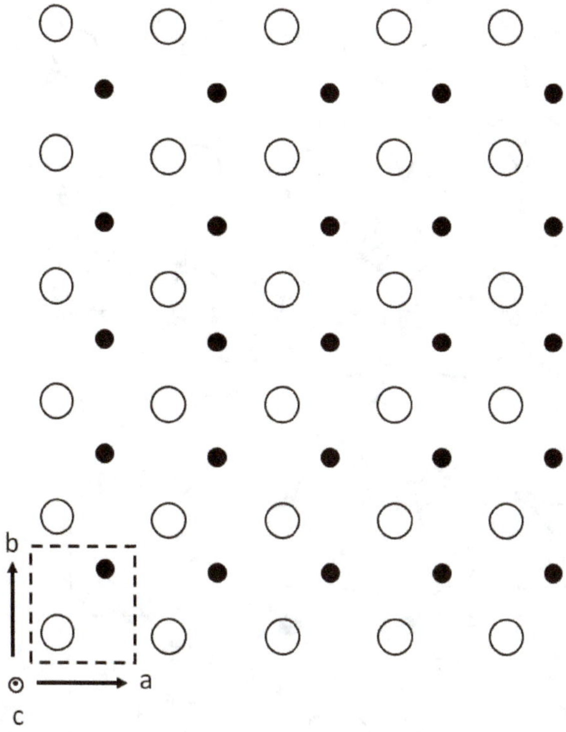

Fig. 1.2: Two-dimensional crystal. The overall crystal is constructed by displacing a basis, a group of atoms encircled, by constant distances three-dimensionally.

by a alone to show explicitly that a = b = c. Thus, for instance, a simple tetragonal system is expressed by a,a,c: This means a = b ≠ c.

Strictly speaking, some of the 14 Bravais lattices can be expressed by simpler unit cells. For instance, a face centered cubic (FCC) lattice can be expressed by a simpler unit cell composed of a', b' and c' (this is referred to as rhombohedral lattice) in Fig. 1.7. Notwithstanding, since the FCC is beautiful and has a high symmetry, it is added to the 14 Bravais lattices. Unit cells that exclude these expedient ones are referred to as primitive unit cells.

Exercise 1.1: Let the fundamental vectors of the primitive unit cell of FCC be a', b' and c', then

$$a' = 1/2[\mathbf{a}+\mathbf{b}], \quad b' = 1/2[\mathbf{b}+\mathbf{c}], \quad c' = 1/2[\mathbf{a}+\mathbf{c}].$$

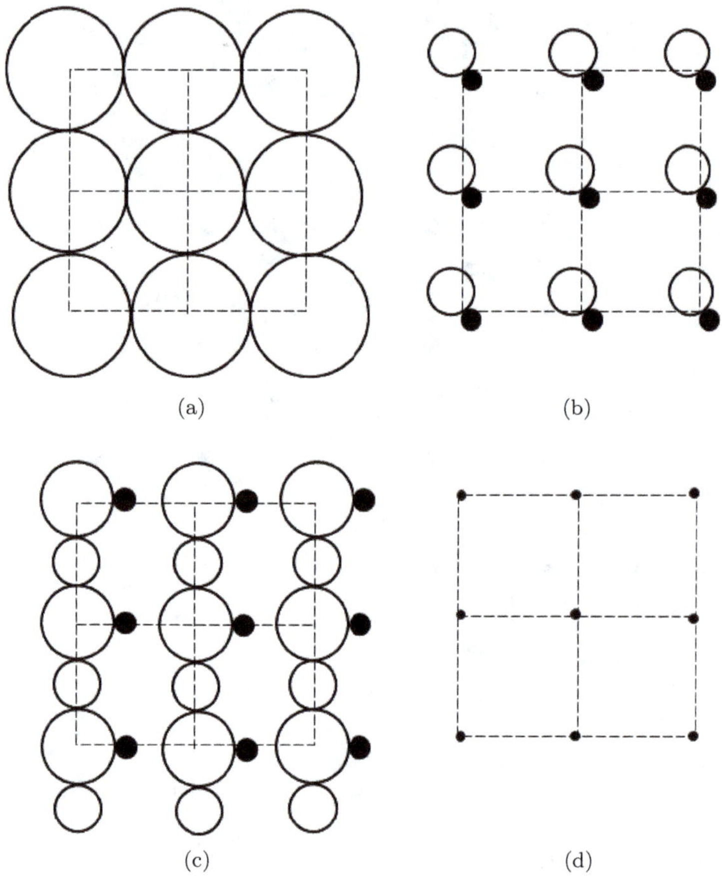

Fig. 1.3: Different crystal structures such as those shown in (a), (b) and (c) are represented by a single lattice space shown in (d).

Similarly, the fundamental vectors of the primitive unit cell of BCC $\mathbf{a}'$, $\mathbf{b}'$ and $\mathbf{c}'$ are given by

$$\mathbf{a}' = 1/2[\mathbf{a} + \mathbf{b} - \mathbf{c}], \quad \mathbf{b}' = 1/2[-\mathbf{a} + \mathbf{b} + \mathbf{c}], \quad \mathbf{c}' = 1/2[\mathbf{a} - \mathbf{b} + \mathbf{c}].$$

Here **a**, **b**, and **c** are fundamental vectors of FCC and BCC lattices. Prove this.

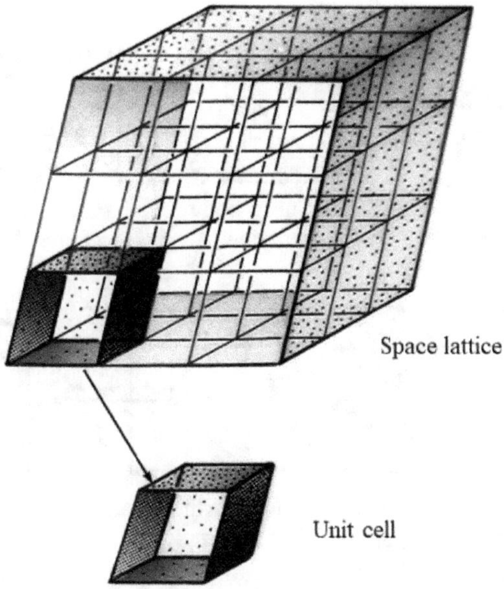

Fig. 1.4: A lattice space and a unit cell.

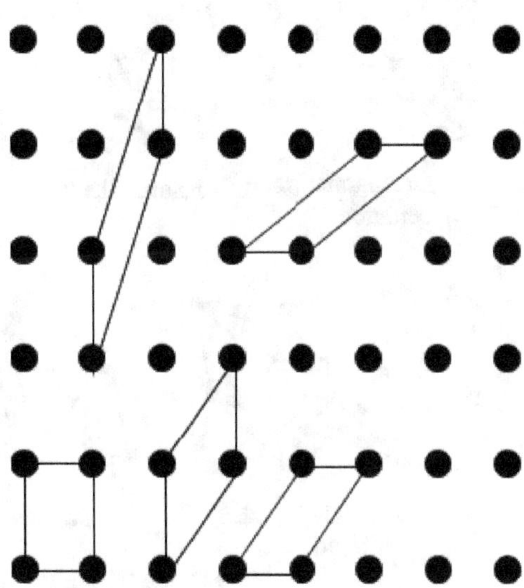

Fig. 1.5: Choice of unit cell.

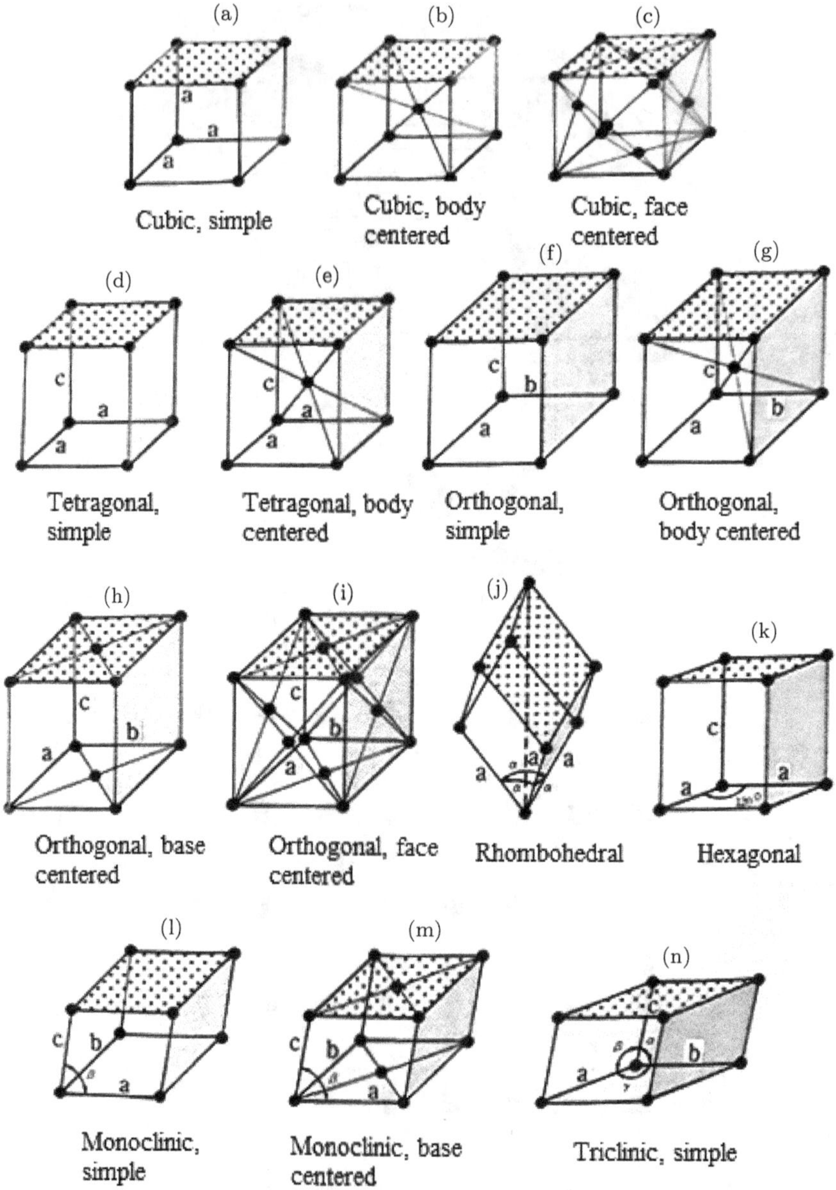

Fig. 1.6: Bravais lattices.

Table 1.1: Bravais lattices.

Crystal system	Axial ratio, axial angle	Bravais lattices	Fig. 1.6
Cubic	a = b = c	Simple cubic	(a)
	$\alpha = \beta = \gamma = 90°$	Body-centered cubic	(b)
		Face-centered cubic	(c)
Tetragonal	a = b ≠ c	Simple tetragonal	(d)
	$\alpha = \beta = \gamma = 90°$	Body-centered tetragonal	(e)
Orthorhombic	a ≠ b ≠ c	Simple orthorhombic	(f)
	$\alpha = \beta = \gamma = 90°$	Body-centered orthorhombic	(g)
		Base-centered orthorhombic	(h)
		Base-centered orthorhombic	(i)
Trigonal or	a = b = c	Simple	(j)
rhombohedral	$\alpha = \beta = \gamma \neq 90°$		(n)
Hexagonal	a = b ≠ c	Simple	(k)
	$\alpha = \beta = 90°$,		
	$\gamma = 120°$		
Monoclinic	a ≠ b ≠ c	Simple monoclinic	(l)
	$\alpha = \gamma = 90° \neq \beta$	Base-centered monoclinic	
Triclinic	a ≠ b ≠ c	Simple	(m)
	$\alpha \neq \beta \neq \gamma \neq 90°$		

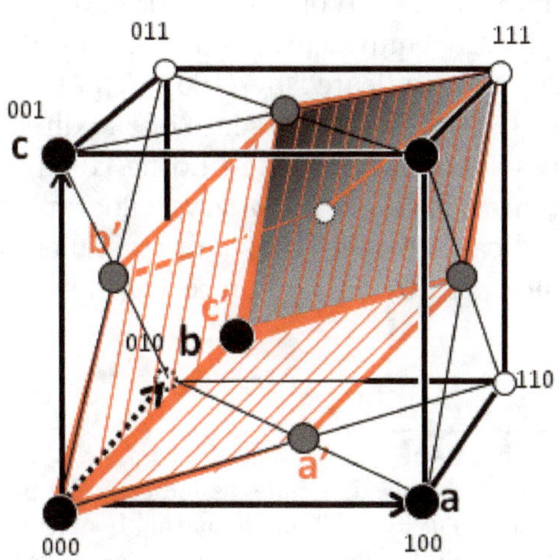

Fig. 1.7: Face centered cubic lattice (**a,b,c**) can be expressed by simple rhombohedral lattice (**a**′, **b**′, **c**′).

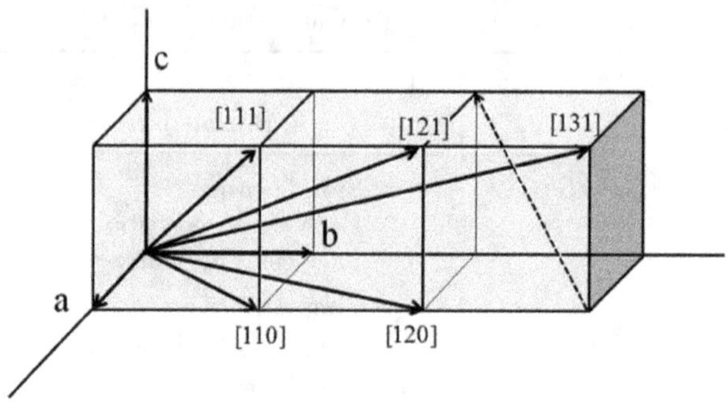

Fig. 1.8: Definition of directions.

1.2 Miller Indices

A crystallographic direction is denoted by a vector such as that shown in Fig. 1.8.

Definition 1.2:	$[\mathrm{uvw}] = \mathrm{u}\mathbf{a} + \mathrm{v}\mathbf{b} + \mathrm{w}\mathbf{c}$	(1.2)

For instance [100] stands for $1 \cdot \mathbf{a} + 0 \cdot \mathbf{b} + 0 \cdot \mathbf{c}$, so that it indicates **a**-axis. In a cubic system, $[100], [010], [001], [\bar{1}00]$,[1] $[0\bar{1}0]$ and $[00\bar{1}]$ are crystallographically equivalent. Such a set of crystallographically equivalent directions is indicated by $\langle 100 \rangle$. A particular direction is indicated by [uvw]. To indicate the sense of the direction, a notation like $[\mathrm{uvw}\rangle$ or $\langle\mathrm{uvw}]$ is used (Hirth and Lothe, 1968).

Miller indices are used to stand for crystallographic planes. A particular plane is indicated by (hkl). A set of crystallographically equivalent planes is indicated by $\{hkl\}$. Figure 1.9(a) shows the definition of (hkl) plane.[2]

[1] $[\bar{1}00] = -1 \cdot \mathbf{a} + 0 \cdot \mathbf{b} + 0 \cdot \mathbf{c}$.

It is customary that the negative index is indicated by placing a minus sign above the index: that is, $-h = \bar{h}$. Throughout this book, the negative index is also indicated by underlying the index, i.e., $-h = \underline{h}$, as described later.

For instance, $\bar{1}$ indicates -1. Similarly $\underline{h}$ indicates $-h$. $(\bar{h}kl) = (-hkl) = (\underline{h}kl)$.

[2] Throughout this book, Miller indices of the directions are denoted by roman type, while those for planes are denoted by *italic* types.

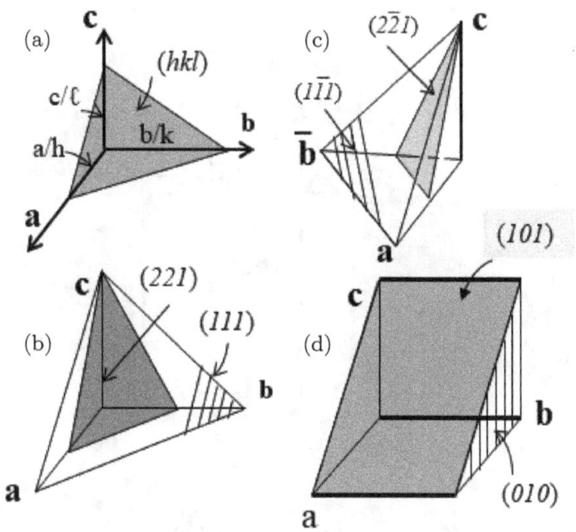

Fig. 1.9: (a)–(d) Miller indices of some important planes.

> **Definition 1.3:**
> The intercepts of the (hkl) plane with the **a**, **b** and **c** axes are $\mathbf{a}/h$, $\mathbf{b}/k$ and $\mathbf{c}/l$, respectively.

For instance, $(111), (221), (2\bar{2}1), (1\bar{1}1), (101)$ and (010) are shown in Figs. 1.9(a)–1.9(d).

Exercise 1.2: Draw $(112), (123), (100), (1\bar{1}0)$ and $(1\bar{2}3)$.

A set of crystallographically equivalent planes such as (100), $(010), (001), (\bar{1}00), (0\bar{1}0)$ and $(00\bar{1})$ is expressed by $\{100\}$.

Exercise 1.3: Name all the planes indicated by $\{110\}$ and $\{111\}$.

Exercise 1.4: In cubic crystals the normal to (hkl) plane is coincident with the direction $[hk\ell]$. This does not necessarily hold for crystals with low symmetry. Prove this by taking (101) plane and [101] in a tetragonal crystal as an example.

(*Hint*) View along **b**-axis.

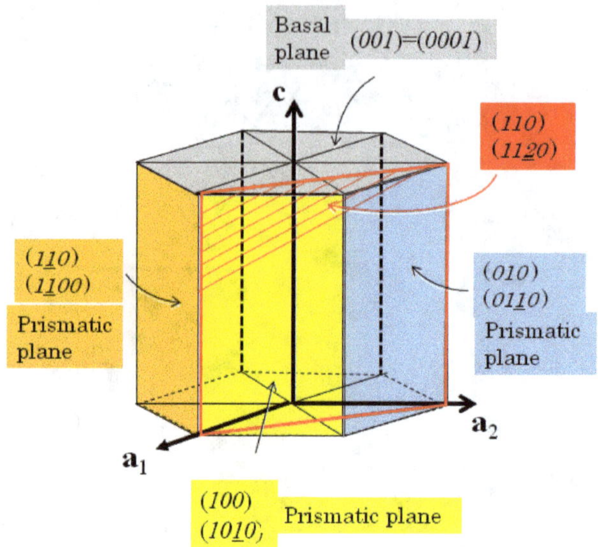

(a)

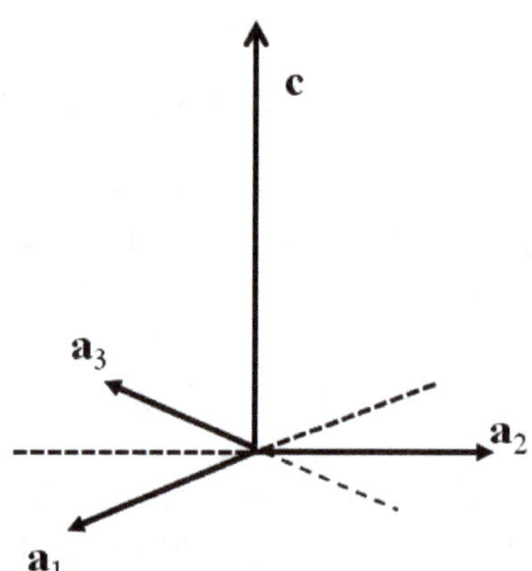

(b)

Fig. 1.10: (a) Indices of some important planes in a hexagonal system. (b) Coordinate system for a hexagonal system.

1.3 Hexagonal Indices

Miller indices can be applied to a hexagonal crystal. However, they are open to the objection that equivalent planes do not have similar indices. For instance, the planes (*100*) and ($\bar{1}$*10*) in Fig. 1.10(a) are

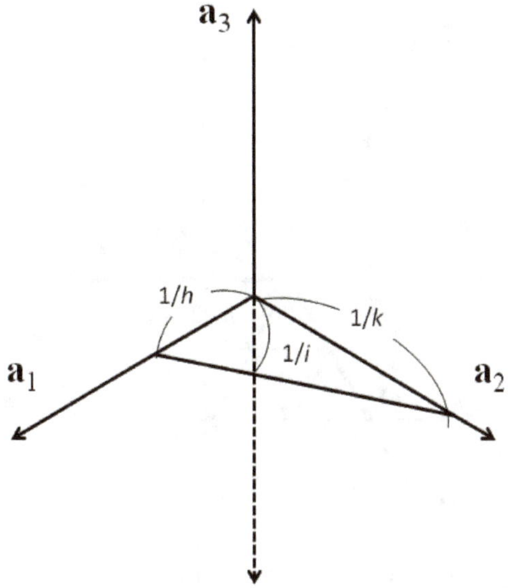

Fig. 1.11: In hexagonal indices $i = -(h + k)$.

both prismatic planes and are equivalent. Similar objection is applicable to directions. For these reasons the hexagonal indices such as shown in Fig. 1.10(b) are used.

(1) In addition to the usual $\mathbf{a}_1(=\mathbf{a})$, $\mathbf{a}_2(=\mathbf{b})$ and $\mathbf{c}$ axes, a fourth axis $\mathbf{a}_3$ is introduced.
(2) Apply Miller indices to $\mathbf{a}_1$, $\mathbf{a}_2$, $\mathbf{a}_3$ and $\mathbf{c}$ axes.
(3) The plane is denoted by $(hkil)$, where h, k, i are indices corresponding to the $\mathbf{a}_1$, $\mathbf{a}_2$ and $\mathbf{a}_3$ axes. l is index corresponding to the $\mathbf{c}$ axis. h, k, i are related by the equations

$$h + k = -i.$$

Exercise 1.2: Prove that $h + k = -i$.

(*Hint*) Figure 1.11.

Directions can be expressed by three-indices [UVW] employing $\mathbf{a}_1$, $\mathbf{a}_2$ and $\mathbf{c}$ axes and four-indices [uvtw] employing $\mathbf{a}_1$, $\mathbf{a}_2$, $\mathbf{a}_3$ and $\mathbf{c}$ axes.

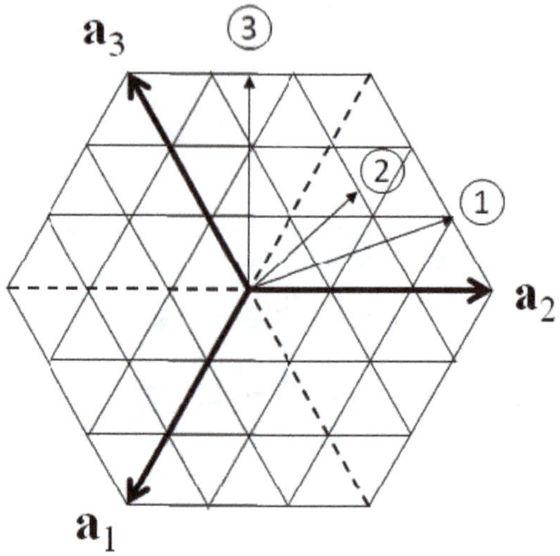

Fig. 1.12: Exercise 1.7.

Conversion between the three and four-indices, i.e., between (HKL) and ($hkil$) and between [UVW] and [uvtw] is given by

Theorem 1.1:

$$H = h, \quad K = k, \quad L = \ell, \quad i = -(h+k) = -(H+K)$$

$$\mathrm{U} = 2\mathrm{u} + \mathrm{v}, \ \mathrm{V} = \mathrm{u} + 2\mathrm{v}, \ \mathrm{W} = \mathrm{w} \tag{1.3}$$

$$\mathrm{u} = 1/3(2\mathrm{U} - \mathrm{V}), \quad \mathrm{v} = 1/3(2\mathrm{V} - \mathrm{U}), \quad \mathrm{w} = \mathrm{W}, \quad \mathrm{t} = -(\mathrm{u} + \mathrm{v}). \tag{1.3'}$$

Exercise 1.6: Prove Eqs. (1.3) and (1.3′).

(*Hint*) $\mathbf{R} = \mathrm{u}\mathbf{a}_1 + \mathrm{v}\mathbf{a}_2 + \mathrm{t}\mathbf{a}_3$, $\mathbf{R} = \mathrm{U}\mathbf{a}_1 + \mathrm{V}\mathbf{a}_2$, $\mathbf{a}_3 = -(\mathbf{a}_1 + \mathbf{a}_2)$.

Exercise 1.7: Express directions ①, ② and ③ in Fig. 1.12 in both [UVW] and [uvtw].

(*Hint*) First determine [UVW].

1.4 Typical Crystal Structures

Crystal structures of pure elements

Face-centered cubic (FCC) structure

Figure 1.13 illustrates FCC structure. Atoms are located at the corners and centers of faces. Any atom at the corners is shared by 8 unit cells. The number of the corners of a unit cell is 8, so that the net number of atoms at the corner is $(1/8) \times 8 = 1$. The co-ordinate of the atom is represented by the origin, that is, $x = 0\ y = 0\ z = 0$ (hereinafter denoted by 000). Similarly the number of atoms at the face centers is $(1/2) \times 6 = 3$. The co-ordinates of these atoms are represented by $x = 0\ y = 1/2\ z = 1/2$, $x = 1/2\ y = 0\ z = 1/2$, $x = 1/2\ y = 1/2\ z = 0$. They are summarized as follows. FCC includes Al, Cu, Au, Ag, Pr, Ni and γ-Fe.

	Origin	Face centers		
x	0	0	1/2	1/2
y	0	1/2	0	1/2
z	0	1/2	1/2	0

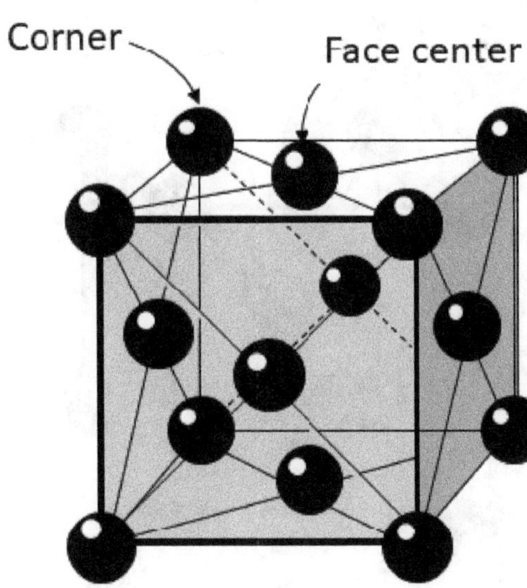

Fig. 1.13: Face-centered cubic structure.

Exercise 1.8: The lattice constant of Cu is 0.36147 nm. The atomic weight is 63.54. Calculate the density of Cu.

Exercise 1.9: Let us suppose the atoms be hard spheres. The (*111*) plane of FCC is a most closely packed plane. Prove this.

Body-centered cubic (BCC) structure

Figure 1.14 illustrates BCC structure. Atoms are located at the center of body and corners. Similarly to FCC structure, the number of atoms in a unit cell is 2, their co-ordinates being represented as follows:

	Origin	Body center
x	0	1/2
y	0	1/2
z	0	1/2

BCC includes α-Fe, Nb, Cr, W, Mo, Na and Li.

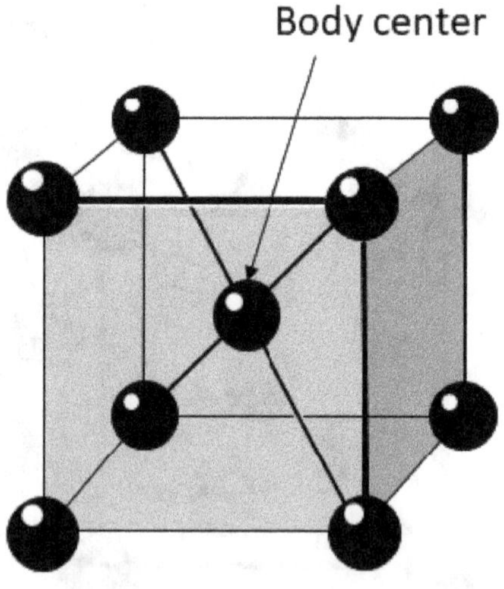

Fig. 1.14: Body-centered cubic structure.

Diamond structure

Figure 1.15 illustrates diamond structure. Two FCC lattices, each of which is located at the origin 000 and 1/4 1/4 1/4, respectively, are interpenetrated. The number of atoms in a unit cell is 8 as summarized in the following. Diamond structure includes diamond, Si, Ge and α-Sn.

	FCC lattice with a corner at 000				FCC lattice with a corner at 1/4 1/4 1/4			
x	0	0	1/2	1/2	1/4	1/4	3/4	3/4
y	0	1/2	0	1/2	1/4	3/4	1/4	3/4
z	0	1/2	1/2	0	1/4	3/4	3/4	1/4

Hexagonal closed packed (HCP) structure

Figure 1.16 illustrates HCP structure. Unit cell of HCP is indicated by red lines. Here, three unit cells are packed face-to-face, making up a hexagonal prism. The number of atoms in a unit cell is 2, with the co-ordinates as follows. HCP includes Cd, Zn, Mg, Be and α-Ti.

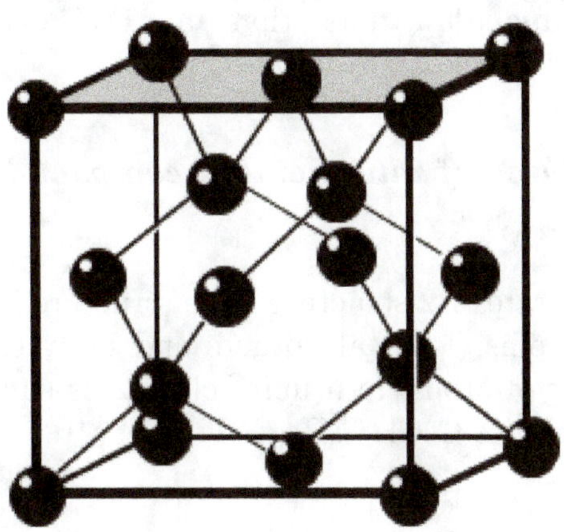

Fig. 1.15: Diamond structure.

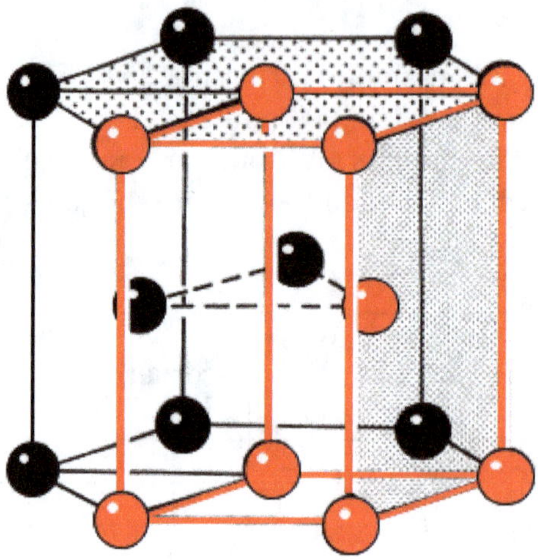

Fig. 1.16: Hexagonal closed packed structure.

	Origin	Others
x	0	2/3
y	0	1/3
z	0	1/3

Exercise 1.10: Stacking of close-packed planes in HCP is expressed by ***ababab***, while that in FCC is expressed by ***abcabcabc***... Prove this.

Crystal structures of intermetallic compounds

B2 (CsCl) structure

Figure 1.17 illustrates B2 structure. The structure is based on BCC structure, but atoms (ions) at corners and body center are different. The number of atoms in a unit cell is 2 as summarized in the following. B2 includes CsCl, CsBr, CsI, etc.

	Cs^+	Cl^-
x	0	1/2
y	0	1/2
z	0	1/2

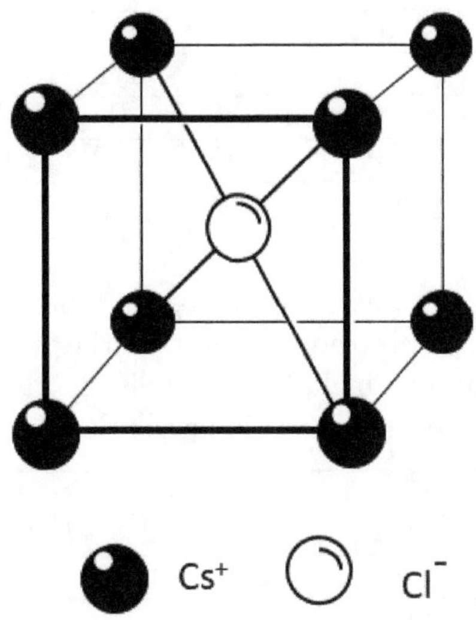

Fig. 1.17: B2(CsCl) structure.

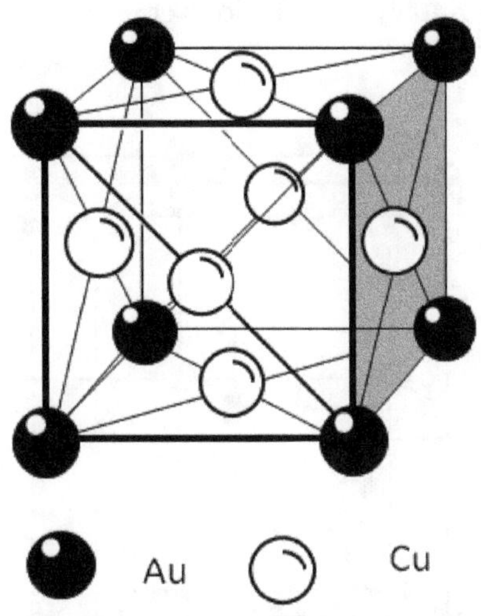

Fig. 1.18: $L1_2$ structure.

$L1_2$ (Cu_3Au) structure

Figure 1.18 illustrates $L1_2$ structure. This structure is based on FCC, but atoms at corners and face centers are different. This includes Cu_3Au, Ni_3Al, Ni_3Ga etc. The co-ordinates of atoms are as follows:

	A atom	B atom		
x	0	0	1/2	1/2
y	0	1/2	0	1/2
z	0	1/2	1/2	0

NaCl structure

Figure 1.19 illustrates NaCl structure. Na^+ occupies lattice points of FCC and Cl^- occupies body center of the unit cell and centers of edges. This structure includes NaCl, KCl, MgO, FeO, NiO etc. The co-ordinates of atoms are as follows:

	Na^+				Cl^-			
x	0	0	1/2	1/2	1/2	1/2	0	0
y	0	1/2	0	1/2	1/2	0	1/2	0
z	0	1/2	1/2	0	1/2	0	0	1/2

Exercise 1.11: Show that Cl^- ions construct an FCC lattice with its origin at $\frac{1}{2}\frac{1}{2}\frac{1}{2}$.

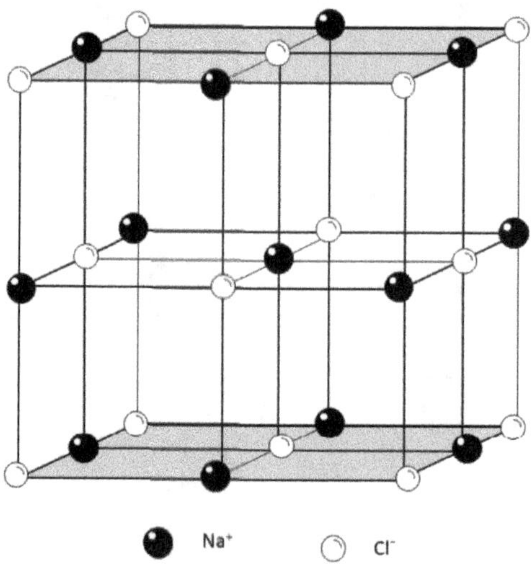

Fig. 1.19: NaCl structure.

Zincblende structure

Figure 1.20 illustrates zincblende structure. This is an ordered form of the diamond structure. This includes ZnS, CdS, GaAs, InSb etc. The co-ordinates of atoms are as follows:

	Zn				S			
x	0	0	1/2	1/2	1/4	1/4	3/4	3/4
y	0	1/2	0	1/2	1/4	3/4	1/4	3/4
z	0	1/2	1/2	0	1/4	3/4	3/4	1/4

Wurtzite structure

Figure 1.21 illustrates wurtzite structure. This is an ordered form of HCP structure. It includes ZnO, CdSe, GaN etc. The co-ordinates of atoms are as follows:

	Zn		O	
x	0	1/3	0	1/3
y	0	2/3	0	2/3
z	0	1/2	z	1/2+z

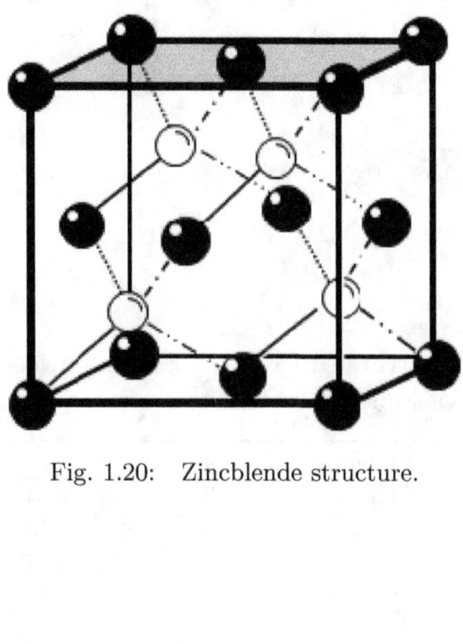

Fig. 1.20: Zincblende structure.

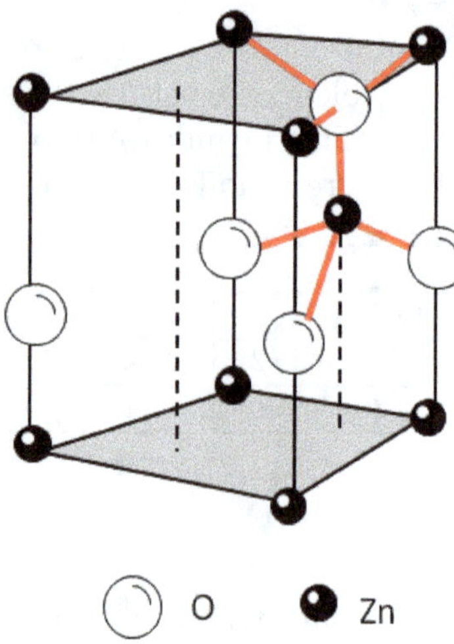

Fig. 1.21: Wurtzite structure.

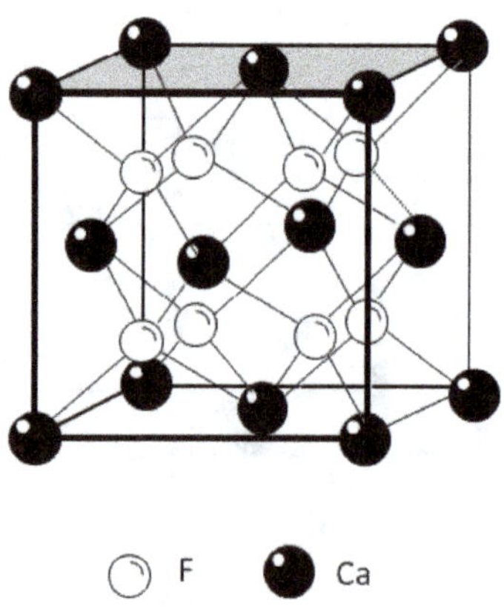

Fig. 1.22: Fluorite structure.

Fluorite structure

The structure is illustrated in Fig. 1.22. It includes CaF_2, ZrO_2 etc. The co-ordinates of ions are as follows:

	Ca^+				F^-							
x	0	0	1/2	1/2	1/4	1/4	3/4	3/4	3/4	3/4	1/4	1/4
y	0	1/2	0	1/2	1/4	3/4	1/4	3/4	3/4	1/4	3/4	1/4
z	0	1/2	1/2	0	1/4	3/4	3/4	1/4	3/4	1/4	1/4	3/4

Model of BCC lattice in Brussel. Inside the lattice points are museums.

Chapter 2

Stereographic Projection and the Reciprocal Lattice

2.1 Space Projection

Reference sphere

Beginners may think that crystallographic planes and axes are represented most conveniently by a three-dimensional model. Appendix A shows an opened-up cubic crystal bounded by {*100*} and {*111*} planes. If readers construct a three-dimensional model using this opened-up model, they will most probably realize that expressing many crystallographic planes and axes using a three-dimensional model is not practical. Crystallographic axes, planes and angles are described conveniently by a sphere, known as a reference sphere or polar sphere.

Definition 2.1:
Locate a tiny crystal at the center of the globe. The position at which the extension of crystallographic axis pierces the surface of the sphere is referred to as a pole. Intersection of a crystallographic plane passing through the center of crystal globe with the surface of globe is referred to as a great circle (Fig. 2.1).

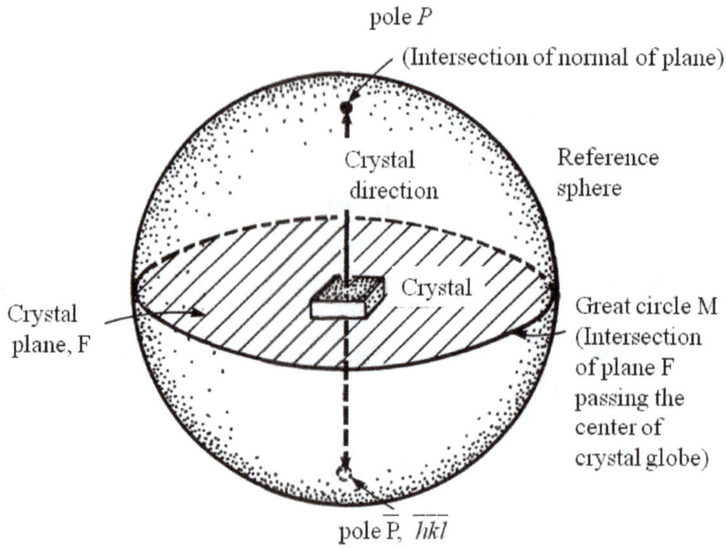

Fig. 2.1: Projection of crystal plane (hkl) on the reference sphere. On a crystal sphere, plane P is represented by the great circle M or the pole P.

A crystallographic plane is expressed by the normal to the plane. Therefore,

Definition 2.2:
Poles represent (1) directions and (2) planes the perpendiculars of which are represented by the poles.

The position of a pole is expressed with a polar co-ordinate (with the radius $r = 1$) as is shown by (ϕ, θ) in Fig. 2.2, where ϕ stands for the longitude and θ stands for the latitude.

Point 2.1.
All the longitudes (ϕ = constant) are great circles. By contrast, the latitudes (θ = constant) are small circles except the equator.

Exercise 2.1: In Fig. 2.3, some of typical planes and their perpendiculars in a cubic crystal are illustrated. Indicate indices in parenthesis in Fig. 2.3.

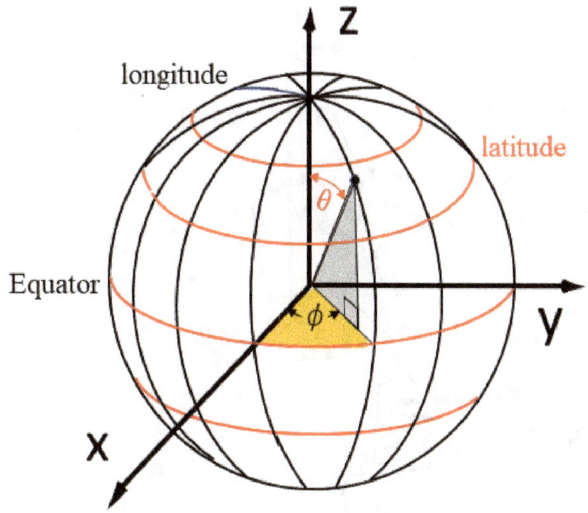

Fig. 2.2: Co-ordinate for pole (P).

Stereographic projection

While the crystal sphere is useful, it is not convenient to describe the sphere on a sheet two-dimensionally. Converting a point on the crystal sphere, which is represented by a spherical co-ordinate, onto a two-dimensional sheet is equivalent to converting a location on the Earth onto a two-dimensional map.[1] There are different methods to convert three-dimensional distribution of locations on the Earth into a two-dimensional map. In order to measure an angle between any two poles the *stereographic projection* is most useful. In this method, a light source is located at a point on the surface of crystal globe (for instance A in Fig. 2.4), and then a screen is located at a point which

[1]The greatest difference between the terrestrial globe and the crystal globe is that while Z and Z' axes on the terrestrial globe point to the north and the south poles, respectively, and the longitude at $\phi = 0°$ in the terrestrial globe passes through Royal Greenwich Observatory in London, choice of the co-ordinates in the crystal globe is at your disposal. The reason for the choice of the co-ordinates stems from the rotation of the Earth, but the choice of the longitude at $\phi = 0°$ so as to pass through London is of no scientific significance. The choice of co-ordinates in crystal globe is compared to a spaceman arriving near the Earth to set the co-ordinates neglecting the rotation of the Earth.

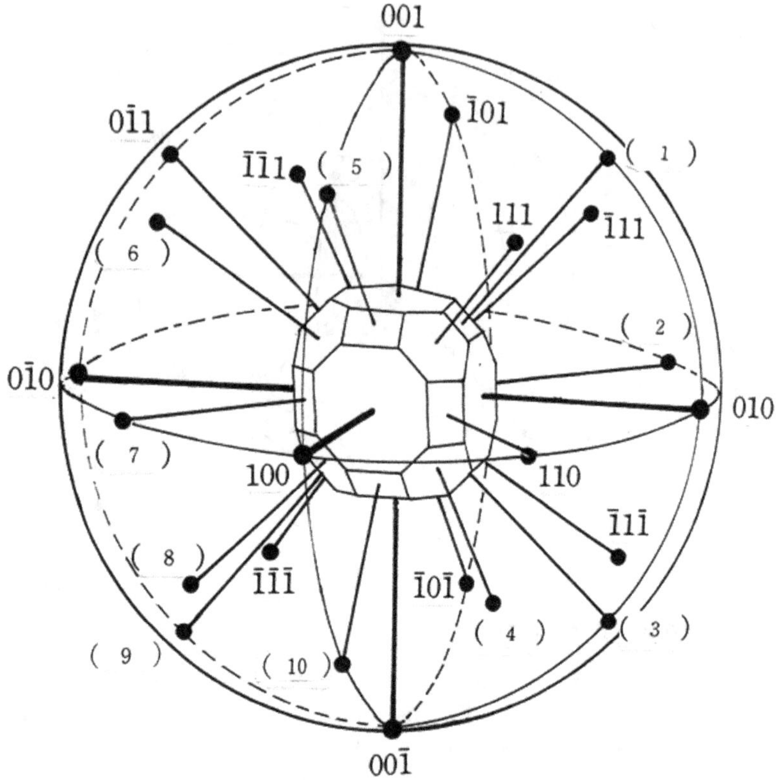

Fig. 2.3: Some important planes and their perpendiculars in a cubic crystal.

is opposite to A (A') in such a way that AA' is perpendicular to the screen (Fig. 2.4).

If a light source is located at $\theta = 180°$ as shown in Fig. 2.5 and longitudes and latitudes on the semi-sphere of $0 < \theta < 0°(Z > 0)$ are projected, the projection is referred to as a polar net. If a light source is located at a position of $\theta = 90°$ and $\phi = 0°$ and longitudes and latitudes on the semi-sphere are projected, the projection is referred to as a Wulff net. In practice, the Wulff net is more frequently used. Appendices B.1 and B.2 show the polar net and the Wulff net.

Exercise 2.2 (NA means north latitude and EL and WL mean east longitude and west longitude, respectively). Tokyo (NA 38°, EL 140°), London (NA 52°, 0°), Anchorage (NA 62°, EL

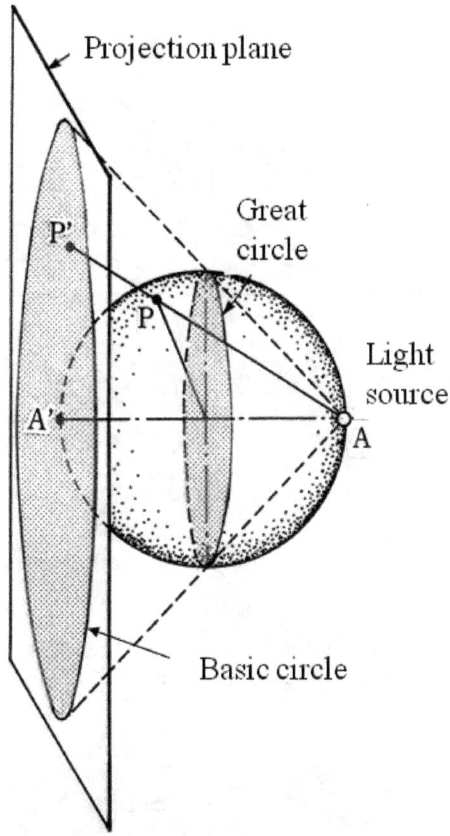

Fig. 2.4: Stereographic projection.

150°), Moscow (NA 56°, EL, 38°), Honolulu (NA 20°, WL 160°), San Francisco (NA 36°, WL 125°), New York (NA 49°, WL 74°), Beijing (NA 40°, EL 110°), Mumbai (NA 29°, NL 73°), Bagdad (NA 33°, EL 45°), Lilly hammer (NA 61°, EL 11°). Locate these cities on the polar net centered at the north pole.

Standard projection

Let us locate a principal pole (either orientations or normal to planes) at the center of stereographic projection, and project other important poles, then we obtain a standard projection. Usually 001, 110 and 111 are used as the principal poles. Figure 2.6 shows a 001 standard

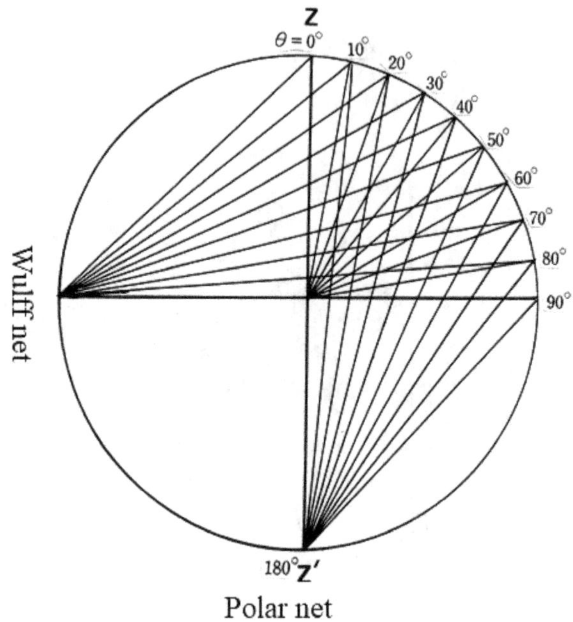

Fig. 2.5: Relationship between the Wulff net and the polar net. In the polar net a light source is located at Z', while in the Wulff net a light source is located on the equator.

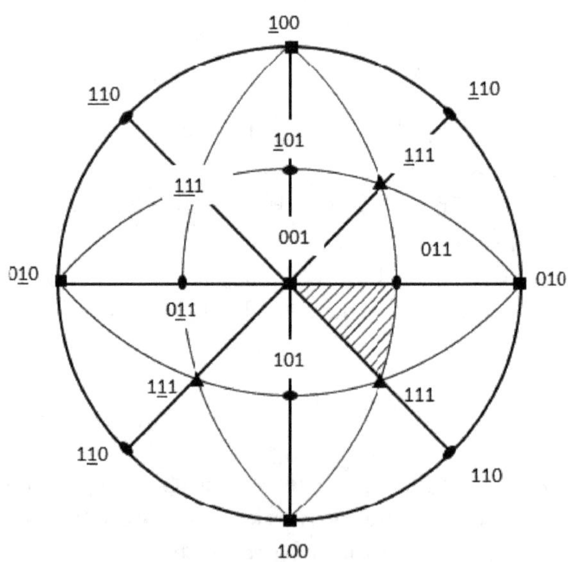

Fig. 2.6: Standard 001 projection of a cubic crystal. Hatched area is the standard triangle.

projection of a cubic crystal. In a cubic crystal (*hkl*) plane and [hkℓ] directions are coincident, so that the 001 standard projection illustrates both (*001*) standard projection and [001] standard projection. Here, 001, 110, 111 and their equivalents are denoted by ■, ⬬, ▲. The spherical triangle connecting 001, 011 and 111 is referred to as a standard triangle (indicated by a hatched triangle in Fig. 2.6). Appendix C shows the standard 001 projection of a cubic crystal.

2.2 Application of Stereographic Projection

Rotation of crystal

When a crystal is rotated, the rotation axis should be the Z-axis of the spherical co-ordinate. This operation brings P_1 to P'_1 and P_2 to P'_2 etc. (Fig. 2.7(a)). Since the rotation axis is the Z-axis, points remain on the same latitude, i.e., θ = constant, crossing the longitudes by the number corresponding to the rotation angle ($\Delta\phi$). This operation is represented on the Wulff net as shown in Fig. 2.7(b).

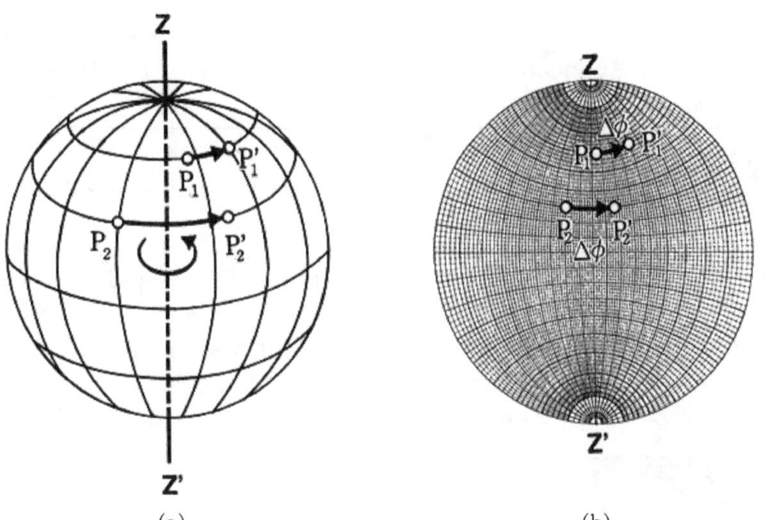

(a) (b)

Fig. 2.7: Rotation of a crystal. Rotation should always be made around Z–Z' axis.

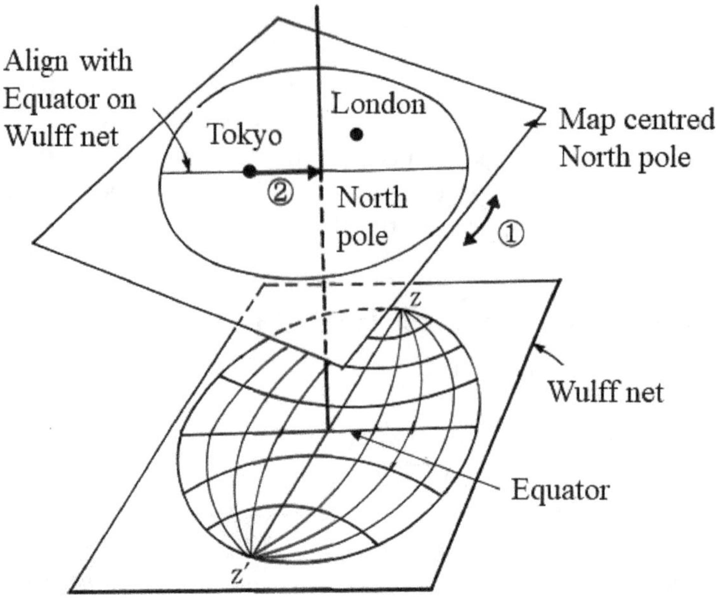

Fig. 2.8: Hint to Exercise 2.4. How to locate Tokyo at the center of the Wulff net.

> **Theorem 2.1:**
> If you want to bring any pole to the center of a stereographic projection, first locate the pole on the equator of the Wulff net, and then rotate a necessary angle around Z–Z' axis.

Exercise 2.3: Using the result of Exercise 2.2, draw a map where Tokyo is centered.

(*Hint*) As shown in Fig. 2.8, overlap a map where N-pole (see Exercise 2.2) is centered on the Wulff net. Then, rotate this N-pole-centered map around the center of the Wulff net in such a way that Tokyo is located on the equator of the Wulff net (①). Then, rotate it around Z–Z' axis so as to bring Tokyo to the center of the Wulff net.

Exercise 2.4: Using the result of Exercise 2.2, draw a conventional map of the eastern hemisphere of the Earth. Confirm that the locations (co-ordinates) of each of cities are coincident with those given in Exercise 2.2. Name the cities which disappear from the stereographic projection.

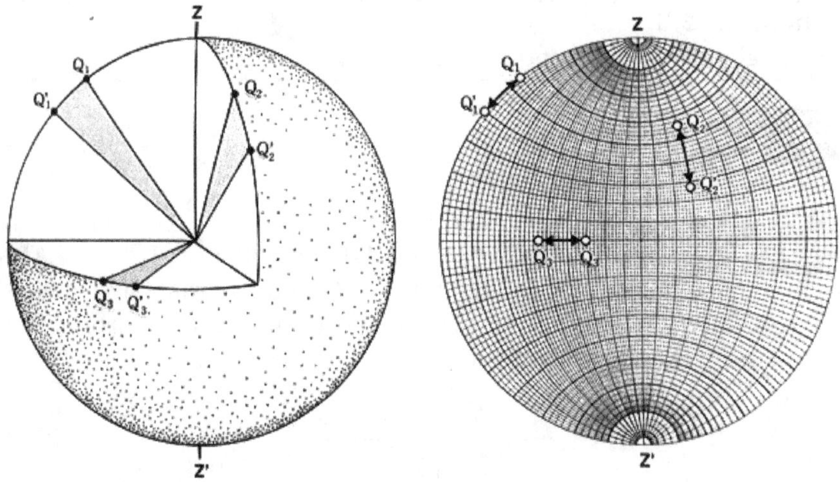

Fig. 2.9: (a) Measuring an angle between poles. Measurement is always to be made on great circles, that is, longitudes or equator. (b) Examples of measurements of angles between two poles using the Wulff net.

(*Hint*) Draw a stereographic projection whose center is NA 0° and EL 90°.

Thanks to symmetry of crystal, the following theorem holds.

Theorem 2.2:
When a pole $hk\ell$ disappears as a result of rotation, the corresponding opposite pole $\bar{h}\bar{k}\bar{\ell}$ appears at the opposite site of the diameter. Therefore, thereafter pay attention to $\bar{h}\bar{k}\bar{\ell}$.

Measuring an angle between two poles

An angle between two poles is measured by locating these two poles on a same great circle (Fig. 2.9(a)). Since in the Wulff net the longitudes are always great circles, this operation is to locate the two poles on a single longitude (for instance Q_1Q_1', Q_2Q_2' in Fig. 2.9(b)). If latitude is used, only the equator is a great circle, so that it is necessary to rotate the Wulff net in such a way that the two poles are on the equator.

Theorem 2.3:
An angle between two poles is measured on a great circle.

Exercise 2.5: Calculate the distance (angle) between Tokyo and London using the results of Exercises 2.2–2.4.

Exercise 2.6: Measure the angles $011^\wedge 111$ and $011^\wedge \bar{1}\bar{1}1$ using the standard 001 projection and the Wulff net. Confirm that these angles are coincident with the values calculated from the inner product of the vectors. Here, for a cubic crystal,

$$\cos\theta = \frac{h_1h_2 + k_1k_2 + l_1l_2}{(h_1^2 + k_1^2 + l_1^2)^{1/2}(h_2^2 + k_2^2 + l_2^2)^{1/2}} \tag{2.1}$$

Appendix D lists up angles between principal poles in a cubic crystal.

Exercise 2.7: Draw the standard 011 and 111 projections using the standard 001 projection in Fig. 2.6.

Exercise 2.8: Locate poles which make 45° with respect to 101 and confirm that the trajectory of these points lies on a circle.

Exercise 2.9: Trajectory of poles which make an angle less than 90° with respect to P is a circle on a stereographic projection (this circle is referred to as a small circle). Prove this.

(*Hint*) OBB′ which passes through the origin of the crystal globe shown in Fig. 2.10(a) is a cone (since its section BB′ is a circle). On the other hand, ACC′ (and therefore ADD′ also) is oblique cone with its section BB′ being a circle. Consider the condition under which the stereographic projection of BB′ (i.e., DD′) becomes a circle (see Fig. 2.10(b)).

Theorem 2.4:
A small circle BB′ is projected to DD′ on the stereographic projection. The center of the projected small circle (indicated by • in Fig. 2.10(c) is not coincident with the stereographic projection of P (*i.e.*, P′ indicated by ∘).

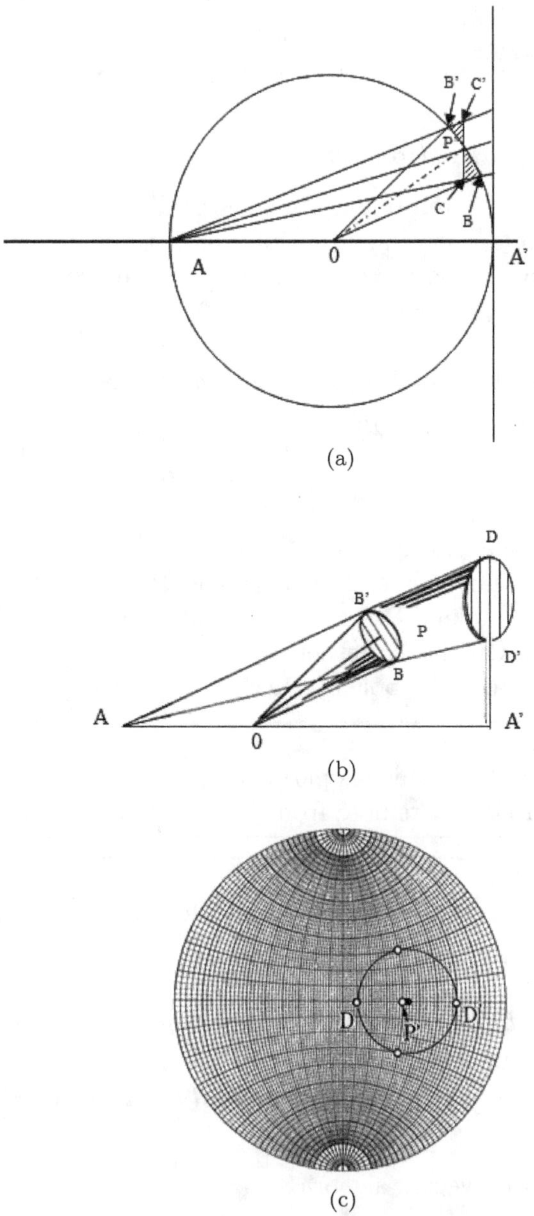

Fig. 2.10: Proof that a small circle on the crystal sphere projected onto a stereographic projection is also a circle.

Exercise 2.10: Locate a pole on the standard 001 projection which is 36° remote from 001, 19° from 011 and 22° from 111.

Determining direction and plane

It is frequently desired to determine crystallographic directions and/or planes inside a crystal from information on the crystal surface. In what follows some typical examples will be presented for a cubic crystal, which is simplest.

Determining [hkℓ] *which is perpendicular to* [uvw]

Putting $\theta = 90°$ in Eq. (2.1), $\cos 90° = 0$. Therefore,

$$\mathrm{uh} + \mathrm{vk} + \mathrm{w}\ell = 0 \tag{2.2}$$

Consider the section of a watermelon cut right into two (Fig. 2.11(a)). Let this section be plane **P**, then all the directions (**p**) lying on the plane **P** are perpendicular to the normal of **P** (**P′**) (Fig. 2.11(a)). On the Wulff net, **p** is located on a large circle which is 90° remote from the pole of **P** (**P′**).

Theorem 2.5:
Any direction [hkℓ] which is perpendicular to P′ = [uvw] is located on a great circle 90° remote from P′ = [uvw] (Fig. 2.11(b)).

Definition 2.3:
Such a set of (*hkl*) planes belongs to the [uvw] zone axis (see also Fig. 2.12(b)).

Determining the direction [uvw] *which is perpendicular to two directions* [HKL] *and*
[hkℓ] *or the zone axis* [uvw] *containing two planes* (*HKL*) *and* (*hkl*)

[uvw] is given by vector algebra

$$\begin{vmatrix} \mathrm{u} & \mathrm{v} & \mathrm{w} \\ \mathrm{H} & \mathrm{K} & \mathrm{L} \\ \mathrm{h} & \mathrm{k} & \ell \end{vmatrix} = \mathrm{u}\begin{vmatrix} \mathrm{K} & \mathrm{L} \\ \mathrm{k} & \ell \end{vmatrix} - \mathrm{v}\begin{vmatrix} \mathrm{H} & \mathrm{L} \\ \mathrm{h} & \ell \end{vmatrix} + \mathrm{w}\begin{vmatrix} \mathrm{H} & \mathrm{K} \\ \mathrm{h} & \mathrm{k} \end{vmatrix}$$

$$= \mathrm{u}(\mathrm{K}\ell - \mathrm{kL}) - \mathrm{v}(\mathrm{H}\ell - \mathrm{hL}) + \mathrm{w}(\mathrm{Hk} - \mathrm{hK})$$

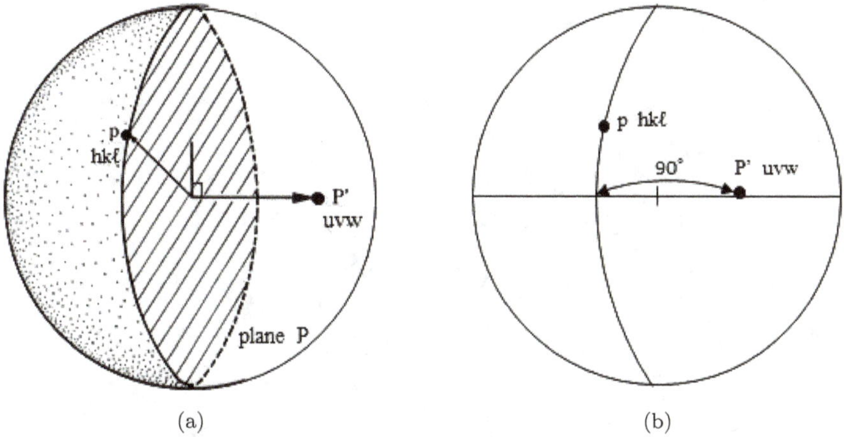

Fig. 2.11: Determining directions perpendicular to pole P.

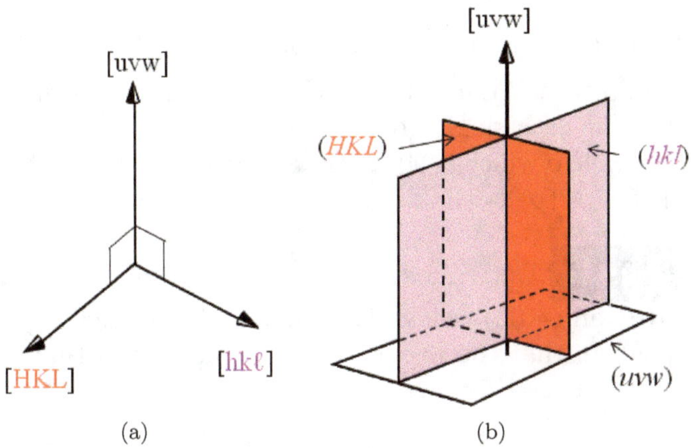

Fig. 2.12: (a) shows how to determine the direction [uvw] which is perpendicular to two directions [HKL] and [hkℓ]. (b) shows how to determine the zone axis [uvw] which contains two planes (*HKL*) and (*hkl*).

That is,

$$u = K\ell - kL, \quad v = hL - H\ell, \quad w = Hk - hK \tag{2.3}$$

There are two ways to determine [uvw].

(1) Using Theorem 2.4 determine the great circle **a** which is perpendicular to [HKL]. Similarly determine the great circle **b** which is perpendicular to [hkℓ]. [uvw] is the intersection of **a** and **b** (Fig. 2.13(a)).

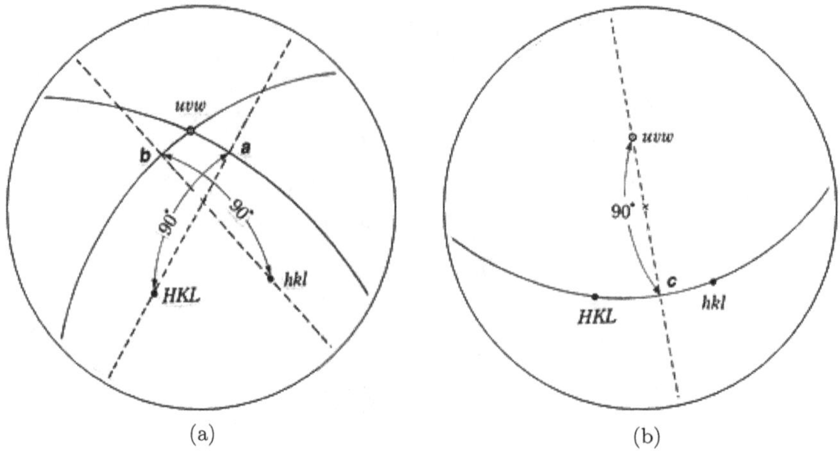

Fig. 2.13: How to locate the pole uvw which is perpendicular to HKL and hkℓ.

(2) Inverse calculation of Theorem 2.5.

(i) Rotate the Wulff net in such a way that both [HKL] and [hkℓ] lie on a single great circle.

(ii) The pole **c** which is 90° remote from this great circle is [uvw] (Fig. 2.13(b)).

Exercise 2.11: Determine the direction which is perpendicular to both [101] and [011] using the aforementioned methods (i) and (ii), and confirm that the result is in agreement with that given by Eq. (2.3).

Determining an intersection [uvw] *on the crystal surface* (*HKL*) *with* (*hkl*) *plane inside a crystal*

This corresponds to (*HKL*) lying on the surface in the foregoing (ii). Let us consider a book opened halfway. The halfway-opened pages correspond to (*hkl*) planes and the back-cover on the desk corresponds to the crystal surface (*HKL*). [uvw] is the direction of the spine (Fig. 2.14(a)). Let (*HKL*) stereographic projection be placed at center. Directions lying on (*HKL*) are located on the circumstance of the stereographic projection. Among them, the direction along which planes $(hkl)_a$, $(hkl)_b$, $(hkl)_c$, i.e., $\boldsymbol{a}$, $\boldsymbol{b}$, $\boldsymbol{c}$ intersect with circumstance is [uvw] (indicated by star in the figure), as shown in Fig. 2.14(b)).

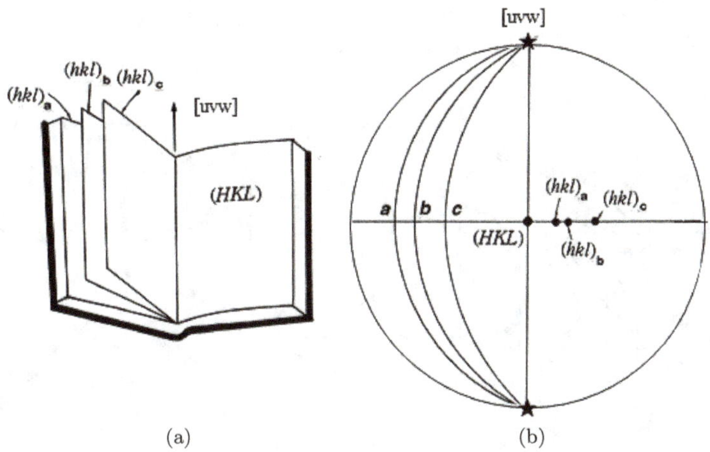

Fig. 2.14: How to determine traces. Traces of $(hkl)_\mathbf{a}$, $(hkl)_\mathbf{b}$ and $(hkl)_\mathbf{c}$ are [uvw].

A line along which a crystallographic plane intersects with surface is referred to as a trace.

Exercise 2.12: Let us consider a crystal the surface of which is (*001*) and contains planar precipitates (with a crystal structure different from the matrix) lying on {*111*}. Determine traces on the crystal surface of the precipitates.

Exercise 2.13: What if the surface of the crystal is (*011*)?

Determining a plane inside a crystal from its traces on two different surfaces

As mentioned in foregoing section, a plane inside a crystal cannot be determined from traces on one surface only. In order to determine the plane inside a crystal, it is necessary to observe traces on at least two surfaces.

Now let us assume that a trace makes an angle of 26° on (*010*) surface and on (*001*) surface 45°, as shown in Fig. 2.15(a). Let us determine the plane P inside crystal. The normal of P is perpendicular to A and B, so that P can be determined by the method described in foregoing section.

On stereographic projections, trace A is indicated by pole A in Fig. 2.15(b), and trace B by pole B in Fig. 2.15(c). Problem here

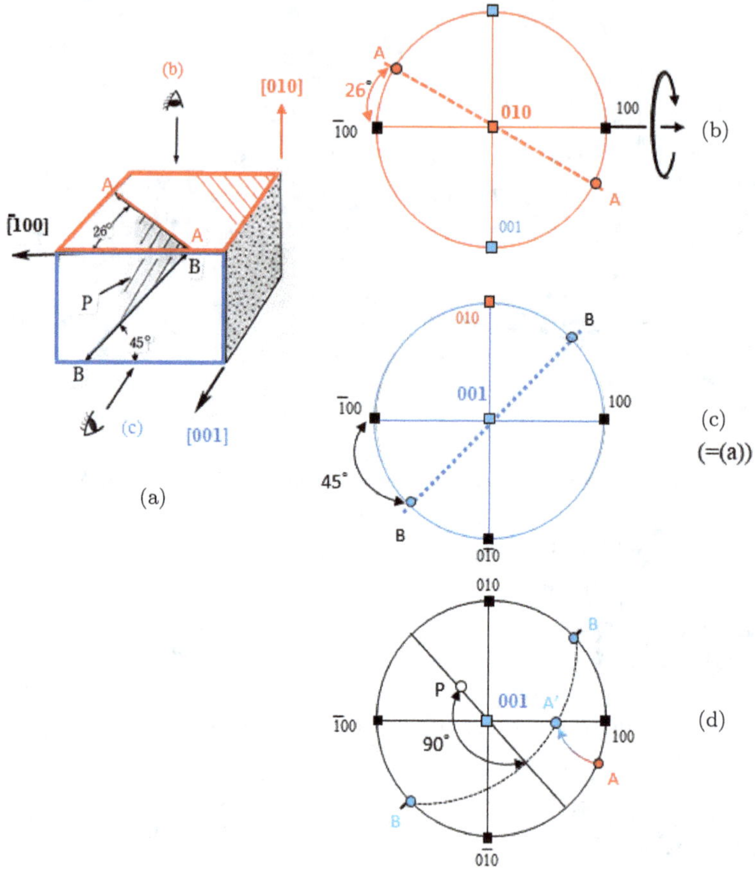

Fig. 2.15: How to determine a plane inside a crystal from traces on two surfaces.

is that while the trace A in Fig. 2.15(b) is on a 010 stereographic projection, the trace B is on a 001 stereographic projection, so that it is impossible to determine P straightforwardly. It is necessary to rotate the stereographic projection (b) 010 to (c) 001 or vice versa. For instance, let the 010 stereographic projection (Fig. 2.15(b)) be rotated around [100] axis to the 001 stereographic projection. In this operation, pole A is also rotated 90° along AA′ to A′ as shown in Fig. 2.15(d). In other words, pole A on the 010 stereographic projection is now indicated by A′ on the 001 stereographic projection (Fig. 2.15(d)). Then, it is easy to determine P, which is perpendicular to both A′ and B.

Determining plane inside a crystal from traces on many surfaces

Although a plane inside a crystal can be determined from their traces on at least two surfaces, it is needless to say that if the plane is determined from many more traces, the accuracy in the determination is improved. Suppose that a parallelepiped of crystal is shaped into poly-piped crystal covered with surfaces ***a***, ***b***, ***c***, ***d*** and ***e***, as shown in Fig. 2.16(a) to which an operation similar to the foregoing section is applied. Normally traces are observed under an optical microscope. When the crystal is rotated by 66° anticlockwise around the axis of column (in this case [100]), plane ***d*** (*021*) comes right beneath the microscope. The trace D makes an angle $\beta = 36°$ with respect to the rotation angle [100] (running to the lower right), so that it is plotted as D on the 021 stereographic projection, as shown in Fig. 2.16(b). In order to rotate this 021 stereographic projection back to the 001 stereographic projection, it should be rotated clockwise by 66°. Needless to say, pole D is also rotated 66° along a latitude of 36° back to D′. In summary,

Theorem 2.6:
Suppose that when a crystal is rotated by $\alpha°$ anticlockwise, trace D is inclined by $\beta°$ with respect to the rotation axis, then pole of trace D is located at D′ on the original (initial) stereographic projection with $\alpha = 0°$ (Fig. 2.16(c)).

The result is such that traces (directions) of P on surfaces ***a***, ***b***, ***c***, ***d*** and ***e*** are expressed by poles A, B. C, D and E in Fig. 2.16(d), so that plane P is located at pole P.

Exercise 2.14: When a single crystal is deformed plastically in tension or compression, the crystal slides (or slips) preferentially on a crystallographic plane (see Section 3.1). This plane is referred to as a slip plane. Slip is the elementary process of plastic deformation of crystal, so that its determination is a very important operation in crystal plasticity. Figure 2.17 shows slip lines on a single crystal deformed in tension. Figure 2.18 shows a tensile axis of a columnar single crystal. Around the surface two sets of slip lines (① and ②) are observed. The angular information of the respective slip lines is

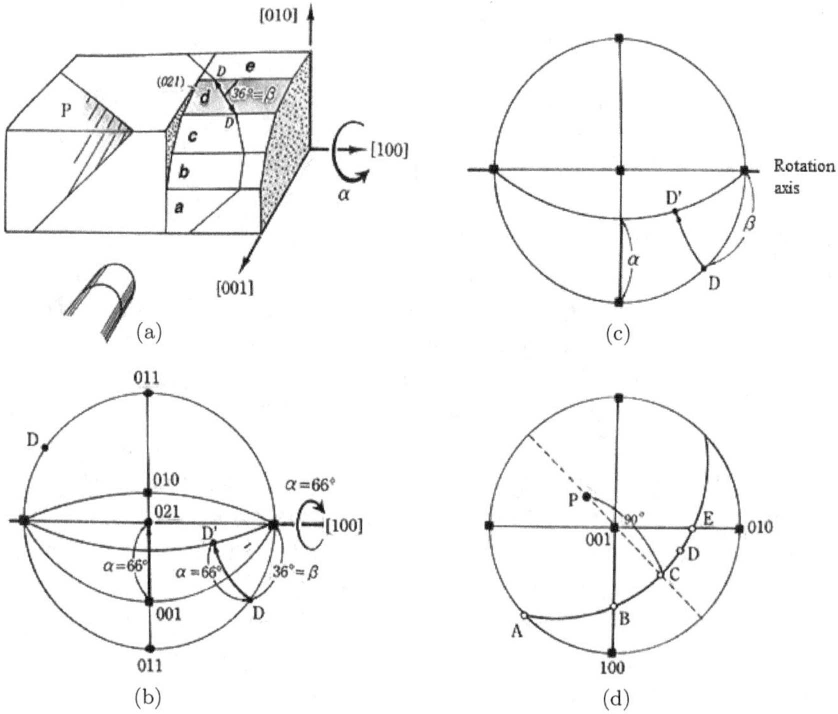

Fig. 2.16: How to determine a plane inside a crystal from many traces on the cylindrical surface.

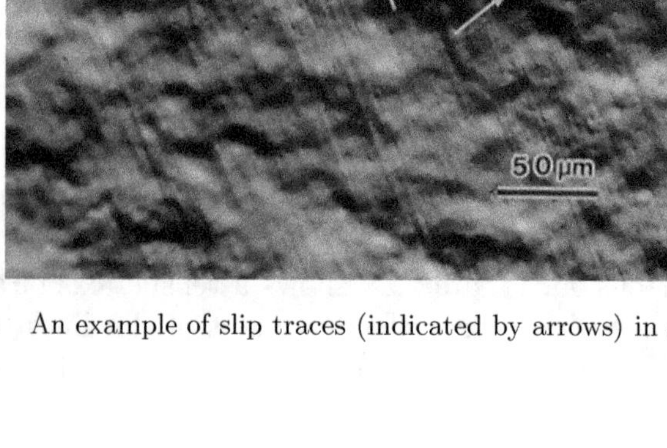

Fig. 2.17: An example of slip traces (indicated by arrows) in a β-CuZn.

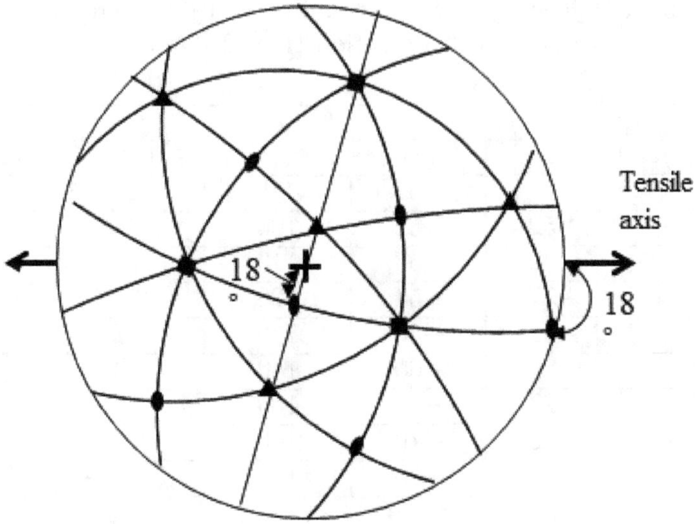

(a)

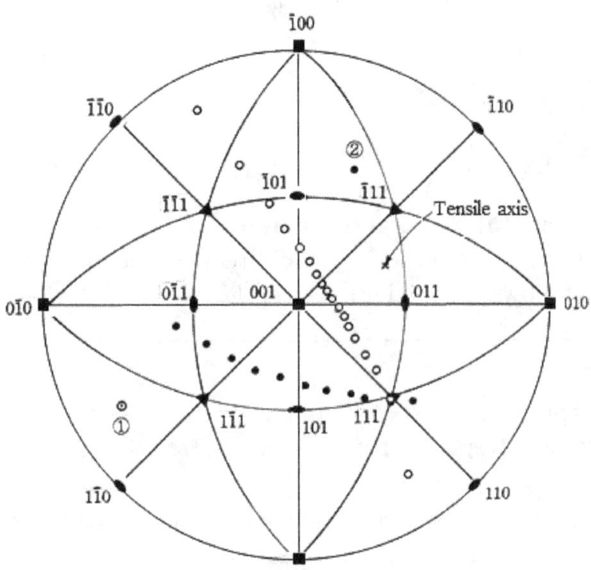

(b)

Fig. 2.18: Exercise 2.14.

shown in the following table. Determine planes ① and ②, and plot them on the 001 standard projection (Fig. 2.18(b)).

α (°)	β (°)		α (°)	β (°)	
	(1)	(2)		(1)	(2)
0	−74	−52	90	−26	−82
10	−52	−52	100	−27	−90
20	−42	−53	110	−29	84
30	−37	−54	120	−31	n.d.
40	−32	−56	130	−37	n.d.
50	−29	−60	140	−43	n.d.
60	−27	−64	150	−52	n.d.
70	−26	−70	160	−68	n.d.
80	−26	−76	170	−86	n.d.

Note: Refer to Figs. 2.16(a)–2.16(c) for α and β. Minus (–) means that the trace points to the lower half and n.d. stands for "not determined".

2.3 Reciprocal Lattice

Reciprocal vector

In defining the crystal globe and its stereographic projection, crystallographic planes are denoted by the directions of their perpendiculars. In what follows let us define a vector $\mathbf{g}^*_{hkl}$ as follows.

Definition 2.4:

(1) $\mathbf{g}^*_{hkl}$ is perpendicular to (hkl) plane.
(2) The length is inversely proportional to its lattice spacing.

$$|\mathbf{g}^*_{hkl}| = \frac{1}{d_{hkl}}, \text{ where } d_{hkl} \text{ is spacing of } (hkl) \text{ plane.}$$

Such a vector is referred to as a reciprocal vector. $|\mathbf{g}^*_{hkl}|$ has the dimension of $[L^{-1}]$ (L: dimension standing for length) and this imaginary space with dimension of $[L^{-1}]$ is referred to as a reciprocal space. By contrast the space with [L], where we live, is referred to as a real space.

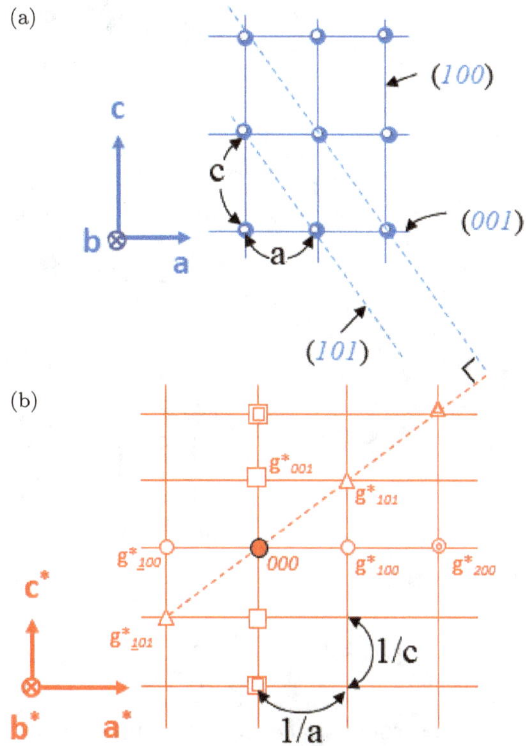

Fig. 2.19: Lattice in real space projected two-dimensionally (a) and its reciprocal lattice (b). **b** and **b*** are out of the plane of paper.

Problem. Figure 2.19(a) illustrates a two-dimensional projection of a three-dimensional crystal (in real space). Here, let **b** be out of the paper. Construct the reciprocal lattice of this crystal.

Solution. The reciprocal vector $\mathbf{g}^*_{100}$ of (100) plane in (a) is a vector drawn from the origin (●) to ○ in (b). If a is the lattice spacing of (100) plane, $1/|\mathbf{g}^*_{100}| = 1/a$. Similarly, the reciprocal vector $\mathbf{g}^*_{001}$ of (001) plane in (a) is a vector drawn from the origin to □. The plane shown by a broken line in (a) is (101) plane and its reciprocal vector $\mathbf{g}^*_{101}$ (indicated by △) is the sum of $\mathbf{g}^*_{100}$ and $\mathbf{g}^*_{101}$.

Exercise 2.15: Construct the reciprocal lattice of a crystal shown in Fig. 2.20, where **b** is out of the paper.

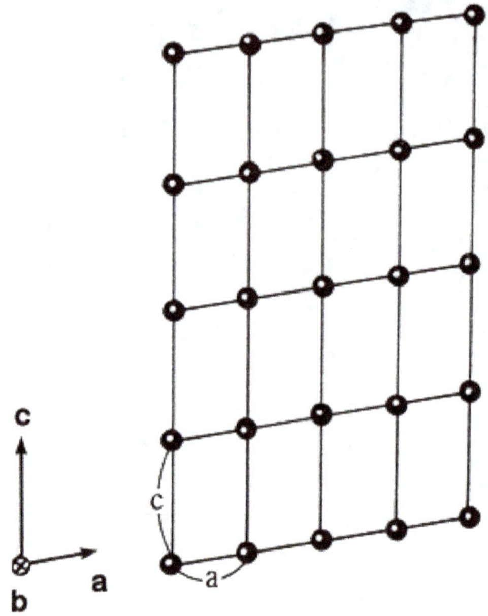

Fig. 2.20: Exercise 2.15.

A vector in the reciprocal space $\mathbf{g}^*_{hkl}$ is defined by the fundamental vectors in the reciprocal space ($\mathbf{a}^*$, $\mathbf{b}^*$, $\mathbf{c}^*$), which are different from the fundamental vectors in the real space (a, b, c). In other words,

$$\mathbf{g}^*_{hkl} = h\mathbf{a}^* + k\mathbf{b}^* + l\mathbf{c}^* \tag{2.4}$$

where ($\mathbf{a}^*, \mathbf{b}^*, \mathbf{c}^*$) are referred to as the fundamental vectors of reciprocal space.

> **Definition 2.5:**
> **a**, **b**, **c**: fundamental vectors in the real space,
> **a***, **b***, **c***: fundamental vectors in the reciprocal space.

Here,

$$\mathbf{a}^* = \frac{[\mathrm{b} \times \mathrm{c}]}{\mathrm{V}}, \quad \mathbf{b}^* = \frac{[\mathrm{c} \times \mathrm{a}]}{\mathrm{V}}, \quad \mathbf{c}^* = \frac{[\mathrm{a} \times \mathrm{b}]}{\mathrm{V}} \tag{2.5}$$

where $\mathrm{V} = (\mathbf{a} \cdot [\mathbf{b} \times \mathbf{c}])$ is the volume of the unit cell (Fig. 2.21).

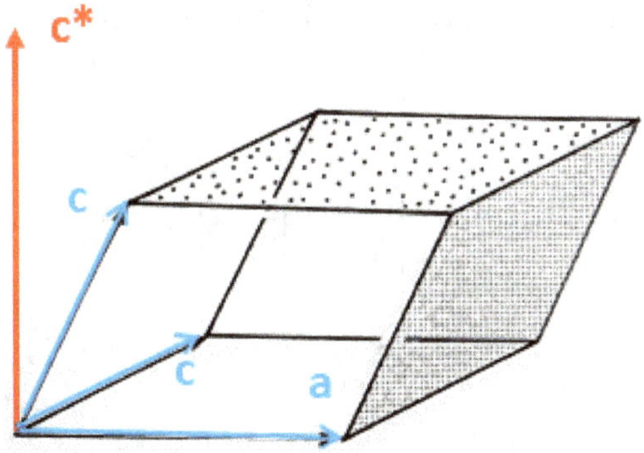

Fig. 2.21: $V = (\mathbf{a} \cdot [\mathbf{b} \times \mathbf{c}])$.

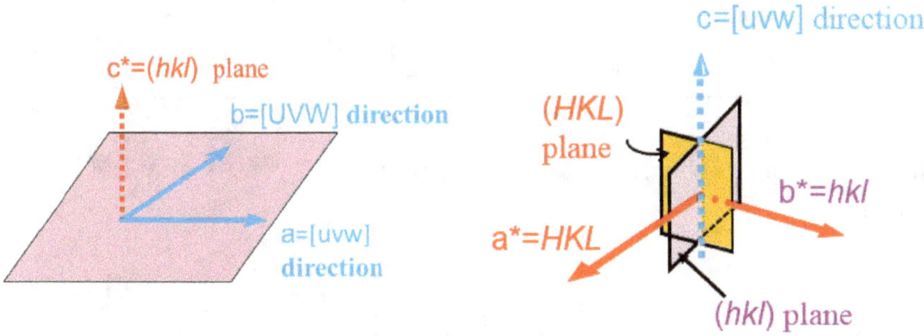

Fig. 2.22: Relation between the real space and the reciprocal space. Vectors in the real space stand for directions, while vectors in the reciprocal space (reciprocal vectors) stand for perpendiculars to planes. Perpendicular to the plane determined by vectors **a** and **b** in real space is $\mathbf{c}^*$ in reciprocal space. Direction common to the crossing planes in the real space **c** is a zone axis.

The following relationships hold.

$$\mathrm{a}^* \cdot \mathrm{b} = \mathrm{a}^* \cdot \mathrm{c} = \mathrm{b}^* \cdot \mathrm{a} = \mathrm{b}^* \cdot \mathrm{c} = \mathrm{c}^* \cdot \mathrm{a} = \mathrm{c}^* \cdot \mathrm{b} = 0 \tag{2.6}$$

$$\mathrm{a}^* \cdot \mathrm{a} = \mathrm{b}^* \cdot \mathrm{b} = \mathrm{c}^* \cdot \mathrm{c} = 1. \tag{2.7}$$

Let us apply the concept of reciprocal lattice to Fig. 2.12. If we pick up only a and b in Fig. 2.21, the result is as shown in Fig. 2.22. From this it is evident that a reciprocal vector c^* is normal of the plane which is defined by two vectors a[uvw] and b[UVW] in the real space (shaded in Fig. 2.22(a)). According to Eq. (2.5),

$$\mathbf{c}^* = \frac{[\mathbf{a}\times\mathbf{b}]}{V} = [\mathbf{a}\times\mathbf{b}]/\mathbf{c}\cdot[\mathbf{a}\times\mathbf{b}]$$

$$= \begin{vmatrix} h & k & l \\ U & V & W \\ u & v & w \end{vmatrix} = k\begin{vmatrix} V & W \\ v & w \end{vmatrix} - k\begin{vmatrix} U & W \\ u & w \end{vmatrix} + l\begin{vmatrix} U & V \\ u & v \end{vmatrix}$$

$$= h(Vw - vW) - k(Uw - uW) + l(Uv - uV).$$

That is,

$$h = Vw - vW, \quad k = uW - Uw, \quad l = Uv - uV$$

in keeping with Eq. (2.3).

Inversely, since

$$\mathbf{a} = \frac{[\mathbf{b}* \times \mathbf{c}*]}{V*}, \quad \mathbf{b} = \frac{[\mathbf{c}* \times \mathbf{a}*]}{V*}, \quad \mathbf{c} = \frac{[\mathbf{a}* \times \mathbf{b}*]}{V*},$$

the direction $\mathbf{c}$ which contains two planes $\mathbf{a}^*(HKL)$ and $\mathbf{b}^*(hkl)$ is given by

$$\mathbf{c} = \begin{vmatrix} u & v & w \\ H & K & L \\ h & k & l \end{vmatrix} = u\begin{vmatrix} K & L \\ k & l \end{vmatrix} - v\begin{vmatrix} H & L \\ h & l \end{vmatrix} + w\begin{vmatrix} H & K \\ h & k \end{vmatrix}$$

$$u = Kl - kL, \quad v = hL - Hl, \quad w = Hk - hK.$$

Exercise 2.16: Prove that $\mathbf{g}^*_{hkl}$, which is defined by Eqs. (2.4) and (2.5), is (1) perpendicular to plane (hkl) and that (2) its length is $1/d_{hkl}$.

(*Hint*) See Fig. 2.23. (1) Prove $\left(\frac{\mathbf{b}}{k} - \frac{\mathbf{a}}{h}\right)\cdot\mathbf{g}^*_{\mathrm{hkl}} = 0$. (2) Use the relation $d_{hkl} = \mathbf{n}^*\frac{\mathbf{a}}{h}$, where n^* is the unit vector of $\mathbf{g}^*_{hkl}$.

Exercise 2.17: Prove Exercise 1.4 (Chapter 1) using the reciprocal lattice.

Exercise 2.18: Prove that the reciprocal lattice of a simple cubic lattice is a simple cubic lattice, the reciprocal lattice of a simple tetragonal lattice is a simple tetragonal lattice and the reciprocal lattice of a hexagonal lattice is a hexagonal lattice.

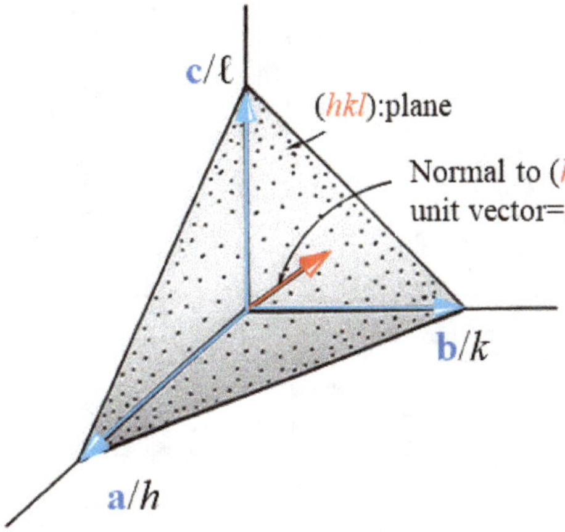

Fig. 2.23: Exercise 2.16.

Theorem 2.7:
The reciprocal lattice of *FCC* is *BCC*.
The reciprocal lattice of *BCC* is *FCC*.

Exercise 2.19: Prove Theorem 2.7.

(*Hint*) The fundamental unit cell of FCC is expressed by **a**=1/2[110], **b** = 1/2[011] and **c** = 1/2[101]. The fundamental unit cell of BCC is expressed by **a** = 1/2[11$\bar{1}$], **b** = 1/2[$\bar{1}$11] and **c** = 1/2[111] (see Exercise 1.1).

Calculating lattice spacing d_{hkl}

Spacings of lattice planes can be calculated using reciprocal lattices. Appendix F summarizes lattice spacings of different unit cells.

Exercise 2.20: Prove Appendix F.

Chapter 3

Crystal Dislocations

3.1 Definition of Dislocation

So far, it has been assumed that a crystal is faultless and misarrangement of atoms (referred to as a lattice defect) does not exist. However, in real crystals a variety of crystal lattice defects are present. Since this book is intended to analyze lattice defects by electron microscopy, we will discuss briefly the fundamentals of dislocations, which is the most important lattice defect.

A crystal is deformed when an external force exerts on it. Figure 3.1 shows a stress–strain curve when a single crystal is deformed in tension. In tension and compression, stress and strain are defined as follows.

Definition 3.1:
Stress = external force/cross-sectional area of a specimen: dimension = $[F][L^{-2}]$.
Strain = change in length/initial length: dimensionless = $[L^0]$.

Deformation of a crystal is categorized into two; elastic deformation and plastic deformation. In elastic deformation Hooke's law holds where stress is proportional to strain and on the removal of external stress the crystal returns to its original (initial) shape. By contrast in plastic deformation, the crystal does not return to the original (initial) shape after the external stress is removed, producing a permanent strain. Now let us consider this process on atomic level.

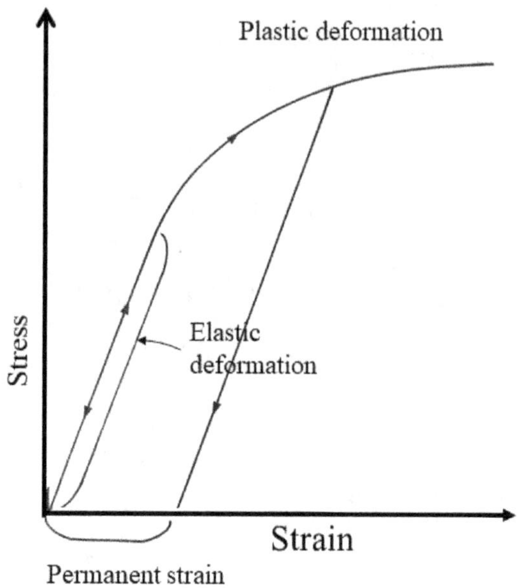

Fig. 3.1: Stress (= external force/cross-sectional area of sample) versus strain (= change in length/original length) curve.

For the sake of simplicity let us consider a single crystal as schematically illustrated in Fig. 3.2(a). In elastic deformation, the vertical distance between atoms is slightly elongated as compared to before deformation (Fig. 3.2(b)). On further increasing an external force, the upper crystal on a (crystallographic) plane tilted by approximately 45° around the tensile direction slides over the lower (Fig. 3.2(c)). This is plastic deformation. The stress required for the slide (or slip) to take place simultaneously over the slip plane is calculated to be $10^{-1}\mu \sim 3 \times 10^{-2}\mu$ (μ is the shear modulus[1] and for steel $\sim 8 \times 10^4$ MPa[2]), while the experimental value is 10^{-3}–$10^{-4}\mu$, much smaller than the theoretical value.

The mechanism by which this gap between theory and experiment is bridged is a dislocation. As shown in Fig. 3.3, the slip does not take

---

[1]One of the proportional constants of Hooke's law. Dimension is $[F][L^{-2}]$.

[2]Unit of pressure (and hence stress). $1\,\text{MPa} = 10^6\,\text{Pa} = 10^6\,\text{N/m}^2 = 0.10197$ kg/mm^2.

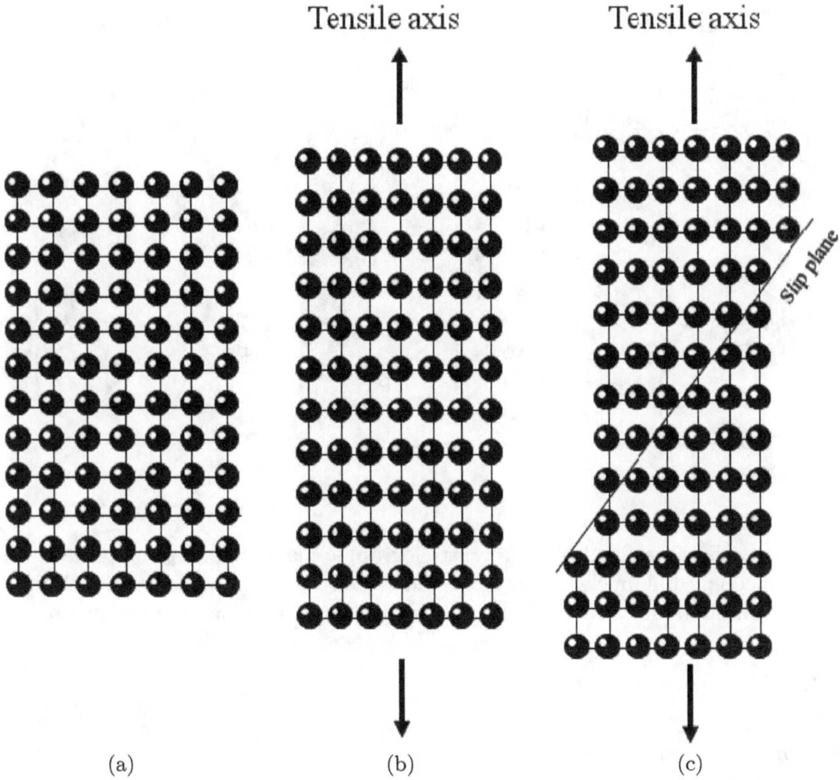

Fig. 3.2: Behavior of atoms in an extended crystal. (a) Before extension, (b) elastic deformation, (c) plastic deformation.

place simultaneously, but initiates at some region of the crystal and extends its region. The stress required for this to take place is much smaller than the aforementioned theoretical value. The boundary between the slipped and the un-slipped regions is a dislocation.

Definition 3.2:
A dislocation is a boundary between slipped and un-slipped regions.

The boundary is a linear defect, so that it is referred to as a dislocation line, or simply a dislocation. The dislocation must close as a loop inside a crystal or must emerge at the crystal surfaces as shown in Fig. 3.3. Such a loop is referred to as a dislocation loop.

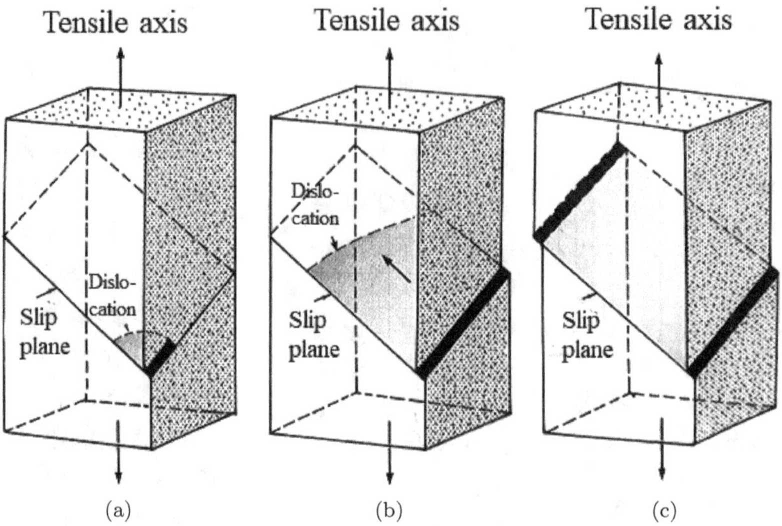

Fig. 3.3: Slip above and below a crystallographic plane. Slip does not take place simultaneously but nucleates at a limited region and then expands the slipped region. The boundary between slipped and un-slipped regions is a dislocation.

3.2 Burgers Vector

Burgers circuit

Slip is specified by length and direction: It is a vector. This vector is referred to as Burgers vector and denoted by **b** (The strength or length is denoted by b). The Burgers vector is defined by a Burgers circuit.

Definition 3.3:

(1) Define a unit vector ζ to assign the sense of direction of the dislocation under investigation.[3] ζ is either +1 or 1. Which of the two is chosen is arbitrary.
(2) Straddle the dislocation line looking toward the direction ζ, and draw a clockwise closed circuit C_1 (Fig. 3.4(a)).

[3]In order to distinguish between arrows indicating the direction of dislocation line ζ and Burgers vector itself **b**, ζ is denoted by a double headed arrow ($\twoheadrightarrow$) and **b** is denoted by a single headed arrow ($\rightarrow$).

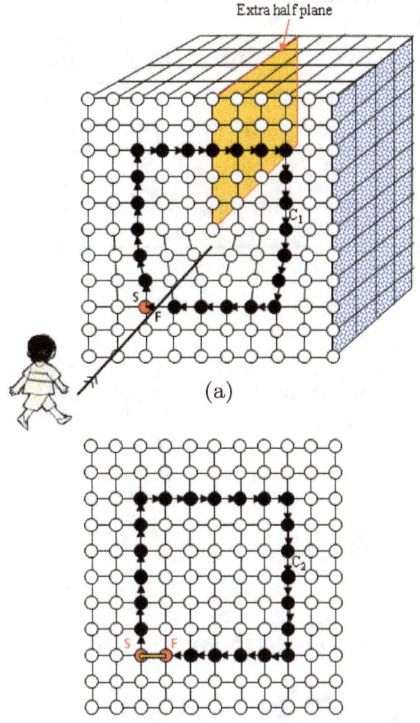

Fig. 3.4: Definition of Burgers vector **b** by Burgers circuit (FS/RH(Perfect) convention).

(3) Do the same operation in a perfect crystal. This circuit C_2 is not closed (Fig. 3.4(b)).
(4) To close the C_2 draw an additional vector from the finishing point (**F**) to the starting point (**S**) of the circuit C_2, $\underset{\mathbf{FS}}{\longrightarrow}$. The vector $\underset{\mathbf{FS}}{\longrightarrow}$ is the Burgers vector.

This method is referred to as FS/RH(Perfect)[4] convention.

[4]If one of FS, RH, Perfect is reversed to SF, LH (left hand), Imperfect (imperfect crystal containing a dislocation), then the sign of **b** is reversed. In standard textbooks on dislocation the definition of **b** is different, so that readers should be careful. For instance, Hirth and Lothe (1968) employs FS/RH (Perfect), while Friedel (1964) employs (Imperfect).

Theorem 3.1:
The first thing to do in determining Burgers vector is to define ζ. It is arbitrary to assign the sign (+or−) to ζ. It is noted that reversing the sign of ζ results in reversing the sign of **b**.

Exercise 3.1: Confirm Theorem 3.1 by changing ζ in Fig. 3.4.

Theorem 3.2:
In FS/RH(Perfect) convention,

(1) looking toward ζ,
(2) viewing the slipped region at the right hand, and
(3) fixing the lower half crystal,

then the upper half crystal is displaced by **b**.

Exercise 3.2: Prove Theorem 3.2.

Edge dislocation, screw dislocation and mixed dislocation

When the Burgers vector **b** is perpendicular to a dislocation line, the dislocation is referred to as an edge dislocation: When the Burgers vector **b** is parallel to the dislocation line, it is referred to as a screw dislocation. The dislocation shown in Fig. 3.4 is an edge dislocation. An edge dislocation has an extra half plane. By contrast

Fig. 3.5: Screw dislocation.

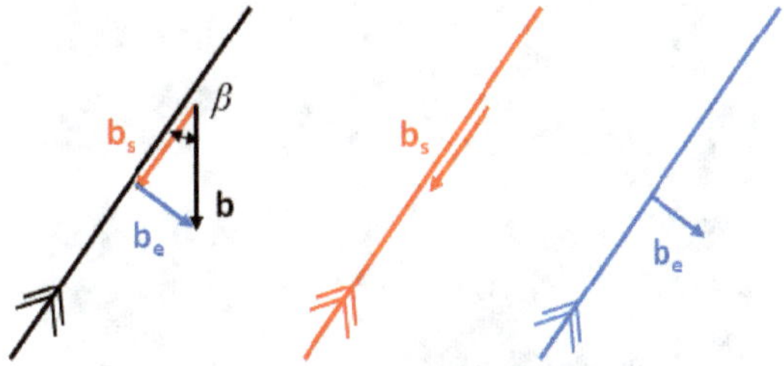

Fig. 3.6: Mixed dislocation = Screw dislocation + Edge dislocation.

a screw dislocation, where $\mathbf{b}//\zeta$, does not have an extra plane. As shown in Fig. 3.5, if you move around along a Burgers circuit, you move forward or backward by **b** as if tightening a screw. Appendix G illustrates a model which shows displacement of lattice planes around a screw dislocation.

When **b** is inclined by an arbitrary angle β ($\beta \neq 0$) with respect to ζ, the dislocation is referred to as a mixed dislocation (Fig. 3.6). The Burgers vector of a mixed dislocation is resolved into $\mathbf{b}_e$ (edge component) and $\mathbf{b}_s$ (screw component). It can be considered that two dislocations having $\mathbf{b}_e$ and $\mathbf{b}_s$ are present independently.

$$\mathbf{b}_e = \mathbf{b}\sin\beta \tag{3.1}$$

$$\mathbf{b}_s = \mathbf{b}\cos\beta \tag{3.1$'$}$$

Kirchhoff's law

In a real crystal, dislocations often compose a network. Figure 3.7 shows an example of dislocation network. Burgers vectors (**b**'s) of dislocations making up a node follow the following rule.

If all of the directions of dislocation lines (ζ_i) point to the same direction as shown in Fig. 3.8,

$$\mathbf{b}_1 = \mathbf{b}_2 + \mathbf{b}_3. \tag{3.2}$$

Theorem 3.3:
When all of the ζ_i are taken as flowing away from (or flowing into) the node

$$\sum_i^N \mathbf{b}_i = 0 \tag{3.2$'$}$$

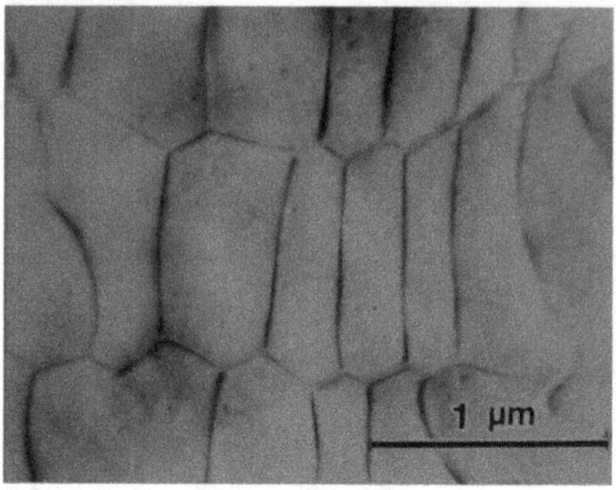

Fig. 3.7: A dislocation network (by courtesy of Dr. Enokido).

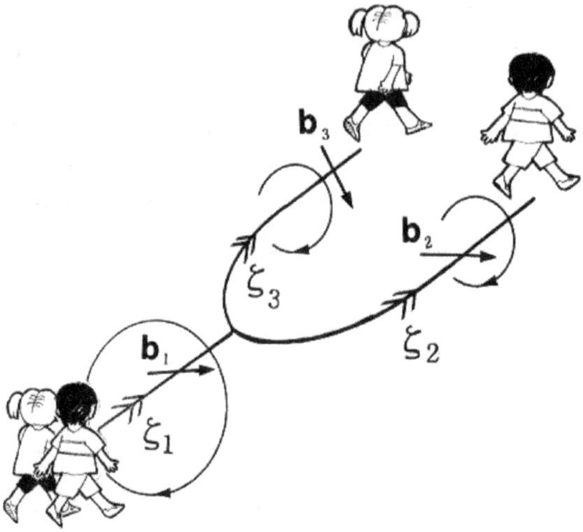

Fig. 3.8: Kirchhoff's law.

Since this rule is in a close analogy with the Kirchhoff's law in electrical circuit, it is referred to as Kirchhoff's law.

Exercise 3.3: Determine **b** of a screw dislocation depicted in Fig. 3.5. Show that when $\mathbf{b}\cdot\zeta = \mathrm{b}$, the dislocation is of a right-handed screw, while when $\mathbf{b}\cdot\zeta = -\mathrm{b}$, the dislocation is of a left-handed screw.

(*Hint*) See Appendix G.

In a crystal some atoms which should occupy their normal sites are missing or some extra atoms are tightly squeezed in confined spaces between atoms on normal sites. The former is referred to as a vacancy (V) and the latter as an interstitial atom (I). They are generically referred to as point defects. These point defects are present in a crystal as an individual point defect or as a cluster. Especially point defects with the identical nature (V or I) can form a disk with thickness of a single atom. When interstitials form such a disk (I-type), one atomic layer is added into the matrix (Fig. 3.9(a)). In the case of vacancies (V-type), one atomic layer is removed from the matrix (Fig. 3.9(b)). The periphery of such a disk is an edge

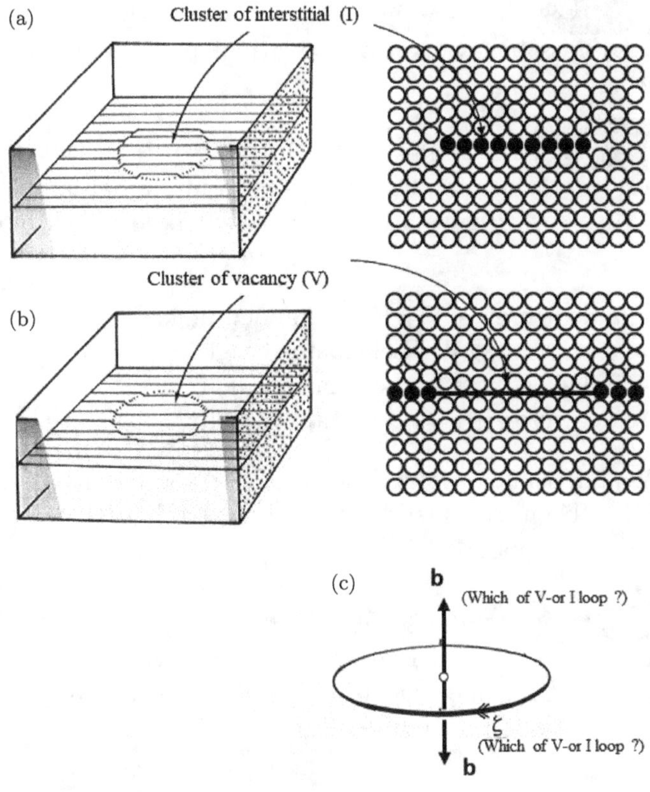

Fig. 3.9: (a) Interstitial (I)-type dislocation loop. (b) Vacancy (V)-type dislocation loop. Atomic planes indicated by ● are extra-half planes of edge dislocations. When ζ is taken as shown in (c), Burgers vectors **b** of I-type loop and V-type loop point either upward or downward.

dislocation: The dislocation forms a closed loop. Diagrams at right sides in Figs. 3.9(a) and 3.9(b) illustrate cross-sections of I-type and V-type dislocation loop, respectively. The Burgers vector **b** is perpendicular to the plane on which the loop lies.[5]

Exercise 3.4: Relationship between ζ and **b** for I-type and V-type loops is either **b** is upward or downward (see Fig. 3.9(c)). Determine which is I-type or V-type.

Exercise 3.5: When two dislocations with opposite signs meet, they annihilate. Prove this using Kirchhoff's law.

Exercise 3.6: Suppose that point defects react with an edge dislocation. Discuss behavior of an extra half plane under such a circumstance.

Exercise 3.7: When point defects react with a screw dislocation, the screw dislocation changes its form into a helical dislocation (helix). Prove this. Figure 3.10 shows an example of helical dislocations.

- Right-handed screw + V = Right-handed helix
- Left-handed screw + V = Left-handed helix
- Right-handed screw + I = Left-handed helix
- Left-handed screw + I = Right-handed helix

(*Hint*) Assume that a screw dislocation is locally bent as shown in Fig. 3.10(b). Then, consider whether the extra half is located above or below the slip plane.

Exercise 3.8: Assuming that instead of point defect itself a dislocation loop as shown in Exercise 3.4 reacts with a screw dislocation, show that the same conclusion is drawn (see Fig. 3.10(c) where, $\mathbf{b}_1 = -\mathbf{b}_2$).

[5]This type of a dislocation loop is referred to as a prismatic loop. By contrast, when **b** is in the plane of the dislocation loop, the loop is referred to as a glide dislocation loop.

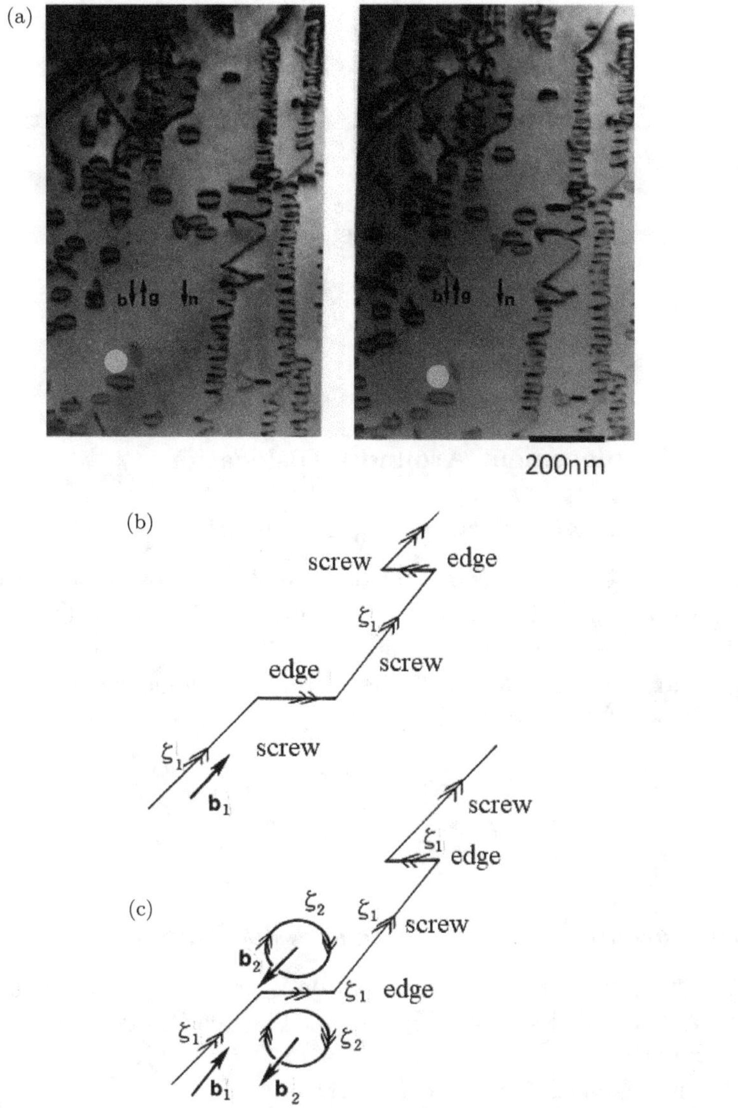

Fig. 3.10: (a) A pair of stereo-micrographs of helical dislocations. Place a partition between the two micrographs and see the right-sided picture with right eye and the left-sided picture wit left eye. The images appear in 3D (by courtesy of Dr. Nakada). (b) An edge segment formed on a bent screw dislocation absorbs point defects, leading to formation of a helix. (c) An edge segment on a bent screw dislocation reacts with a dislocation loop as shown in Fig. 3.9(c), leading to the same configuration as (b). The configuration of the helix depends on (1) whether the screw dislocation is right-handed or left-handed, and (2) whether the dislocation loop is I-type or V-type.

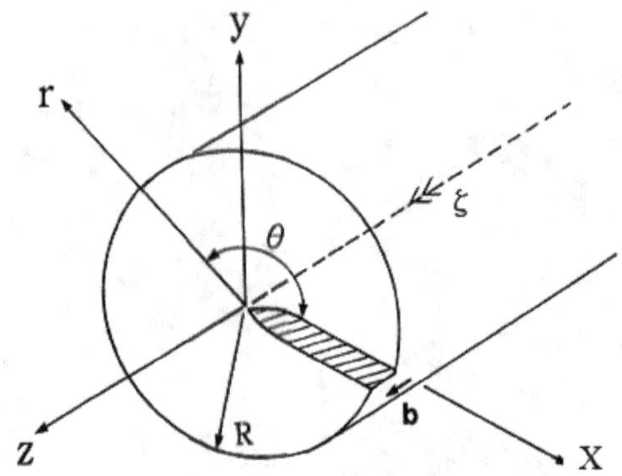

Fig. 3.11: Displacement around a screw dislocation.

3.3 Displacement Around a Dislocation

Displacement around a screw dislocation

Figure 3.11 shows a screw dislocation sheared by b along the z-axis across $x - y$ plane. Cylindrical coordinate is useful. Going around the z-axis anticlockwise completely ($\theta = 2\pi$), the displacement in the direction of z-axis is b. Therefore, the displacement at a rotation angle of θ from the x-axis is given by

$$u_z(r, \theta) = \frac{b\theta}{2\pi} = \frac{b}{2\pi}\tan^{-1}\frac{y}{x} \tag{3.3}$$

$$u_x = u_y = 0.$$

Displacement around an edge dislocation

The situation for an edge dislocation is much more complicated than for a screw dislocation (Fig. 3.12). The readers refer to Hirth and Lothe (1968).

The displacements are given by

$$u_x = \frac{b}{2\pi}\left[\tan^{-1}\frac{y}{x} + \frac{xy}{2(1-\nu)(x^2+y^2)}\right] \tag{3.4}$$

$$u_y = -\frac{b}{2\pi}\left[\frac{1-2\nu}{4(1-\nu)}\ln(x^2+y^2) + \frac{x^2-y^2}{4(1-\nu)(x^2+y^2)}\right] \tag{3.4'}$$

where ν is the Poisson ratio.

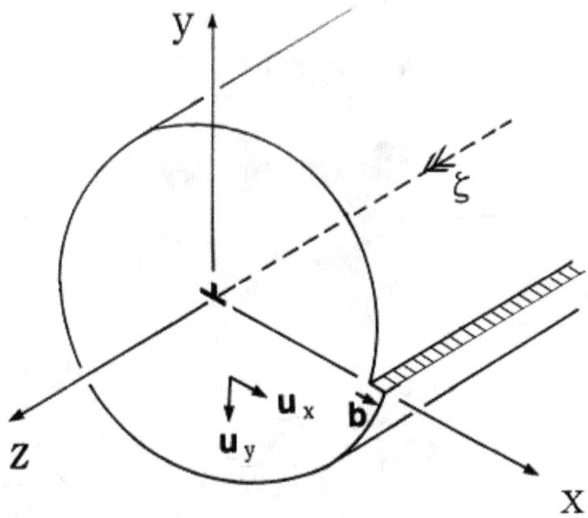

Fig. 3.12: Displacement around an edge dislocation. ⊥ indicates an edge dislocation. Note the direction of $\mathbf{u}_y$.

3.4 Dissociated Dislocations in FCC

Thompson tetrahedron

As shown in Fig. 3.13(a) four $\{111\}$ planes are present in the unit cell of FCC structure, that is, DAB, DAC, DBC and ABC. They are labeled (c), (b), (a) and (d) on the outside surface, respectively. These four planes (a), (b), (c) and (d) planes make up a tetrahedron ABCD, which is referred to as Thompson tetrahedron. Figure 3.13(b) shows a Thompson tetrahedron opened-up at corner D. Details of the Thompson tetrahedron are shown in Appendix H. Readers are recommended to build the Thompson tetrahedron and master how to use it.

Planes $\{111\}$ are closed packed planes where atoms are packed together most tightly. Among them, (d) plane (ΔABC) has a Miller index (111). Viewing this plane along [111] direction towards the origin (Fig. 3.14(a)), there are three possible (equivalent) positions of atoms indicated by ○, ◉ and ●. Let us label by ◉ ***b***, ○ by ***a***, and ● by ***c***. The stacking sequence of the (111) planes is, from the origin (D), ***abcabc*** (Fig. 3.14b)). (By contrast, the sequence ***ababab***, where ***c*** is missing, corresponds to HCP structure (cf. Exercise 1.10)).

Plastic deformation of a crystal results from motion of a dislocation. In FCC crystals, a dislocation moves on (111) plane. For instance, motion of a dislocation shifts an atom ***c*** in Fig. 3.14 to the

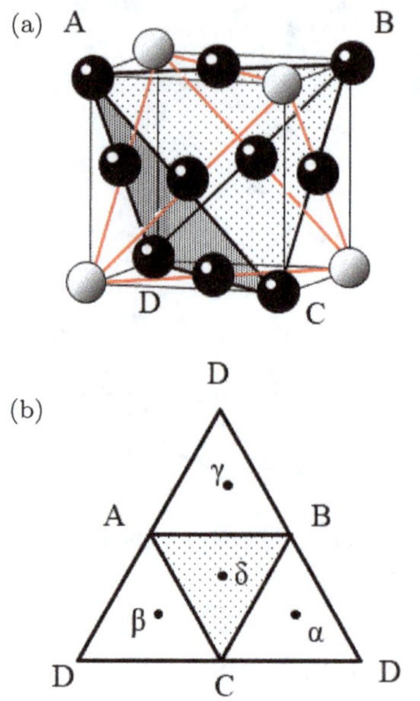

Fig. 3.13: (a) Thompson tetrahedron (ABCD) in FCC structure. There is another tetrahedron drawn in red. (b) Thompson tetrahedron opened-up at corner D.

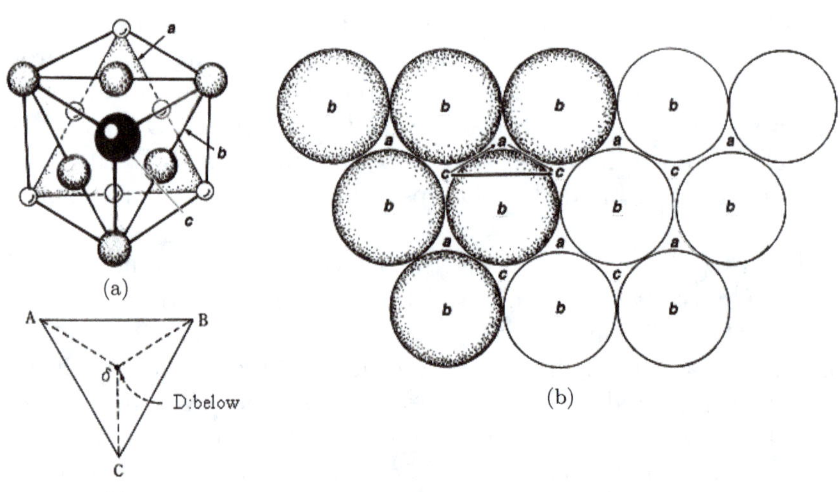

Fig. 3.14: Stacking of (*111*) planes in FCC crystal. The (*111*) planes are stacked in the order of ***abc***.... Atoms shaded in (b) correspond to atoms on plane ***b*** in (a). (c) shows the corresponding Thompson tetrahedron.

right-sided neighboring position of $\boldsymbol{c}$ on the plane of $\boldsymbol{b}$. (A dislocation is defined as a boundary between slipped and un-slipped regions.)

Now let us suppose that $\boldsymbol{c} \to \boldsymbol{c}$ takes place. In this process the atom $\boldsymbol{c}$ must travel near position $\boldsymbol{a}$ on the way. The position of $\boldsymbol{a}$ is lower. In other words, $\boldsymbol{c}$ crosses the hillside while viewing the valley on its left side. Thus, instead of moving directly $\boldsymbol{c} \to \boldsymbol{c}$, $\boldsymbol{c}$ may take a rest at $\boldsymbol{a}$ and then move to $\boldsymbol{c}$ stepwise like $\boldsymbol{c} \to \boldsymbol{a} \to \boldsymbol{c}$. That is, the first step $\boldsymbol{c} \to \boldsymbol{a}$ followed by the second step $\boldsymbol{a} \to \boldsymbol{c}$ completes the motion of $\boldsymbol{c} \to \boldsymbol{c}$ (Fig. 3.15(a)). Dislocations corresponding to $\boldsymbol{c} \to \boldsymbol{a}$ and $\boldsymbol{c} \to \boldsymbol{c}$ are referred to as Shockley's partial dislocations, or imperfect dislocations and the total dislocation $\boldsymbol{c} \to \boldsymbol{a} \to \boldsymbol{c}$ is referred to as an extended dislocation or a dissociated dislocation.

Figure 3.16 shows an example of dissociated dislocations in a Cu-13.43at.%Al alloy. A pair of nearly parallel partial dislocations (p_1, p_2) are observed. $\mathbf{b}_T$ shows the Burgers vector of

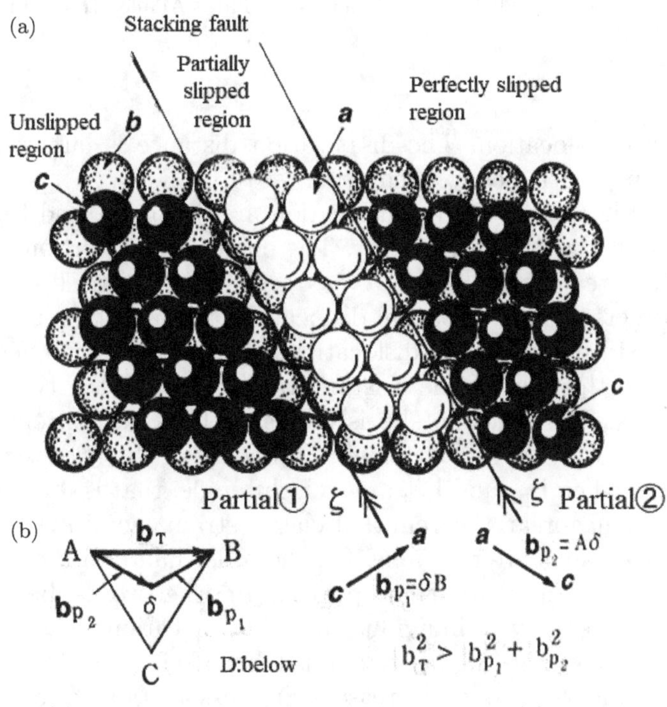

Fig. 3.15: (a) Structure of a dissociated dislocation. (b) The corresponding Thompson tetrahedron.

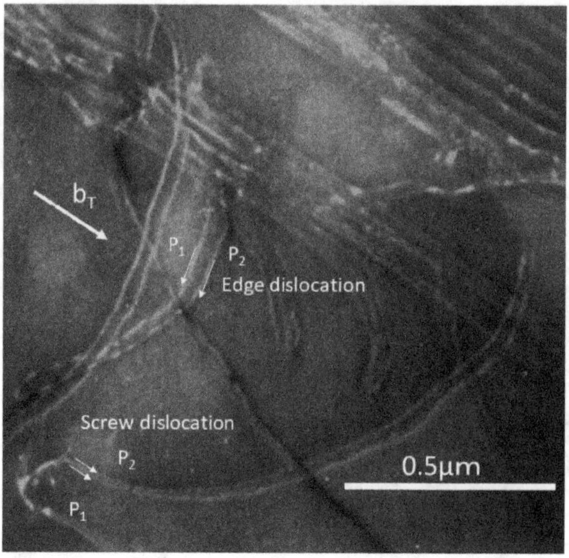

Fig. 3.16: Dissociated dislocations in a Cu-13.43 at %Al alloy (reproduced from Saka *et al.* (1978)).

the perfect dislocation. The dissociation distance is larger in edge orientation than in screw orientation (see Eq. (3.5)).

Crystal lattices around a dislocation are distorted, and therefore produces an elastic strain energy. The energy is proportional to b^2. In the two-step motion ($\boldsymbol{c} \to \boldsymbol{a} \to \boldsymbol{c}$), the amount of slip in either step (Burgers vectors of partial dislocations $\mathbf{b}_{\mathrm{p1}}$ and $\mathbf{b}_{\mathrm{p2}}$) is smaller. However, since two partial dislocations are involved, it is necessary to compare the sum of the elastic strain energy of the two partial dislocations. The two-step slip is indicated by a vector representation in Fig. 3.15(b): The elastic strain energy of the two-step slip is lower. It is when $\mathbf{b}_{\mathrm{p1}}$ and $\mathbf{b}_{\mathrm{p2}}$ are at right angle, that is, $\mathbf{b}_{\mathrm{p1}}$ and $\mathbf{b}_{\mathrm{p2}}$ satisfies Pythagorean theorem that elastic strain energies of one-step and two-step slips are the same. On the other hand, in the two-step slip two partial dislocations are present. In other words, besides the elastic strain energy of individual partial dislocations, elastic interaction energy of the two partials must be taken into account. The elastic interaction energy decreases with dissociation distance (Δ_{d}) of the two partials. From the viewpoint of elastic energy, the two-step slip $\boldsymbol{c} \to \boldsymbol{a} \to \boldsymbol{c}$ is favored the more, when Δ_{d} is the larger.

Stacking fault

In a dissociated dislocation, the region where the slip in the first step has already taken place, yet the slip in the second step has not taken place, that is, a region bounded by the two partial dislocations, atoms which should sit on ***c*** are now sitting on ***a***. In other words, the stacking of atoms is disturbed. This disturbance is referred to as the stacking fault (SF). Indeed, in this region the stacking sequence of atoms is represented by ***abcababc***..., where underlined portion assumes the stacking sequence of HCP. In other words,

> **Point 3.1.**
> Stacking fault in FCC is a very thin crystal with HCP structure embedded in the matrix of FCC.

Therefore, the energy of a stacking fault is higher than that of FCC matrix. This extra energy is referred to as the stacking-fault energy[6] (SFE, usually denoted by γ). From the viewpoint of SFE, the narrower dissociation distance has the lower energy. To summarize, the dissociation distance is determined by the sum of elastic strain energy of individual partial dislocations, elastic interaction energy between the two partials (denoted by **a** in Fig. 3.17) and SFE

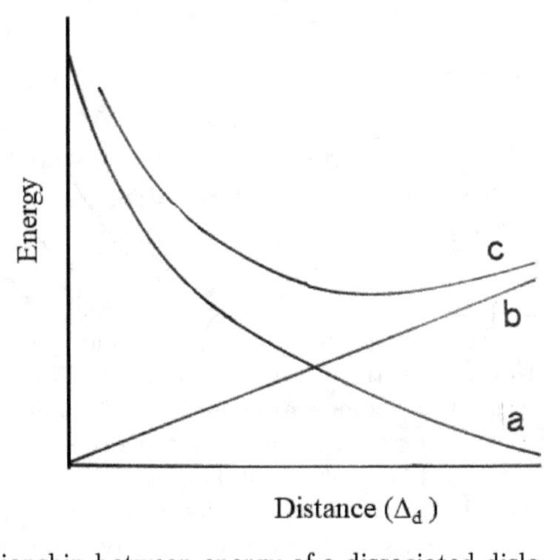

Fig. 3.17: Relationship between energy of a dissociated dislocation and the dissociation distance (Δ_d).

[6]The SFE is defined as an energy per unit area, the unit being J/m^2.

(denoted by **b** in Fig. 3.17). That is, the sum of **a** and **b** is a minimum at the equilibrium.

The dissociation distance Δ_d between the two partials at equilibrium is given by

$$\Delta_e = \frac{\mu b_p^2(2-\nu)}{8\pi\gamma(1-\nu)}\left(1 - \frac{2\nu\cos 2\beta}{2-\nu}\right) \tag{3.5}$$

where β is an angle between the total Burgers vector $\mathbf{b}_T$ and the dislocation line, b_p is the strength of the partial dislocations and ν is the Poisson's ratio ($\doteqdot 1/3$).

Burgers vectors of a dissociated dislocation

Let us suppose that atoms $\boldsymbol{c}$, which lie one layer above $\boldsymbol{a}$, move to the adjacent $\boldsymbol{c}$. This corresponds to gliding in the direction of **AB** on a Thompson tetrahedron. Therefore, the total Burgers vector is $\mathbf{b}_T = \mathbf{AB}$. We are looking down the Thompson tetrahedron from above.[7]

The dissociation mode in this case is represented using the Thompson tetrahedron by

$$\mathbf{AB} = \mathbf{A}\delta + \delta\mathbf{B} \tag{3.6}$$

The glide in the first step is ($\boldsymbol{c} \to \boldsymbol{a}$), as can be seen from Fig. 3.14(b). This direction corresponds to $\delta\mathbf{B}$ on the Thompson tetrahedron. Similarly, the glide in the second step ($\boldsymbol{a} \to \boldsymbol{c}$) corresponds to $\mathbf{A}\delta$. Taking the direction of the dislocation in Fig. 3.15(a) upward, the right side is the slipped region. In order for this slip

[7]Thompson tetrahedra with D pointing above and below are shown in Figs. 3.18(a) and 3.18(b), respectively. Suppose that you shrink to such an extent that you can enter inside the Thompson tetrahedron or climbing the outer wall. If you succeed in getting into the Thompson tetrahedron like a grave-robber of a pyramid, you are viewing {*111*} planes of Thompson tetrahedron from inside. A boy in black in Fig. 3.18(a) is standing on the inside of plane ABC, and a skier in black in Fig. 3.18(b) is skiing on the inside of plane ADC. Those who are unsuccessful in getting into the inside view {*111*} planes from the outside. A skier in white in Fig. 3.18(a) is skiing on the outside of plane ABD and a girl in white in Fig. 3.18(b) is standing on the outside of plane ABC and a climber is trying to move from outside of ADB to ABC.

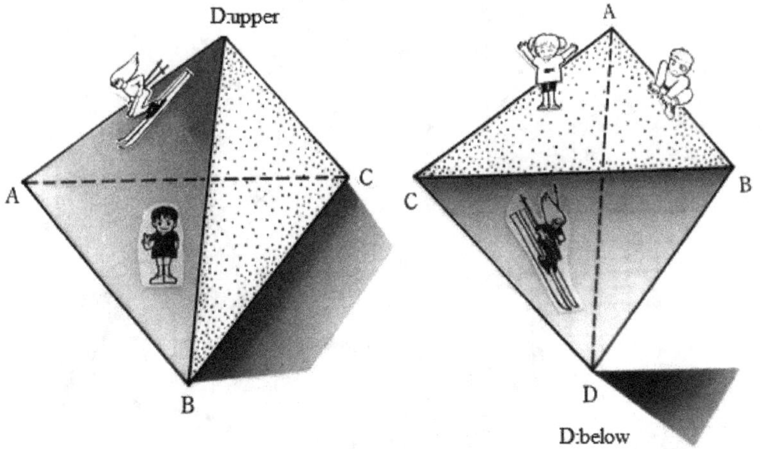

(a) (b)

Fig. 3.18: Viewing Thompson tetrahedron. (a) D: above and (b) D: below.

to proceed, the dislocation must move to the left. In other words, partial ① is the leading partial which corresponds to slip in the first step ($\delta\mathbf{B}$) and partial ② is the trailing partial which corresponds to slip in the second step ($\mathbf{A}\delta$). It follows that:

> **Theorem 3.4:** [8]
> When viewing a dissociated dislocation from outside of the Thompson tetrahedron, Greeks are outside along the dislocation line (Fig. 3.19(a)). When viewing a dissociated dislocation from inside of the Thompson tetrahedron, Greeks are inside along the dislocation line (Fig. 3.19(b)).

In Fig. 3.19(a), ΔABC of the Thompson tetrahedron is viewed from outside. In Fig. 3.19(b), it is rotated anticlockwise around an axis AB. Then, ΔABC of the Thompson tetrahedron is now viewed from inside. In Fig. 3.19(a), partial ② is located to the right of the viewer and partial ① to the left, while in Fig. 3.19(b), the reverse is true, that is, partial ① is located to the right and partial ① is to the left.

[8]This theorem holds when the SF is intrinsic. When the SF is extrinsic, the reverse is true (see Section 3.4.4).

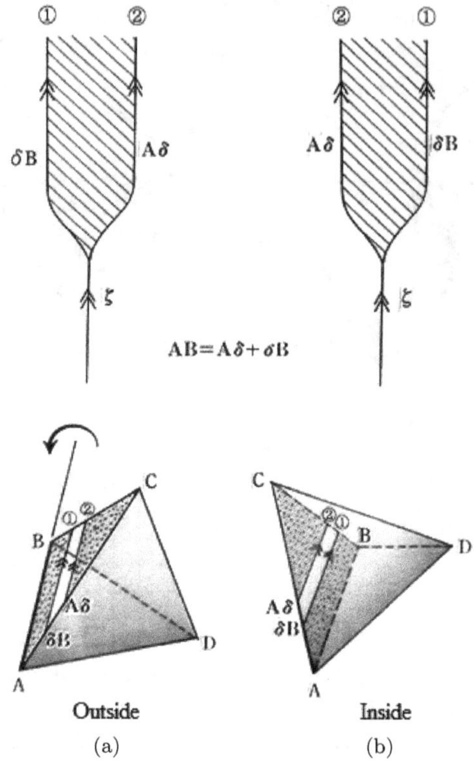

Fig. 3.19: (a) When a dissociated dislocation ($\mathbf{AB} = \mathbf{A}\delta + \delta\mathbf{B}$) is viewed from outside the Thompson tetrahedron, the Greek (in this case δ) is located outer side along the positive sense of the dislocation line (indicated by double-headed arrows ($\twoheadrightarrow$)). (b) However, when the same dislocation is viewed from inside, the Greek (δ) is located inside, where (b) is obtained by rotating (a) anticlockwise around AB.

Exercise 3.9: In Eq. (3.6), the direction of slip is represented by **AB**. The actual motion of atoms corresponding to this slip is $\boldsymbol{c} \rightarrow \boldsymbol{c}$, as shown in Fig. 3.14(b), which is represented by 1/2[110]. Represent Eq. (3.6) in this way using vector representation.

Frank's partial dislocation and intrinsic or extrinsic stacking fault

Partial dislocations described so far result from dissociation of a perfect (slip) dislocation, and expressed by a combination of Roman and Greek with different alphabet such as $\mathbf{A}\delta$ in the Thompson nomenclature. This type of a partial dislocation is referred to as a Shockley partial dislocation. The same stacking fault as that caused by a Shockley partial is formed by removing one layer of $\{111\}$ plane.

Among stacking of $\{111\}$ planes, $\boldsymbol{a} \to \boldsymbol{b}$, $\boldsymbol{b} \to \boldsymbol{c}$ and $\boldsymbol{c} \to \boldsymbol{a}$ are denoted by Δ and $\boldsymbol{a} \to \boldsymbol{c}$, $\boldsymbol{c} \to \boldsymbol{b}$ and $\boldsymbol{b} \to \boldsymbol{a}$ by ∇. This nomenclature is referred to as Frank's symbol. In this symbol, FCC structure ***abcabcabcabc*** is denoted by $\Delta\Delta\Delta\Delta\Delta\Delta\Delta$. Structure of HCP is ***abababab***, ***bcbcbcbc***, ***cacacacaca*** and all of them are denoted by $\Delta\nabla\Delta\nabla\Delta\nabla\Delta\nabla\Delta\nabla$. Now let us remove one layer of (111) plane, say $\boldsymbol{c}$, then the stacking sequence of (111) planes becomes ***abcab*** $\downarrow$ ***abcabc*** ($\boldsymbol{c}$ is missing at the position of $\downarrow$), which is denoted by $\Delta\Delta\Delta\nabla\Delta\Delta\Delta$ by Frank's symbol: The stacking fault is formed. Inversely, when one layer of (111) $\boldsymbol{c}$ is inserted, the stacking sequence becomes $\boldsymbol{abca\overset{\downarrow}{c}bca}$ (c is inserted at the position of $\downarrow$), which is expressed by $\Delta\Delta\Delta\nabla\nabla\Delta\Delta\Delta$ in Frank's symbol. The stacking sequence $\Delta\Delta\Delta\nabla\Delta\Delta\Delta$ containing one ∇ is referred to as an intrinsic stacking fault and the stacking sequence $\Delta\Delta\Delta\nabla\nabla\Delta\Delta\Delta$ containing two $\nabla\nabla$ is referred to as an extrinsic stacking fault.

Next, let us consider the case where the stacking sequence is disturbed by a Shockley partial; sequence ***abcab*** $\downarrow$ $\boxed{\mathbf{c}}$***abc*** is displaced at the position of $\downarrow$. Atom just above $\downarrow$ (encircled by square) is displaced to $\boldsymbol{a}$ (not displaced to $\boldsymbol{b}$ (see Fig. 3.14(b)). Atom above $\boxed{\mathbf{c}}$, i.e., $\boldsymbol{a}$ is displaced to $\boldsymbol{b}$ and so on, resulting in

$$\begin{array}{ccccc|cccc} \boldsymbol{a} & \boldsymbol{b} & \boldsymbol{c} & \boldsymbol{a} & \boldsymbol{b} & \boxed{\mathbf{c}} & \boldsymbol{a} & \boldsymbol{b} & \boldsymbol{c} \\ & & & & & \downarrow & \downarrow & \downarrow & \downarrow \\ \boldsymbol{a} & \boldsymbol{b} & \boldsymbol{c} & \boldsymbol{a} & \boldsymbol{b} & \boldsymbol{a} & \boldsymbol{b} & \boldsymbol{c} & \boldsymbol{a}. \end{array}$$

This sequence is expressed in Frank's symbol as $\Delta\Delta\Delta\Delta\nabla\Delta\Delta\Delta$

Here, let us introduce a new symbol for stacking of close-packed planes, $\{111\}$ planes or equivalently (0001) planes. Let $\Delta\nabla$ and $\nabla\Delta$ be denoted by ⬡ and ⬢, respectively and $\Delta\Delta$ and $\nabla\nabla$ by □ and ■, respectively.[9] In this notation, FCC is denoted by □□□...

[9]It is noted that Frank symbol indicates the way the two neighboring (111) planes are stacked. In other words, it indicates the environment of the two neighboring (111) planes. By contrast, this new symbol indicates whether the (111) plane under consideration actually assumes the FCC structure or the HCP structure.

or ■■■... and HCP by ⬡⬡⬡... or ⬢⬢⬢. ... An intrinsic SF, ΔΔΔ∇ΔΔΔ, is denoted by □□⬡⬢□□. while an extrinsic SF, ΔΔΔ∇∇ΔΔΔ, is denoted by □□⬡■⬢□□. It is clear that the intrinsic SF contains two layers of HCP alone, while the extrinsic SF contains two layers of HCP plus one layer of FCC, that is, a thin twin.

A dislocation loop formed by removing one layer is V-type, and that formed by inserting one layer is I-type (see Exercise 3.4). In FCC structure the stacking fault is formed inside the dislocation loop. Such a loop is referred to as Frank's partial dislocation. In the Thompson nomenclature, the Frank loop is expressed by combination of the same alphabet of Roman and Greek such as $\mathbf{A}\alpha$, $\mathbf{B}\beta$, $\mathbf{C}\gamma$ and $\mathbf{D}\delta$.

3.5 Super-Dislocations in Ordered Alloys and Intermetallic Compounds

Let us consider a binary alloy with 50:50 composition for the sake of simplicity. In a disordered alloy, A atoms and B atoms cannot be distinguished statistically. By contrast, in an ordered alloy or an intermetallic compound, A atoms and B atoms align themselves regularly (Fig. 3.20(a)). In other words, lattice sites occupied by A and B atoms are predetermined. Now, let the Burgers vector in a disordered alloy be **b**, the length of which is the distance of the neighboring atoms. Now let us suppose that a dislocation with **b** be enforced to be inserted into an ordered alloy. In an ordered alloy, A and B atoms are distinguished, so that after the dislocation moves, bonds of atoms across the slip plane become wrong bonds such as A–A or B–B (Fig. 3.20(b)). The area across which wrong bonds are formed (in this case slip plane) is referred to as an anti-phase boundary (APB). An APB contains wrong bonds and has a higher energy: The extra energy is referred to as the anti-phase boundary energy (APB energy). To avoid an APB, the Burgers vector in an ordered alloy is to be 2**b** (Fig. 3.20(a)). Such a dislocation is referred to as a super-dislocation. Since energy of a dislocation is proportional to b^2, energy of a super-dislocation is four times as large as that of a dislocation in a disordered alloy.

On the other hand, when a dislocation with **b**, which corresponds to the Burgers vector in a disordered alloy, is to continue to move further, the area of APB increases, so that the total energy also

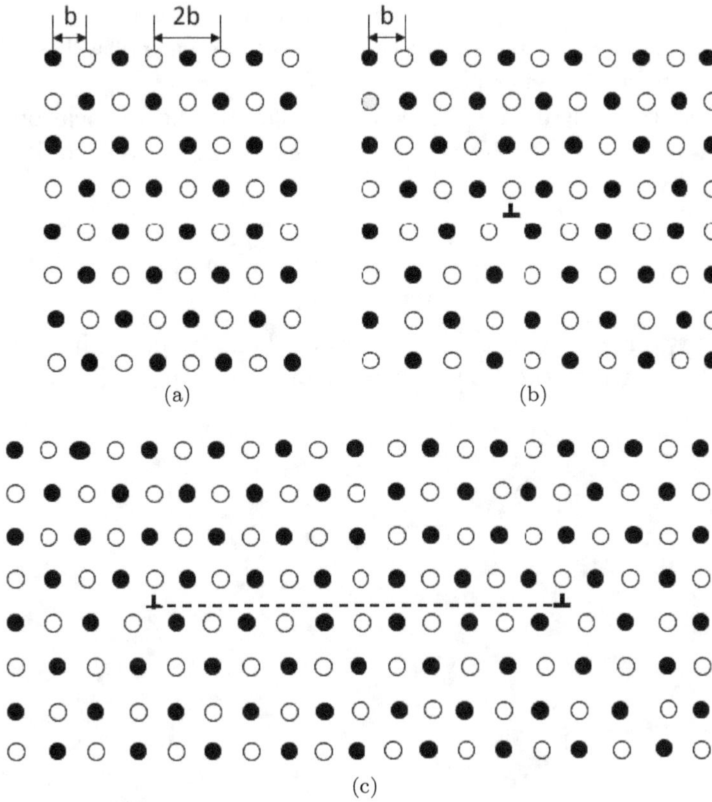

Fig. 3.20: (a) An ordered alloy. (b) Anti-phase boundary (APB) in an ordered alloy. (c) Super-dislocation in an ordered alloy and APB.

increases: The motion is hindered. However, when another dislocation with **b** follows on the same slip plane, wrong bonds A–A and B–B return to the right ones A–B and B–A (Fig. 3.20(c)).

In other words, if a pair of dislocations with **b** corresponding to the Burgers vector in a disordered alloy is present in an ordered alloy, the area of APB is reduced. Therefore, the elastic strain energy is halved roughly as compared with that of a super-dislocation. The individual dislocations with **b** making up a super-dislocation are referred to as super partial dislocations (or super-partials). The Burgers vectors of the two super-partials are identical, so that the super-partials repulse each other. The distance between the two super-partials is determined by the balance of a repulsive force between the two super-partials and an attractive force resulting from APB. This situation is exactly the same as the case of a dissociated dislocation in FCC crystal.

When **b** is inclined by β with respect to the dislocation line, the distance between the two super-partials Δ_e in the equilibrium is given by

$$\Delta_e = \frac{\mu \mathrm{b}_\mathrm{p}^2}{2\pi\gamma_{\mathrm{APB}}}\left(\cos^2\beta + \frac{\sin^2\beta}{1-\nu}\right) \tag{3.7}$$

where γ_{APB} is the APB energy (additional energy per unit area) and b_p is the length of the Burgers vector of a super-partial.

Chapter 4

Diffraction of Electrons by a Crystal

4.1 Plane Wave

Diffraction of electron beams by a crystal has its origin in wave properties of electrons. Let us suppose that a wave propagates from a point source (Fig. 4.1). The wave propagates to all the directions from its source (**O**: origin in Fig. 4.1); the shape of wave front is spherical. Such a wave is referred to as a spherical wave. However, at a point far remote from the origin, the radius of curvature of the spherical wave is so large that its wave front can be approximated by a plane. Such a wave is referred to as a plane wave.

The amplitude (ϕ) of a plane wave propagating in the direction of z is described by a sine or cosine curve. Now let us describe it by a sine curve as

$$\phi = \sin\left(\frac{2\pi z}{\lambda}\right) \tag{4.1}$$

where λ is the wave length. In the case of a plane wave its phase (that is, hill-and-hill or valley-and-valley) is the same in a plane perpendicular to z-axis. Therefore, the phase at a position distant $\mathbf{r}$ from the origin is the same as that at a position $\mathbf{z}$, where $\mathbf{r}$ is projected to the z-axis (Fig. 4.2). Supposing $\mathbf{e_3}$ be a unit vector along z-axis, substituting $z = \mathbf{r} \cdot \mathbf{e_3}$ into Eq. (4.1) gives

$$\phi = \sin\frac{2\pi\mathbf{r} \cdot \mathbf{e_3}}{\lambda} = \sin\left(2\pi\mathbf{r} \cdot \frac{\mathbf{e_3}}{\lambda}\right) \tag{4.1$'$}$$

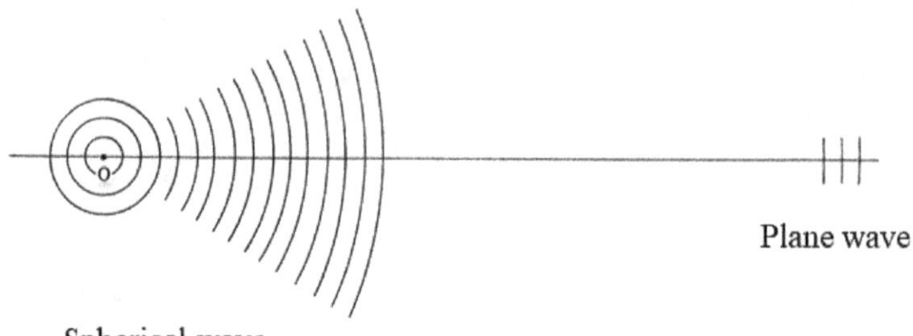

Fig. 4.1: Spherical and plane waves.

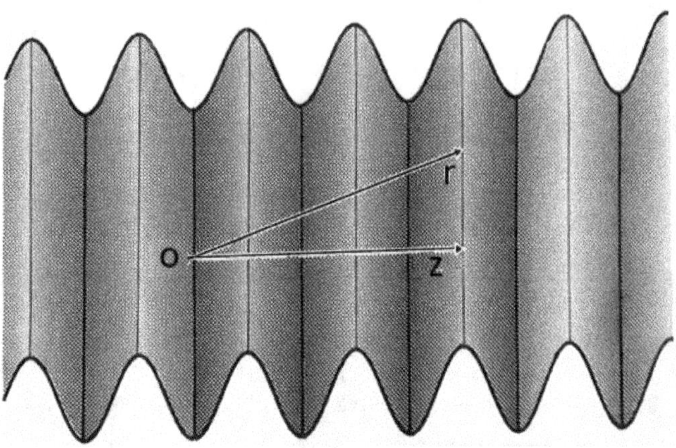

Fig. 4.2: A plane wave propagating in z-direction. The phase in a plane perpendicular to z-axis is the same.

where $\mathbf{e_3}/\lambda$ is a vector whose direction is $\mathbf{e_3}$, i.e., the direction of propagation of the wave and its length is $1/\lambda$. This vector is referred to as a wave number vector and is denoted by $\mathbf{k}^*$.[1] That is,

$$\mathbf{k}^* = \frac{\mathbf{e_3}}{\lambda}, \quad k = |\mathbf{k}^*| = \frac{1}{\lambda} \tag{4.2}$$

The reason why this vector $\mathbf{k}^*$ is referred to as a wave number vector is that it represents how many waves are present per a unit length and its dimension is $[\mathrm{L}^{-1}]$. Thus, $\mathbf{k}^*$ is defined in the reciprocal space (see Section 2.3.1). In summary, Eqs. (4.1) and (4.1′) are

[1]Throughout this book, vectors in the reciprocal space are attached by asterism *.

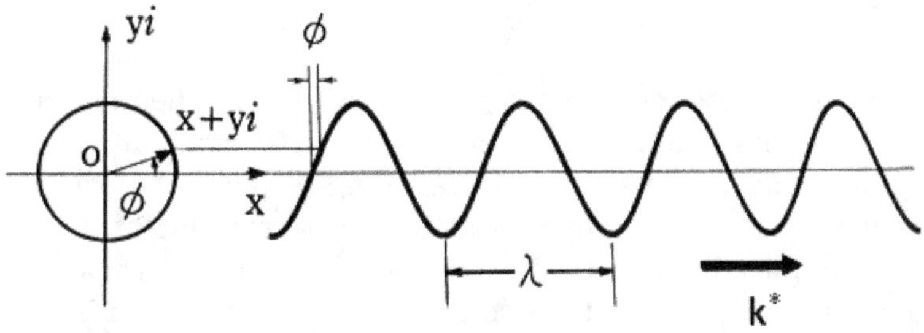

Fig. 4.3: Representation of a plane wave in a complex plane. A plane wave is denoted by a complex vector $\mathrm{x} + \mathrm{y}i$.

expressed as

$$\phi = \sin(2\pi\mathbf{r} \cdot \mathbf{k}^*) \tag{4.1''}$$

Here, let us introduce a complex plane in order to describe a plane wave. A coordinate (x, y) in a conventional (x–y) plane is represented by a complex $\mathrm{x} + \mathrm{y}i$. Therefore, a vector drawn from the origin to a co-ordinate (x, y) is also represented by a complex $\mathrm{x} + \mathrm{y}i$ in a complex plane (see Fig. 4.3). For a unit vector $(\mathrm{x}^2 + \mathrm{y}^2)^{1/2} = 1$,

$$\mathrm{x} + \mathrm{y}i = \frac{\mathrm{x}}{(\mathrm{x}^2 + \mathrm{y}^2)^{1/2}} + \frac{\mathrm{y}i}{(\mathrm{x}^2 + \mathrm{y}^2)^{1/2}} = \cos\phi + i\sin\phi.$$

Using Euler's formula, we obtain

$$\cos\phi + i\sin\phi = \exp(i\phi) \equiv \mathrm{e}^{i\phi} \tag{4.3}$$

Here, ϕ is an angle measured from the x-axis anticlockwise. Suppose that a unit vector expressed by $\exp(i\phi)$ rotates anticlockwise, then its projection onto the x-axis ($\cos\phi$) is the real part and the projection onto y-axis ($\sin\phi$) is the imaginary part.

In conclusion, a plane wave (Eq. (4.1″)) is expressed in a complex plane as

$$\phi = \exp(2\pi i\mathbf{k}^* \cdot \mathbf{r}) \equiv \cos(2\pi i\mathbf{k}^* \cdot \mathbf{r}) + i\sin(2\pi i\mathbf{k}^* \cdot \mathbf{r}) \tag{4.4}$$

What we can observe in reality is not the amplitude of a beam but its intensity, which is expressed by

$$\mathrm{I} = \phi\phi^* \tag{4.5}$$

where

$$\phi^* = \exp(-2\pi i\mathbf{k}^* \cdot \mathbf{r}) \equiv \cos(2\pi i\mathbf{k}^* \cdot \mathbf{r}) - i\sin(2\pi i\mathbf{k}^* \cdot \mathbf{r}) \tag{4.6}$$

4.2 Scattering by a Single Atom

First let us consider scattering by a single atom. When an electron does not lose its kinetic energy, its wave length (and wave number) does not change.[2] This type of scattering is referred to as elastic scattering. Now let us define an atomic scattering factor f to indicate the amplitude of scattering by a single atom. Figure 4.4 shows atomic scattering factors for electrons. The atomic scattering factor f decreases with scattering angle θ and increases with atomic number Z of the scattering atom.[3]

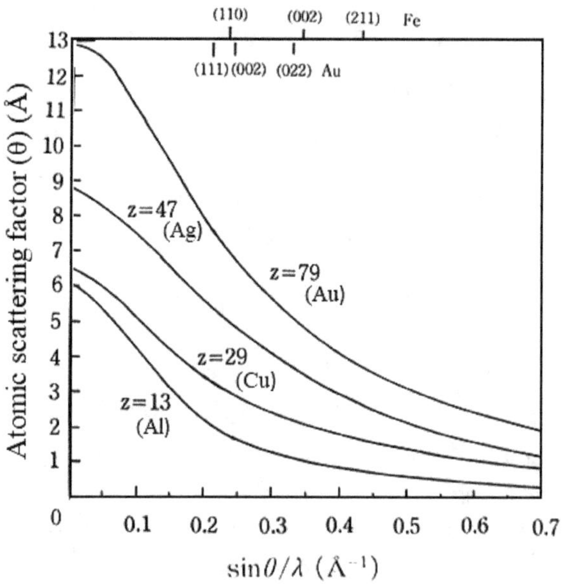

Fig. 4.4: Relationship between the atomic scattering factor $f(\theta)$ and the scattering angle $\sin\theta/\lambda$.

---

[2]Wave vector $\mathbf{k}^*$ and kinetic energy E are related by $k^2h^2/2m = E$, where h is Planck's constant).

[3]Atomic scattering factors for X-ray are similar to those for electrons, however, their absolute values are much larger than those for electrons. By contrast, atomic scattering factors for neutrons do not depend on θ and the dependence on atomic number Z is different from X-ray and electrons.

4.3 Scattering by a Unit Cell: Diffraction

Let us consider a simple unit cell which contains only one atom. Here, two atoms, that is, an atom at the origin (**O**) and the other atom at a position distant **r** (from the origin) are involved in the scattering event (Fig. 4.5). Here,

$$\mathbf{r} = \mathrm{x}\mathbf{a} + \mathrm{y}\mathbf{b} + \mathrm{z}\mathbf{c} \quad \text{(x, y, z are integers)} \tag{4.7}$$

where **a**, **b** and **c** are the fundamental vectors of a simple unit cell in the real space. From now on we treat electrons as waves. The wave scattered at **r** (ϕ_{sr}) is scattered after it travels $\mathbf{r} \cdot \mathbf{k}^*/|\mathbf{k}^*|$ behind the wave scattered at the origin **O** (ϕ_{s0}): At this point, ϕ_{s0} (the wave scattered at the origin **O**) travels already $\mathbf{r} \cdot \mathbf{k}'^*/|\mathbf{k}'^*|$ ahead

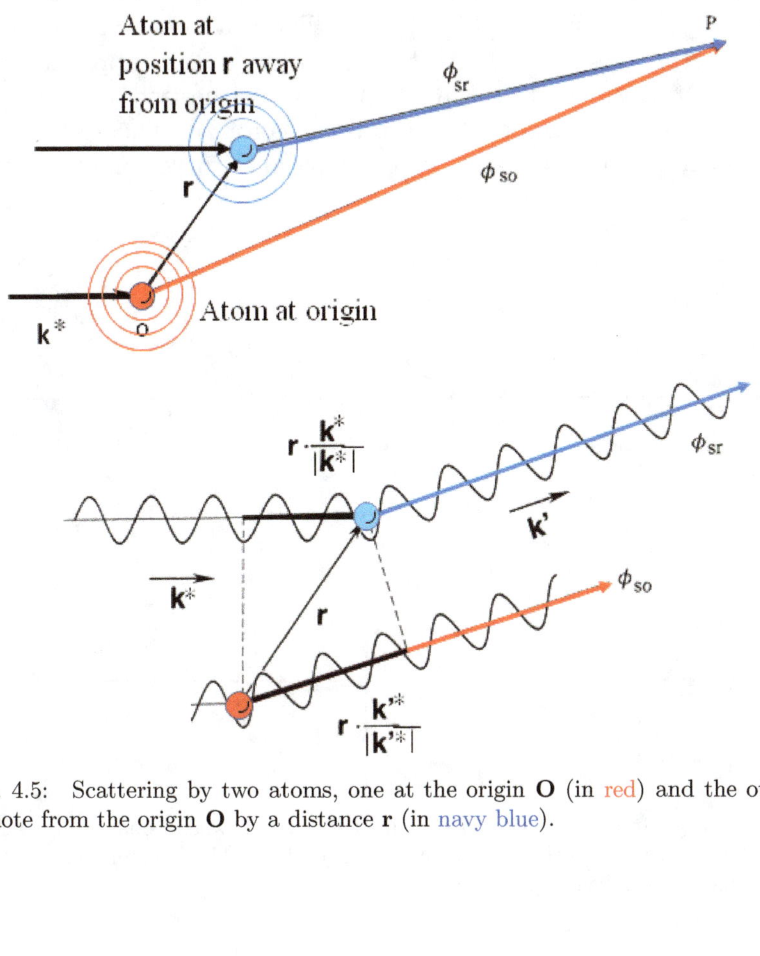

Fig. 4.5: Scattering by two atoms, one at the origin **O** (in red) and the other remote from the origin **O** by a distance **r** (in navy blue).

in the direction of $\mathbf{k}'^*$. In other words, ϕ_{sr} is delayed by a distance $\mathbf{r} \cdot \mathbf{k}'^*/|\mathbf{k}'^*| - \mathbf{r} \cdot \mathbf{k}^*/|\mathbf{k}^*|$ as compared with ϕ_{s0}. This difference in distance is referred to as a path difference. When the path difference is $n\lambda$, where n is an integer and λ is the wave length, the phases of ϕ_{s0} and ϕ_{sr} are identical. This can be compared to a racer being n laps behind. However, when n is not an integer, the phases of ϕ_{s0} and ϕ_{sr} are not identical.

Now let us assume that the phase difference between ϕ_{s0} and ϕ_{sr} is 15° and that the two sinusoidal curves corresponding to ϕ_{s0} and ϕ_{sr} are superimposed. The result is shown in Fig. 4.6(a). The amplitude of the superimposed (or synthesized) wave is slightly smaller than two times of the amplitude of ϕ_{s0} (or ϕ_{sr}) and the phase is just between ϕ_{s0} and ϕ_{sr}. This procedure of superimposing two waves is expressed in a complex plane as follows: Let us consider that the phase of ϕ_{s0} is just zero (Fig. 4.6(b)). The wave of ϕ_{s0} is expressed by a vector ①(=1 + 0i). At this instance, ϕ_{sr} is expressed by a vector ②(=cos(−15°) + $i$sin(−15°). The synthesized wave $\phi_{s(1+2)}$ is expressed by a vector ③(=vector①+vector②). The vector ② is equivalent to the vector ②′ which is parallel to vector ②, so that

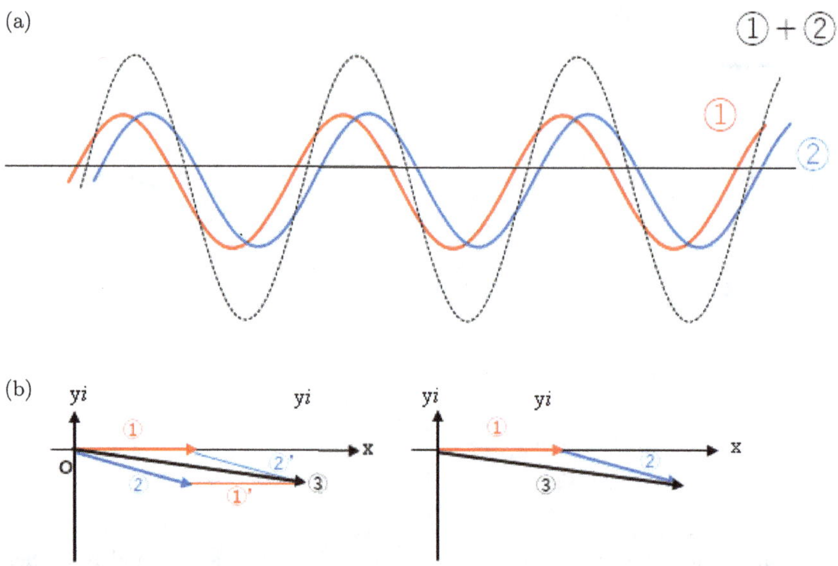

Fig. 4.6: (a) Superimposition of two sinusoidal waves with a phase difference of 15°. The resultant synthesized wave is indicated by a broken line. (b) and (c) Amplitude-phase diagram.

the vector ③ is obtained by a procedure shown in Fig. 4.6(c). Such a figure as Fig. 4.6(c) provides information of amplitudes and phases of the component and the synthesized waves and is referred to as an amplitude-phase diagram.

Definition 4.1:
An amplitude-phase diagram indicates how waves scattered at more than one places are synthesized (or superimposed).

Mathematically:

phase difference between ϕ_{s0} and ϕ_{sr}

$$
\begin{aligned}
&= 2\pi \times (\text{path difference}) \div (\text{wave length})\\
&= 2\pi\left(\frac{\mathbf{r}\cdot\mathbf{k}^*}{|\mathbf{k}^*|} - \frac{\mathbf{r}\cdot\mathbf{k}'^*}{|\mathbf{k}'^*|}\right) \div \lambda\\
&= 2\pi\mathbf{r}(\mathbf{k}^* - \mathbf{k}'^*).
\end{aligned}
$$

Putting $\phi_{s0} = 1$ gives

$$\phi_{sr} = \exp\{2\pi i(\mathbf{k}^* - \mathbf{k}'^*)\cdot\mathbf{r}\} = \exp(-2\pi i\mathbf{K}^*\cdot\mathbf{r}) \tag{4.8}$$

where

$$\mathbf{K}^* = \mathbf{k}'^* - \mathbf{k}^* \tag{4.9}$$

4.4 The Ewald Sphere

Let us suppose that an atom be located at the origin **O** and another atom at a position **r**. For the sake of simplicity, let us assume that these two atoms lie on two separate (*hkl*) planes, which are parallel to each other. Also let us assume that the incident and scattered waves are at the mirror position with respect to the (*hkl*) plane. The geometry is shown in Fig. 4.7, where the incident wave is expressed by its wave number $\mathbf{k}^*$, the scattered wave by its wave number $\mathbf{k}'^*$. $\mathbf{K}^*(=\mathbf{k}'^* - \mathbf{k}^*)$ points to the normal of the (*hkl*) plane, that is, $\mathbf{K}^*$ is parallel to the reciprocal vector of the (*hkl*) plane, $\mathbf{g}^*_{hkl}(\equiv h\mathbf{a}^* + k\mathbf{b}^* + l\mathbf{c}^*)$ (see Eq. (2.4)). Therefore,

$$\mathbf{K}^* = \eta\mathbf{g}^*_{hkl} \tag{4.10}$$

where η is a constant.

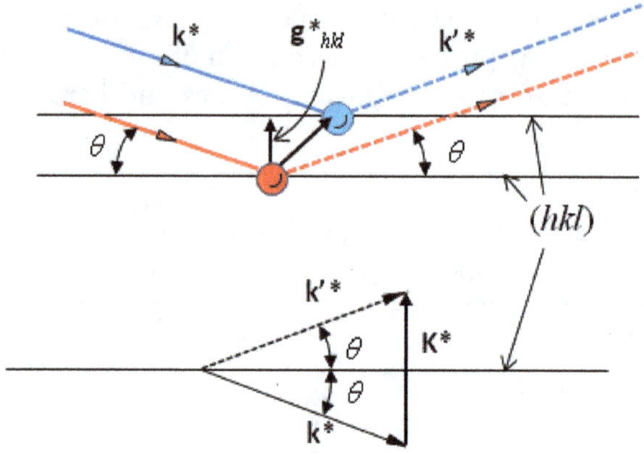

Fig. 4.7: Relationship among (hkl) plane, the incident wave (wave number vector $\mathbf{k}^*$) and the scattered wave (wave number vector $\mathbf{k}'^*$).

Considering Eq. (4.8) and the amplitude-phase diagram,

$$\phi_{\mathrm{s0}} + \phi_{\mathrm{sr}} = 1 + \exp(-2\pi i\mathbf{K}^* \cdot \mathbf{r}) = 1 + \exp\{-2\pi \mathrm{i}\eta(h\mathrm{x} + k\mathrm{y} + l\mathrm{z})\} \tag{4.8$'$}$$

Here, the first term stands for the wave scattered at the origin **O** and the second term the wave scattered at position **r**. A sufficient condition for ϕ_{sr} being a maximum is that the second term is 1, that is,

$$\eta(h\mathrm{x} + k\mathrm{y} + l\mathrm{z}) = \text{integer} \tag{4.11}$$

Since the Miller indices (hkl) are integers and a simple unit cell is under consideration, x, y, and z are also integers (see Eq. (4.7)). Therefore, a sufficient condition for satisfying Eq. (4.11) is that η is an integer, or simply putting $\eta = 1$,

$$\mathbf{K}^* = \mathbf{g}^*_{hkl} \equiv h\mathbf{a}^* + k\mathbf{b}^* + l\mathbf{c}^* \tag{4.12}$$

A construction to find this condition is as follows.

Let us consider a case where a plane wave with a wave length λ (its wave number being $|\mathbf{k}^*| = 1/\lambda$) is incident to a single crystal whose fundamental vectors of the unit cell are **a**, **b** and **c** (Fig. 4.8(a)). In Fig. 4.8(a), vector **c** is out of the paper, so that it is not indicated. There exists a reciprocal lattice corresponding to this unit cell (in the real space), the fundamental of which are $\mathbf{a}^*$, $\mathbf{b}^*$ and $\mathbf{c}^*$ ($\mathbf{c}^*$ is out of the paper and is not shown).

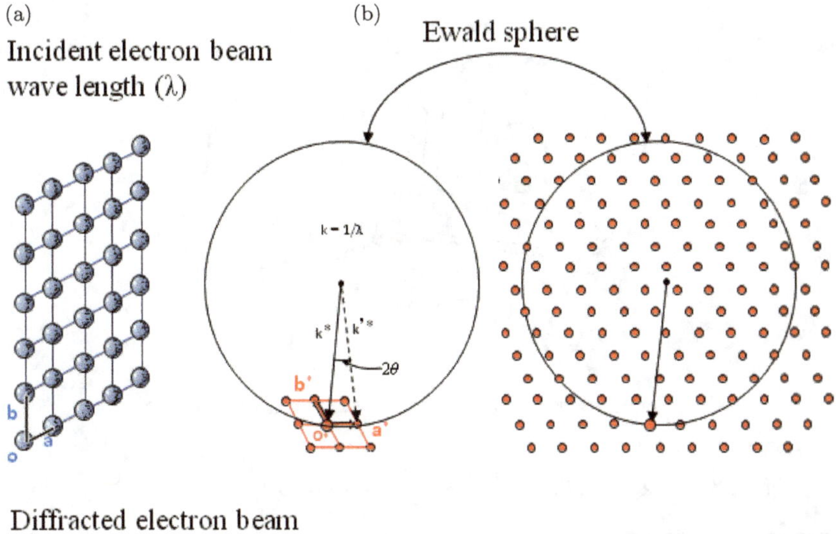

Fig. 4.8: Definition of the Ewald sphere. (a) Relation between a crystal and the electron beams in real space. (b) Relation between reciprocal vectors and the Ewald sphere. (c) Relation between the Ewald sphere and reciprocal spots where high-ordered reflections are taken into account.

Definition 4.2:
The Ewald sphere
Draw a straight line from the origin of the reciprocal lattice in the direction reverse to the incident electron beam (which is in the real space). Mark the point at $\mathbf{k}^* = |\mathbf{k}^*| = 1/\lambda$ and draw a sphere with radius $1/\lambda$ about this end point as center (Fig. 4.8(b)).

This sphere is referred to as a reflection sphere or the Ewald sphere.

Theorem 4.1:
When the Ewald sphere intersects any reciprocal lattice point $\mathbf{g}^*_{hkl}$ (the Miller index of which is hkl), then,

$$\mathbf{K}^*(=\mathbf{k}'^* - \mathrm{k}^*) = \mathbf{g}^*_{hkl} \equiv h\mathbf{a}^* + k\mathbf{b}^* + l\mathbf{c}^* \qquad (4.12')$$

and the (hkl) plane obeys Bragg's condition, i.e.,

$$2\mathrm{d}\sin\theta = \lambda \qquad (4.13)$$

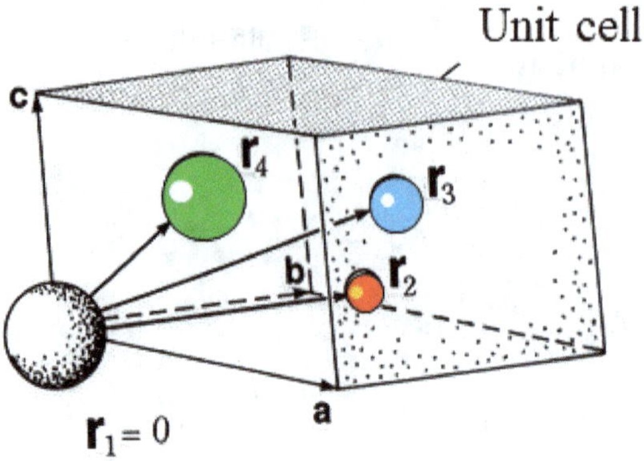

Fig. 4.9: Many atoms are present in a unit cell.

In Fig. 4.8(b), $\mathbf{g}^* = \mathit{100}$ satisfies Bragg's condition.

Exercise 4.1: Prove Theorem 4.1.

4.5 Structure Factor

For those crystals which do not have a simple unit cell, all of x, y and z are not necessarily integers. As shown in Fig. 4.9, when n atoms are present in a unit cell, the amplitude of the total of scattered waves from the unit cell is the summation of Eq. (4.8) and given by

$$\phi_s = \sum_{j=1}^{n} f_j \exp(-2\pi i \mathbf{K}^* \cdot \mathbf{r}_j) \tag{4.14}$$

where $\mathbf{r}_j$ stands for a vector drawn from the origin to the jth atom and given by

$$\mathbf{r}_j = x_j\mathbf{a} + y_j\mathbf{b} + z_j\mathbf{c} \quad (x_j, y_j \text{ and } z_j \text{ are integers or fractions}) \tag{4.7$'$}$$

Each atom making up the unit cell may be different, so that its atomic scattering factor is denoted by f_j in Eq. (4.14).

Figure 4.10 shows an amplitude-phase diagram for the superimposed wave from a unit cell containing four atoms. When this wave

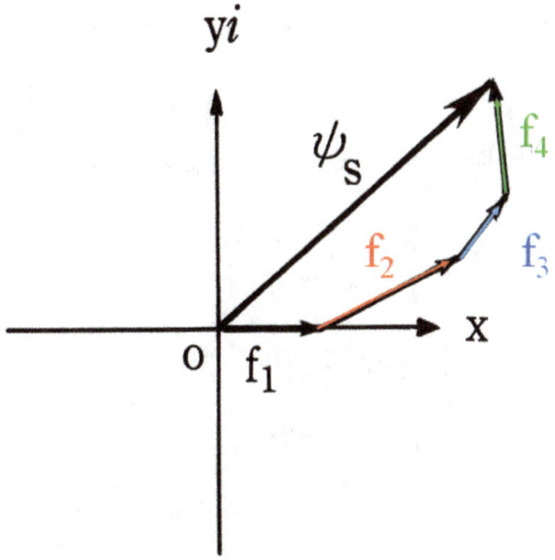

Fig. 4.10: Amplitude-phase diagram of a wave scattered from a unit cell containing many atoms inside.

satisfies Bragg's condition, substituting Eq. (4.12) into Eq. (4.14),

$$\begin{aligned}\mathrm{F}(hkl) &\equiv \phi_s \\ &= \sum_{j=1}^{4} f_j \exp\{-2\pi i(\mathbf{g}^* \cdot \mathbf{r}_j)\} \\ &= \sum_{j=1}^{4} f_j \exp\{-2\pi i(h\mathbf{a}^* + k\mathbf{b}^* + l\mathbf{c}^*)(x_j\mathbf{a} + y_j\mathbf{b} + z_j\mathbf{c})\}.\end{aligned}$$

In summary,

$$\mathrm{F}(hkl) = \sum_{j=1}^{4} f_j \exp\{-2\pi i(hx_j + ky_j + lz_j)\} \tag{4.15}$$

$\mathrm{F}(hkl)$ depends on crystal structure through Eq. (4.7) and is referred to as a structure factor.

Problem.

Calculate the structure factor of BCC.

Solution: Two atoms are present in BCC. Positions of these two atoms are

$$\mathbf{r}_1 = 0\mathbf{a} + 0\mathbf{b} + 0\mathbf{c}, \quad \mathbf{r_2} = \frac{1}{2}\mathbf{a} + \frac{1}{2}\mathbf{b} + \frac{1}{2}\mathbf{c} \quad \text{(see 1.4.1)}$$

so that substituting $x_1 = 0$, $y_1 = 0$, $z_1 = 0$ and $x_2 = 1/2$, $y_2 = 1/2$, $z_2 = 1/2$ into Eq. (4.15), we obtain

$$\begin{aligned} \mathrm{F}(hkl) &= f \exp\{-2\pi i(0h + 0k + 0l)\} \\ &\quad + f \exp\left\{-2\pi i \left(\frac{1}{2}h + \frac{1}{2}k + \frac{1}{2}l\right)\right\} \\ &= f[1 + \exp\{-\pi i(h + k + l)\}]. \end{aligned}$$

Therefore, when $h+k+l = \text{odd}$, $\mathrm{F}(hkl) = 0$. In this case, the phase of wave ϕ_0 scattered at $\mathbf{r}_1 = 0\mathbf{a} + 0\mathbf{b} + 0\mathbf{c}$ is different by π with respect to wave ϕ_r scattered at $\mathbf{r}_2 = 1/2\mathbf{a} + 1/2\mathbf{b} + 1/2\mathbf{c}$, so that they cancel each other. A criterion to judge whether a structure factor is zero or not is referred to as an extinction law.

This can be explained using an amplitude-phase diagram as follows: The phase of ϕ_0 is 0. This means that ϕ_0 is denoted by vector ① in the amplitude-phase diagram (Fig. 4.11(a)). When $h + k + l = \text{odd}$, ϕ_r is denoted by vector ② in Fig. 4.11(a)). As a result, $\phi_0 + \phi_r = 0$, and diffraction does not occur. By contrast, when

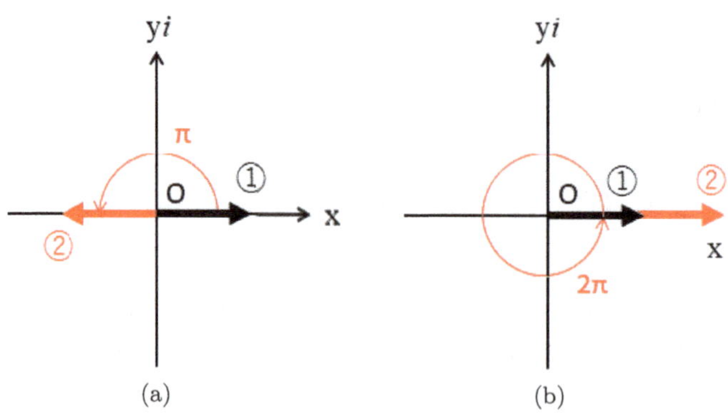

Fig. 4.11: In BCC when $h+k+l = \text{odd}$, the structure factor is zero, while when $h+k+l = \text{even}$, the structure factor is non-zero.

$h+k+l=$ even, $\phi_0+\phi_r=2$f as shown in Fig. 4.11(b), giving rise to diffraction.

Exercise 4.2: In CsCl type structure, which is based on BCC (see Section 1.4.2), the structure factor is non-zero even when $h+k+l=$ odd. Prove this.

When a disordered alloy transforms into an ordered alloy (see Section 3.5), reflections which are absent in the disorder state appear. These reflections are referred to as superlattice reflections or forbidden reflections. On the other hand, those which appear in the disordered state are referred to as fundamental reflections.

Exercise 4.3: Reciprocal points for which the structure factors are non-zero for BCC structure $\mathbf{g}^*_{hkl} = h\mathbf{a}^* + k\mathbf{b}^* + l\mathbf{c}^*$ make up an FCC lattice in the $(\mathbf{a}^*, \mathbf{b}^*, \mathbf{c}^*)$ reciprocal space.

Exercise 4.4: If a similar operation is performed for FCC structure, non-zero reciprocal lattices make up a BCC lattice in the $(\mathbf{a}^*, \mathbf{b}^*, \mathbf{c}^*)$ reciprocal space. (The solutions to Exercises 4.3 and 4.4 are in keeping with solution to Exercise 2.19.)

In summary,

> **Theorem 4.2:**
>
> *BCC*: When $h+k+l=$ odd, the structure factor is 0.
> *FCC*: When h, k, l are mixture of even and odd, the structure factor is 0.

Exercise 4.5: Show that the structure factors for diamond structure $\mathrm{F}_{\text{diamond}}$ is given by

$$\mathrm{F}_{\text{diamond}} = \left[1 + \exp\left\{2\pi i\left(\frac{h}{4} + \frac{k}{4} + \frac{l}{4}\right)\right\}\right] \mathrm{F}_{\text{FCC}}$$

where F_{FCC} is a structure factor for FCC structure. In other words, a diamond structure is superimposition of two FCC lattices, the origins of which are located at 000 and 1/4,1/4,1/4, respectively (see Section 1.4.1).

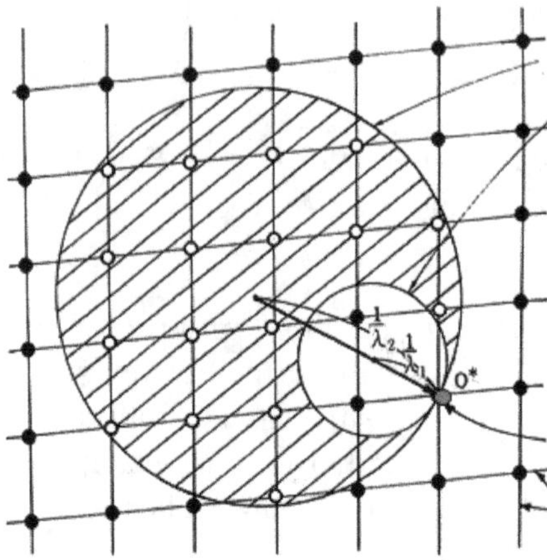

Fig. 4.12: Laue method, where a continuous X-ray is used to change the radius of the Ewald sphere continuously from $1/\lambda_1$ to $1/\lambda_2$. Reciprocal points within a hatched region (open circles) satisfy Bragg's condition.

Exercise 4.6: Calculate the structure factor for NaCl (see Section 1.4.2).

Exercise 4.7: The lattice constant of Cu is 0.362 nm. Wave length of the Cu characteristic X-ray (CuK_α) is 0.154 nm. Draw the Ewald sphere when the incident beam is along [100].

Exercise 4.8: Wave length of 200 kV electron is 0.0025 nm. Perform a procedure similar to Exercise 4.7.

As can be seen from these aforementioned exercises, the Ewald sphere for electrons in a 200 kV electron microscope is almost planar when intersecting reciprocal lattices. By contrast, the radius of Ewald sphere of X-ray is comparable to a reciprocal lattice vector, so that it is rather rare for the Ewald sphere to intersect reciprocal lattices. For this reason, there are some devices to increase an opportunity for the Ewald sphere to intersect reciprocal lattices.

Example 1. Changing the radius of the Ewald sphere continuously.

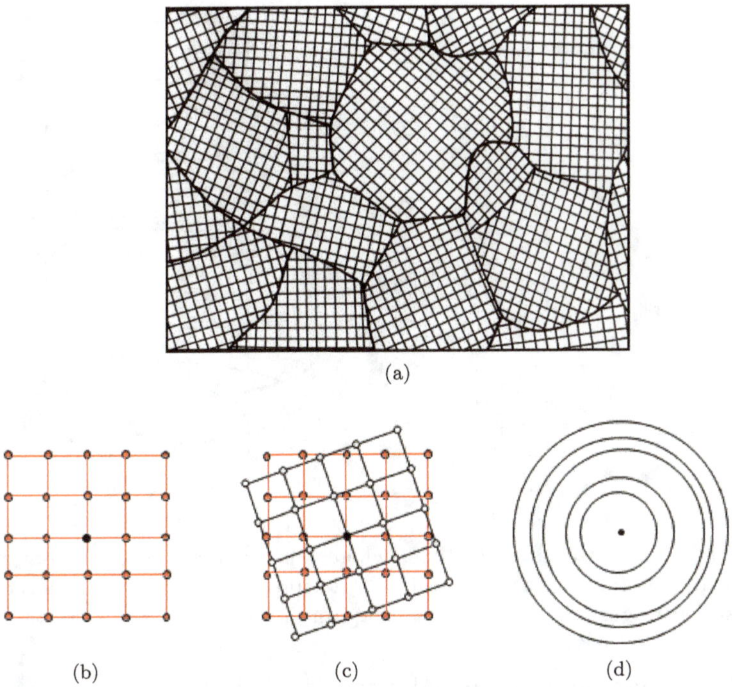

Fig. 4.13: Powder method. (a) Poly-crystal. (b) Reciprocal lattice of a single crystal. (c) Reciprocal lattices of two single crystals. (d) Reciprocal lattices of a poly-crystal. When a poly-crystal or powdered specimen is used, reciprocal points make up concentric spheres, so that they intersect the Ewald sphere.

This corresponds to using continuous X-ray[4] whose wave length changes continuously. This method is referred to as Laue method. As shown in Fig. 4.12, if the wave length of X-ray changes from λ_1 to λ_2 continuously, the radius of the corresponding Ewald sphere also changes from $1/\lambda_1$ to $1/\lambda_2$ continuously. This results in reciprocal lattice points within a hatched region in Fig. 4.12 being intersected by the Ewald sphere.

Example 2. Increasing the number of the reciprocal lattice points for a characteristic X-ray.

A characteristic X-ray has a constant wave length.

[4] Also referred to as white X-ray.

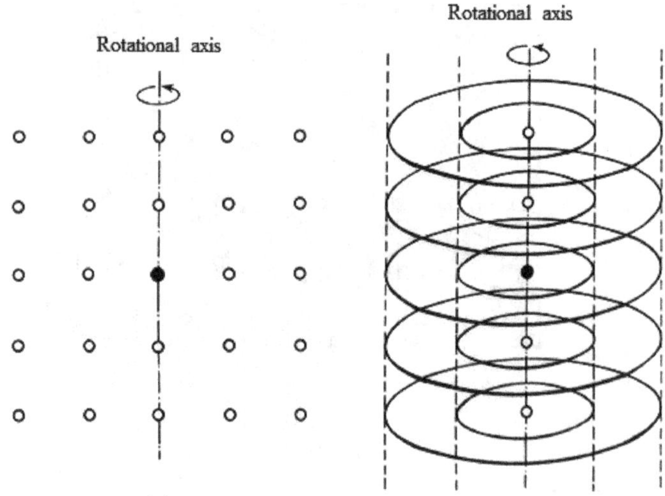

Fig. 4.14: Crystal rotation method, where the reciprocal lattice becomes concentric by rotating a single crystal. (a) Reciprocal lattice of a single crystal. (b) Concentric reciprocal lattice resulting from rotation of a single crystal.

(i) Poly-crystals or powdered crystals are used as a sample. Randomly orientated crystals (grains) are present in these samples (Fig. 4.13(a)). With increasing the number of randomly orientated crystals, the reciprocal lattices increase in number, as shown in Figs. 4.13(b) and 4.13(c) and eventually make up continuous spheres (illustrated by circles in Fig. 4.13(d)).

(ii) In the case where a single crystalline sample is used, when the sample is rotated around an axis, the reciprocal lattice points are also rotated (Fig. 4.14(a)), resulting in increase in opportunity of the Ewald sphere intersecting reciprocal lattice points. This method is referred to as a crystal rotating method.

4.6 Broadening of Reciprocal Lattice Points Due to the Outer Shape of a Crystal

So far, a crystal with an infinite number of unit cells has been treated. In a real crystal, definite number of unit cells, which can be considered as scattering centers having a structure factor F(hkl), are arranged regularly. Here, let us consider a parallelepiped with

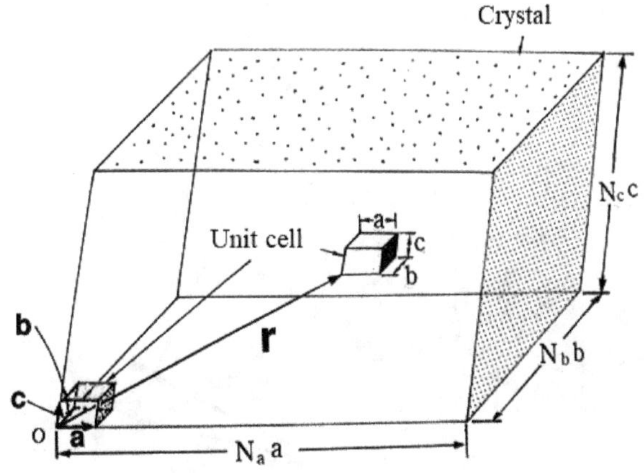

Fig. 4.15: A crystal with a finite size.

$N_a\mathbf{a}$, $N_b\mathbf{b}$ and $N_c\mathbf{c}$ as three edges (Fig. 4.15). Diffraction amplitude ϕ_g from the whole crystal is the summation of diffracted amplitudes from each of the unit cells making up the crystal. Let us denote the position of a unit cell by a vector $\mathbf{r}$ whose components are u, v and w, that is,

$$\mathbf{r}(\mathrm{u}, \mathrm{v}, \mathrm{w}) = \mathrm{u}\mathbf{a} + \mathrm{v}\mathbf{b} + \mathrm{w}\mathbf{c} \tag{4.7''}$$

Then,

$$\phi_{\mathrm{g}} = \mathrm{F} \sum_{\mathrm{u}=0}^{\mathrm{N_a}-1} \sum_{\mathrm{v}=0}^{\mathrm{N_b}-1} \sum_{\mathrm{w}=0}^{\mathrm{N_c}-1} \exp\{-2\pi \mathbf{K}^* \cdot \mathbf{r}(\mathrm{u}, \mathrm{v}, \mathrm{w})\} \tag{4.16}$$

Let us consider a case where $\mathbf{K}^*$ deviates slightly from $\mathbf{g}^*_{hkl}$. The deviation is defined as follows.

Definition 4.3:

$$\begin{aligned} \mathbf{K}^* &= \mathbf{g}^*_{hkl} + \mathbf{s}^*_{\mathrm{g}} \\ &= (h + s_{\mathrm{x}})\mathbf{a}^* + (k + s_{\mathrm{y}})\mathbf{b}^* + (l + s_{\mathrm{z}})\mathbf{c}^* \end{aligned} \tag{4.17}$$

where

$$\mathbf{s}^*_{\mathrm{g}} = s_{\mathrm{x}}\mathbf{a}^* + s_{\mathrm{y}}\mathbf{b}^* + s_{\mathrm{z}}\mathbf{c}^* \tag{4.17'}$$

$\mathbf{s}^*_g$ is referred to as a deviation parameter (Fig. 4.16).

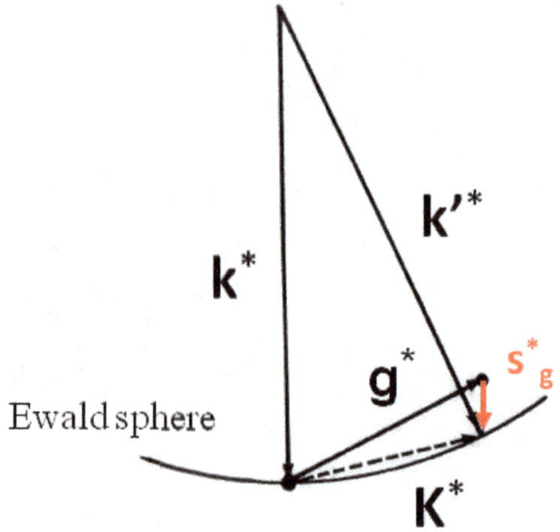

Fig. 4.16: Definition of a deviation parameter $\mathbf{s}_g^*$.

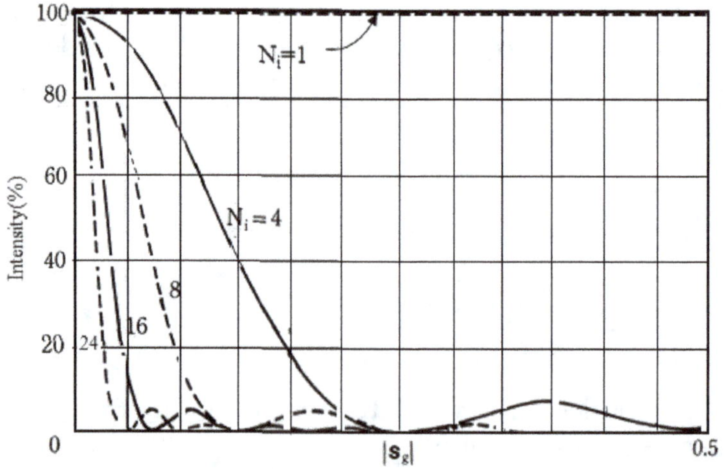

Fig. 4.17: Laue function. N_i is the thickness of a crystal as measured by the number of unit cell. When $0.5 < |\mathbf{s}_g^*| < 1$, the function is symmetrical with respect to $|\mathbf{s}_g^*| = 0.5$. $|\mathbf{s}_g^*| = 0$ indicates the origin of the reciprocal lattice, i.e., direct spot and $|\mathbf{s}_g^*| = 1$ corresponds to {*100*} spot.

Equation (4.16) is rewritten as

$$\phi_g = F \sum_{u=0}^{N_a-1} \sum_{v=0}^{N_b-1} \sum_{w=0}^{N_c-1} \exp[-2\pi\{(h+s_x)u + (k+s_y)v + (l+s_z)w\}] \tag{4.16'}$$

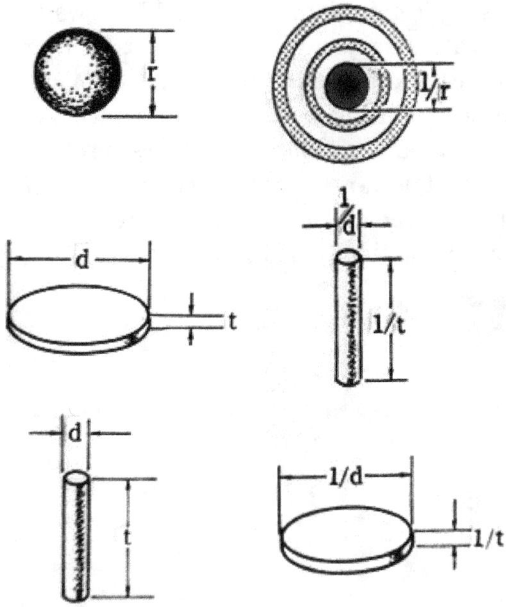

Fig. 4.18: Relationship between the size of a crystal (a) and broadening of the reciprocal point (b). The broadening of a reciprocal point is inversely proportional to the size of the crystal approximately. For the spherical crystal subpeaks of reciprocal points are drawn, while for the others the subpeaks are not drawn.

Equation (4.16′) can be written as

$$\begin{aligned}\phi_g &= F\sum_{u=0}^{N_a-1}\exp\{-2\pi s_x u\}\sum_{v=0}^{N_b-1}\exp\{-2\pi s_y v\}\sum_{w=0}^{N_c-1}\exp\{-2\pi s_z w\}\\ &= F\exp[-\pi i\{(N_a-1)s_x+(N_b-1)s_y+(N_c-1)s_z\}]\\ &\quad\times\frac{\sin\pi N_a s_x}{\sin\pi s_x}\times\frac{\sin\pi N_b s_y}{\sin\pi s_y}\times\frac{\sin\pi N_c s_z}{\sin\pi s_z}\end{aligned}\qquad(4.16'')$$

The diffracted intensity I (*hkl*) from the crystal as a whole is given by I(*hkl*) = F^2G.

Here,

$$G=\frac{\sin^2\pi N_a s_x}{\sin^2\pi s_x}\times\frac{\sin^2\pi N_b s_y}{\sin^2\pi s_y}\times\frac{\sin^2\pi N_c s_z}{\sin^2\pi s_z}\qquad(4.18)$$

is referred to as Laue function. Figure 4.17 shows G. With increasing N_i, that is, with increasing the thickness of a crystal, the peak at $\mathbf{s}_g^* = 0$ becomes sharper and subpeaks diminish. Inversely, with decreasing the size of a crystal, the full width at half maximum (FWHM) becomes larger. The FWHM is proportional to $1/N_i$, roughly speaking. It is noted that when $N_i = 1$, $G = 1$. Relation between the outer shape of a crystal and extension of the reciprocal lattice is shown schematically in Fig. 4.18.

Exercise 4.9: Derive Eq. (4.16″).

Chapter 5

Electron Microscope

5.1 Basic Design of the Electron Microscope

Figure 5.1(b) illustrates a basic design of the transmission electron microscope (TEM). TEM has a close analogy with a biological optical microscope (BOM) (Fig. 5.1(a)). The light source of BOM is replaced with an electron gun from which an electron beam is emitted. Condenser lens and objective lens have the same names. What corresponds to an eyepiece in BOM is the objective lens in TEM.[1] By a condenser lens, electrons which are emitted from an electron gun, are aligned in such a way that parallel electron beams illuminate the sample. The objective lens is the first-stage multiplying lens and the performance of a TEM depends primarily on the performance of the objective lens. An intermediate lens and a projection lens are used to adjust the final magnification of the image. Lens in BOM is a glass and the light is bent because of a difference in refractive index between air and the glass. Electron lens bends electrons by means of electro-magnetism.

A scanning electron microscope (SEM) has quite a different design as shown in Fig. 5.2(a). An electron beam emitted from the electron

[1]In practice, an intermediate lens is inserted between the projection lens and the objective lens.

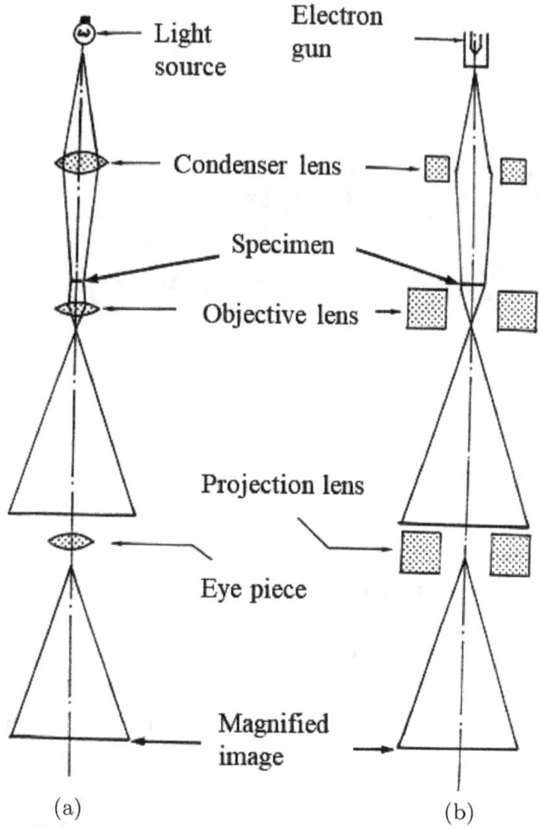

Fig. 5.1: Biological optical microscope (a) and transmission electron microscope (b).

gun is converged by a probe-forming lens to a probe whose diameter is small, which is scanned over the surface of a sample by a reflecting coil. An area illuminated by the probe (electron beam) emits secondary electrons etc., which are collected by a detector placed above the sample. The signals collected by the detector are displayed on a monitor, synchronized with the scanning of the electron probe. The scanned area is enlarged to a size of the monitor.

TEM provides information inside the sample, while SEM provides information on (near) the surface. When the sample is thin enough, electrons scanned on a small area of the sample may transmit the sample and be detected by a detector placed below the sample.

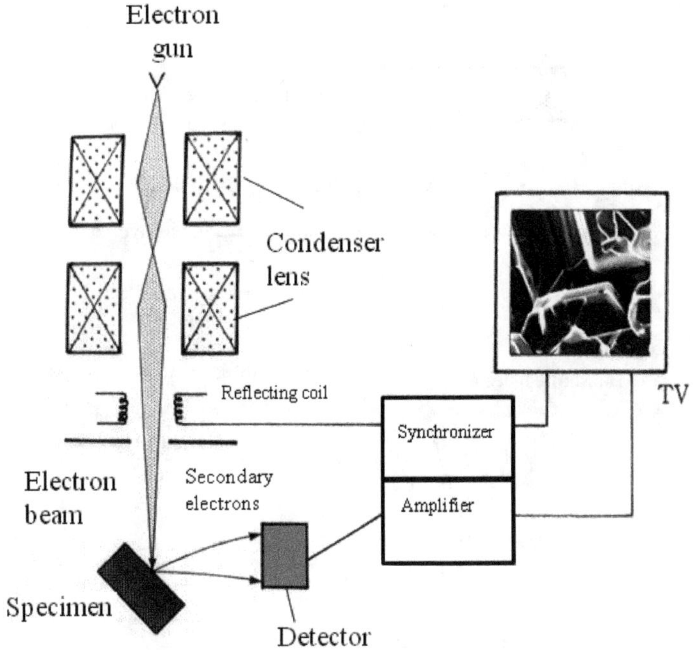

Fig. 5.2: Scanning electron microscope.

This mode is referred to as scanning transmission electron microscopy (STEM).[2] The image of STEM is essentially the same as SEM image except that, in SEM signals emitted in the direction reverse to the incident electron beam are collected, while in STEM an image is formed by electron probe which transmits the (thin) sample. In this chapter, conventional transmission electron microscope (CTEM) will be treated.[3]

5.2 Electron Diffraction and Electron Microscope

Suppose that a parallel electron beam is incident on a single crystal (Fig. 5.3). The electron beam is divided into two: One travels

[2]The mode shown in Fig. 5.1(b) is often referred to as conventional transmission electron microscope (CTEM) in order to distinguish from STEM.

[3]STEM will be treated in Chapter 10.

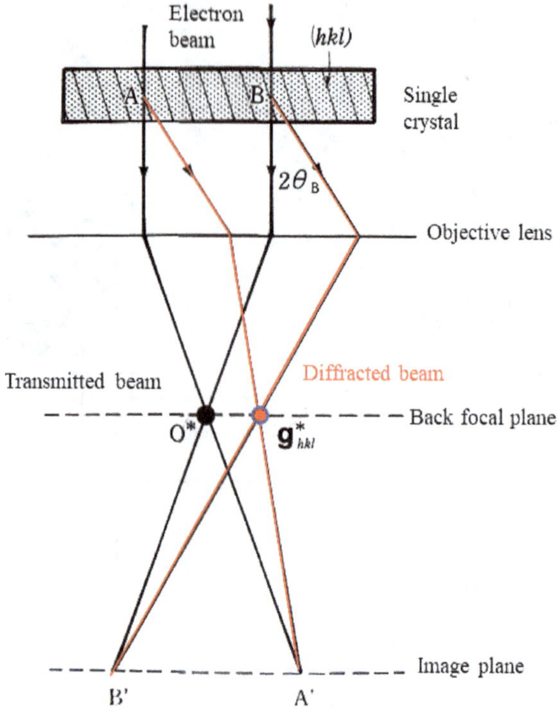

Fig. 5.3: Ray paths of a transmitted beam (which is not bent after entering a crystal: indicated by black line) and a diffracted beam (which is Bragg-reflected in a crystal: indicated by red line. On the back-focal plane the transmitted beam and the diffracted beam focus at $\mathbf{O}^*$ and $\mathbf{g}^*_{hkl}$, respectively. In this case, the electron beams passing through different areas A and B in the crystal cannot be distinguished. On the other hand, on the image plane, enlarged images A' and B', which correspond to areas A and B in the crystal, are obtained.

through the crystal straightly (transmitted beam) and the other is reflected inside the crystal by an angle $2\theta_B$ (θ_B is referred to as Bragg's angle), which is referred to as a diffracted beam. Both transmitted and diffracted beams travel downward through the objective lens and converse at the back-focal plane and then form an image at the image-plane.

If a film is placed at the back-focal plane, an electron diffraction pattern is obtained. If placed at the image-plane, a TEM image is obtained. Practically by adjusting the focal length of an intermediate lens, which is placed just below the objective lens, either the electron diffraction pattern or the TEM image is formed on a film[4] placed below the intermediate and projection lens.

5.3 Electron Diffraction Pattern and its Indexing

As described in Section 4.4, when the Ewald sphere intersects a reciprocal lattice point $\mathbf{g}^*_{hkl}$, the (hkl) plane produces a Bragg diffraction. If a lens is not present between the sample and film, the Bragg-diffracted beam is projected onto the film. This projected spot is referred to as a diffraction spot. The distance from the direct spot (transmitted beam) to the diffraction spot, r_{hkl}, on the film is given by

$$\mathrm{r}_{hkl}\mathrm{d}_{hkl} = \lambda \mathrm{L} \tag{5.1}$$

(Fig. 5.4) where λ is the wave length of electron beam and L is the distance between the sample and the film, which is referred to as a camera length. The right-side term λL does not depend on physical properties of the sample and depends solely on the apparatus and is referred to as a camera constant. In the actual electron microscope an objective lens, an intermediate lens and a projection lens are placed between a sample and a film, so that the physical distance between

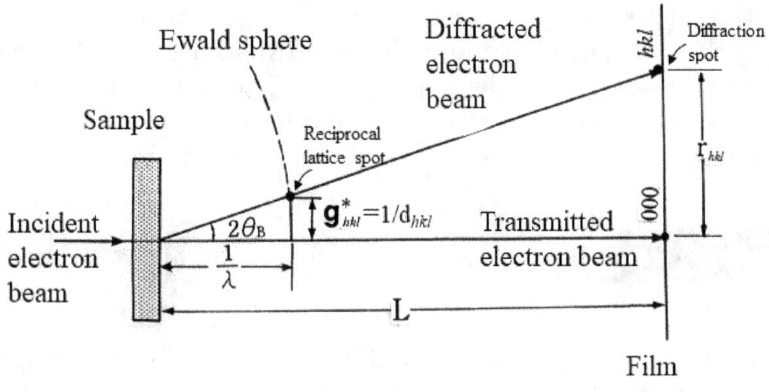

Fig. 5.4: Geometry of a reciprocal lattice spot, diffraction spot and the Ewald sphere.

[4]Instead of a film, a TV camera is used nowadays.

the sample and the film is not L ($= \mathrm{r}_{hkl}\mathrm{d}_{hkl}/\lambda$). However, L is referred to as a camera length, following the suit.

For a cubic crystal as shown in Appendix F,

$$\mathrm{d} = \frac{\mathrm{a}}{(h^2 + k^2 + l^2)^{1/2}} \quad (\mathrm{a} : \text{lattice constant}) \tag{5.2}$$

so that

$$\mathrm{r}_{hkl} = \left(\frac{\lambda \mathrm{L}}{\mathrm{a}}\right)(h^2 + k^2 + l^2)^{1/2} \tag{5.2'}$$

Indexing Debye rings

For a poly-crystal whose grains are randomly orientated, reciprocal lattice points make up concentric spheres, which are referred to as Debye rings. Radius of the Debye ring is given by r_{hkl} in Eq. (5.2). Extinction law is different from one crystal structure to another (see Section 4.5), so that the numerical sequence of $(h^2 + k^2 + l^2)^{1/2}$ in Eq. (5.2′) is characteristic to each of crystal structures. In other words, Debye rings may be regarded as *figure prints* of crystal structure. Figures 5.6(a) and 5.6(b) show schematically Debye rings of FCC and BCC, respectively.

Appendix F summarizes values of $(h^2 + k^2 + l^2)^{1/2}$ for which the structure factors are non-zero for typical cubic structures. By comparing these values with those obtained experimentally the crystal structure of sample can be determined. When the camera constant is known, lattice constants can be determined. Inversely by analyzing Debye rings of a known crystal, the camera constant can be calibrated.

Exercise 5.1: Index Debye rings in Fig. 5.5(a) and determine the camera constant (λL)

The intensity of Debye ring is different from one ring to another. The reason for this is that number of $\{hkl\}$ planes which satisfy Eq. (5.2′) is different (see Exercise 1.3). When distribution of orientations of each grain making up a poly-crystal is not perfectly random and has preferential orientations, parts of the Debye rings are missing (Fig. 5.5(b)).

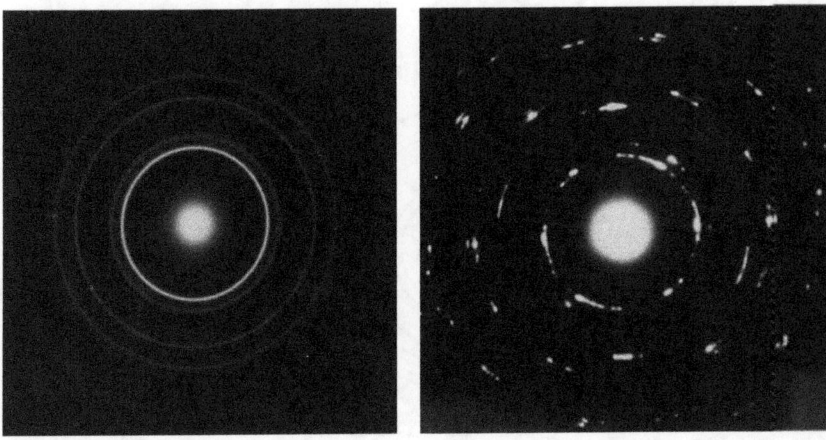

(a) (b)

Fig. 5.5: (a) Debye rings from a poly-crystal containing grains with perfectly random orientations. (b) Debye rings from a poly-crystal containing grains with preferred orientation (sample: Al).

Indexing a diffraction pattern from a single crystal

In the case of electron diffraction, the wavelength (λ) is quite short. Therefore, the radius of Ewald sphere ($1/\lambda$) is so large that it can be approximated by a plane.

> **Theorem 5.1:**
> A diffraction pattern from a single crystal is a cross-section of the reciprocal lattice intersected by a flat plane vertical to the direction of the incident electron beam (**B**).

Exercise 5.2: Suppose that an electron beam is incident in the directions of [001] and [011] of FCC structure. Draw the diffraction patterns.

Let us consider the case where the crystal structure is known but the orientation is unknown. In the diffraction pattern shown in Fig. 5.7, let us select three reciprocal vectors which satisfy

$$\mathbf{g}_3^* = \mathbf{g}_1^* + \mathbf{g}_2^* \tag{5.3}$$

In other words,

$$h_3k_3l_3 = h_1k_1l_1 + h_2k_2l_2 \tag{5.3'}$$

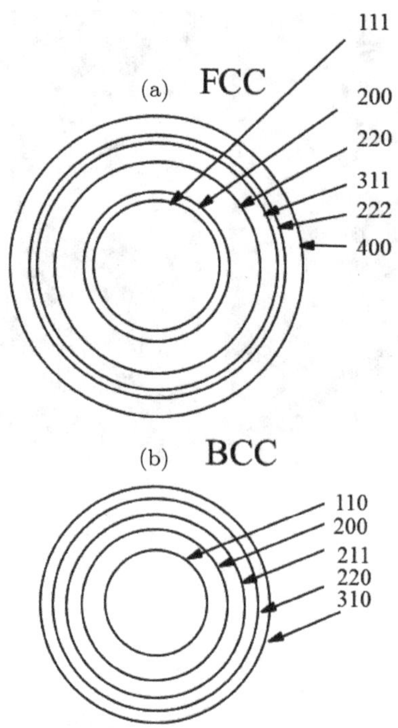

Fig. 5.6: Schematic illustrations of Debye rings from poly-crystals of FCC (a) and BCC (b) structures.

The choice is made by trial-and-error. Needless to say, it is necessary to choose those (hkl) planes whose structure factors are not zero.

Problem. Figure 5.8 shows an electron diffraction pattern from Al (FCC). Index this pattern.

Solution.

(1) When the camera constant is known.

The camera constant in this particular case is 2.48 nm × mm.

Choose OA, OB and OC for $\mathbf{g}_1^*$, $\mathbf{g}_2^*$ and $\mathbf{g}_3^*$. The lengths of OA, OB and OC as measured from the picture are 10.6 mm, 17.2 mm and 20.0 mm, respectively. Then, from Eq. (5.1), d_{hkl} are 0.234 nm, 0.144 nm and 0.124 nm. The lattice constant of Al is a = 0.4049 nm, so that the corresponding values of $(h^2 + k^2 + l^2)^{1/2}$ for OA, OB

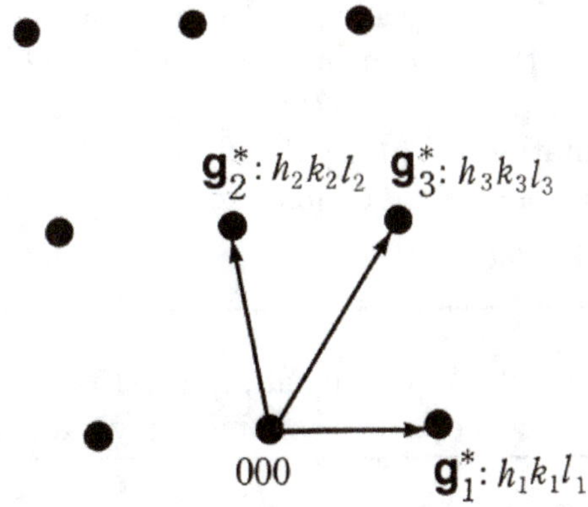

Fig. 5.7: How to index a diffraction pattern of single crystal. The three diffraction spots h_1, k_1, l_1, h_2, k_2, l_2 and h_3, k_3, l_3 satisfy relationship of $h_3, k_3, l_3 = h_1, k_1, l_1 + h_2, k_2, l_2$. The angle between two vectors drawing from the origin to h_i, k_i, l_i and h_j, k_j, l_j, that is, $h_i, k_i, l_i, h_j k_j, l_j$ is equal to the angle between the perpendiculars to the planes (h_i, k_i, l_i) and (h_j, k_j, l_j) in real space.

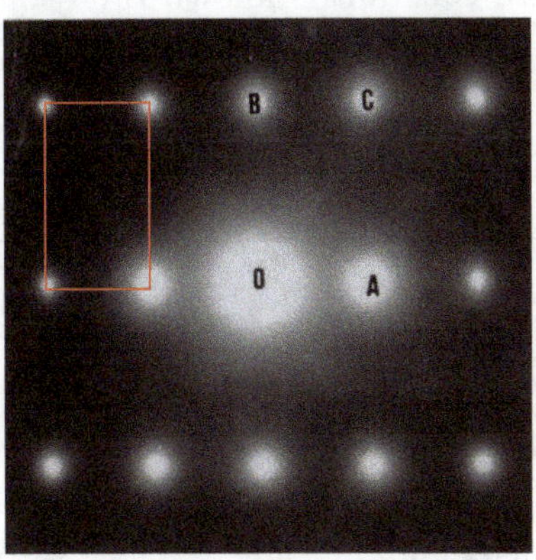

Fig. 5.8: Problem.

and OC are 1.732, 2.812 and 3.265, respectively. Since Al has FCC structure, Theorem 4.2 tells that all of *hkl* are either even or odd. Therefore, candidates of $\{hkl\}$ for OA, OB and OC are $\{111\}$, $\{220\}$ and $\{311\}$ (see Appendix I).

(2) When the camera constant is unknown.

First measure lengths of OA, OB and OC, as in (1). Calculate ratio of $(h^2 + k^2 + l^2)^{1/2}$ for OA, OB and OC from Eq. (5.2′). This is

1:0.615:0.530. This is in agreement with the ratio of $d_{111} : d_{220} : d_{311}$ within an error, so that it can be concluded that $\{hkl\}$ for OA, OB and OC are $\{111\}$, $\{220\}$ and $\{311\}$.

Among combinations of $\{111\}$, $\{220\}$ and $\{311\}$, those which satisfy Eq. (5.3′) are solutions. For instance,

$$1\bar{1}1 + 220 = 311, \quad 111 + 02\bar{2} = 13\bar{1}, \text{ etc.}$$

> **Point 5.1.**
> There are many solutions which satisfy Eq. (5.3′). In other words, indexing of a diffraction pattern of single crystal is not unique.

Exercise 5.3: Measure angles among $h_1k_1l_1$, $h_2k_2l_2$ and $h_3k_3l_3$, and confirm that these are consistent with calculated values (see Eq. (2.1) or Appendix D).

In summary,

(1) When the camera constant is known $L\lambda = 2.48\,\text{nm} \times \text{mm}$.

	r		Experimental value of	Theoretical value of	Solution
	(measured)	$d = (L\lambda)/r$	$\sqrt{h^2+k^2+l^2}$	$\sqrt{h^2+k^2+l^2}$	$\{hkl\}$
OA	10.6 nm	0.234 nm	1.73	1.732	$\{111\}$
OB	17.2 nm	0.144 nm	2.81	2.828	$\{220\}$
OC	20.0 mm	0.124 nm	3.27	3.317	$\{311\}$

(2) When the camera constant is unknown.

	r	$\frac{1}{r} \propto d$	Experimental ratio of d_{hkl}	Theoretical ratio of d_{hkl}	Solution $\{hkl\}$
OA	10.6 nm	0.094 nm	1	1	$\{111\}$
OB	17.2 nm	0.058 nm	0.617	0.6127	$\{220\}$
OC	20.0 mm	0.050 nm	0.530	0.5225	$\{311\}$

The direction of the incident electron beam **B** is given by

$$\begin{aligned}\mathbf{B} &= u\mathbf{a} + v\mathbf{b} + w\mathbf{c} \\ &= \begin{vmatrix} \mathbf{a} & \mathbf{b} & \mathbf{c} \\ h_1 & k_1 & l_1 \\ h_2 & k_2 & l_2 \end{vmatrix} = \mathbf{a}\begin{vmatrix} k_1 & l_1 \\ k_2 & l_2 \end{vmatrix} - \mathbf{b}\begin{vmatrix} h_1 & l_1 \\ h_2 & l_2 \end{vmatrix} + \mathbf{c}\begin{vmatrix} h_1 & k_1 \\ h_2 & k_2 \end{vmatrix} \\ &= \mathbf{a}(k_1l_2 - k_2l_1) + \mathbf{b}(l_1h_2 - l_2h_1) + \mathbf{c}(h_1k_2 - h_2l_1) \qquad (5.4)\end{aligned}$$

In this case, the rotation from $h_1k_1l_1$ to $h_2k_2l_2$ is to be taken anticlockwise.[5]

Exercise 5.4: Determine **B** for the diffraction pattern shown in Fig. 5.8.

Appendices J.1–J.4 show representative diffraction patterns for FCC, BCC, diamond structure and HCP, respectively.

In actual diffraction patterns many diffraction spots (i.e., reciprocal points) are present. These diffraction spots result from higher-order diffractions. This corresponds to putting $\eta = n$ (n: integers) in Eq. (4.10), then

$$\mathbf{K}^* = n\mathbf{g}^*_{hkl} \tag{4.10$'$}$$

Putting $n\mathbf{g}^*_{hkl} = \mathbf{g}'^*_{hkl}$ and applying the Ewald construction formally,

$$2d'_{hkl}\sin\theta_B = \lambda \tag{4.13$'$}$$

is obtained. Here,

$$|\mathbf{g}'^*_{hkl}| = \frac{1}{d'_{hkl}} = n|\mathbf{g}^*_{hkl}| = \frac{n}{d_{hkl}}$$

Rewriting Eq. (4.13′)

$$\begin{aligned} 2\left(\frac{d_{hkl}}{n}\right)\sin\theta_B &= \lambda \\ 2d_{hkl}\sin\theta_B &= n\lambda. \end{aligned} \tag{5.5}$$

In summary, when phases of two waves scattered by two neighboring (hkl) planes are just out of phase by λ, Eq. (4.13) holds. When out of phase by $n \times \lambda$, Eq. (5.5) holds.

[5]Throughout this book all the pictures are positive-printed with an emulsion upward. This is the same as images viewed on the flourescent screen across a viewing window. In other words, this corresponds to viewing a crystal held with hands from outside. For instance, in [001] stereographic projection, [001] is an upward vector. Taking **B** upward corresponds to treating a crystal in a conventional way.

In reciprocal space the reciprocal vector which corresponds to nth order diffraction is $\mathbf{g}'^{*}_{hkl}$, the length of which is $|n\mathbf{g}^{*}_{hkl}| = n|\mathbf{g}^{*}_{hkl}|$. Therefore, if higher-order diffractions are taken into account, the reciprocal space extends infinitely.

Multiple diffraction

Electrons interact with matter much more strongly than an X-ray. As a result, an electron beam, once diffracted and changes its direction, behaves as if the incident beam and is diffracted once again. This phenomenon is referred to as double diffraction (or double reflection) or more generally as multiple diffraction (or multiple reflections). Figure 5.9(a) shows a diffraction pattern obtained on a sample where poly-crystals of Al are deposited on a single crystal of Ag. As is obvious from Fig. 5.9(b), a Bragg-diffracted beam from the Ag single crystal behaves as if the incident beam, producing Debye rings of Al around the Bragg-diffracted spots from Ag.

Because of the double and/or multiple diffractions, an intensity of electron diffraction spots does not reflect exactly what is predicted by Eq. (4.15). This is a drawback in determining crystal structure by electron diffraction, however, here we confine ourselves to effects of double and/or multiple diffraction on the positions of diffraction spots.

Here, a Bragg-diffracted beam behaves as if the incident electron beam, so that the resultant diffraction pattern is obtained by superimposing the original diffraction pattern with its origin displaced to any Bragg-diffracted spot. Let us suppose that $\mathbf{g}^{*}_{h_1k_1l_1}$ is double diffracted, then the resultant (double-diffracted) pattern is obtained by displacing the origin of the non-double-diffracted pattern ($\mathbf{O}$) to the position of $\mathbf{g}^{*}_{h_1k_1l_1}$ and superimposing it. In the case of electron diffraction patterns shown in Fig. 5.8, where symmetry is high, the position of the spots remains unchanged although the intensities change. However, in a diffraction pattern such as shown in Fig. 5.10(a), a diffraction pattern of HCP structure which has low symmetry, this operation brings out spots, which, without double diffraction, should not appear (indicated by $\times$ in Fig. 5.10(b)). Such a spot is referred to as a forbidden spot.

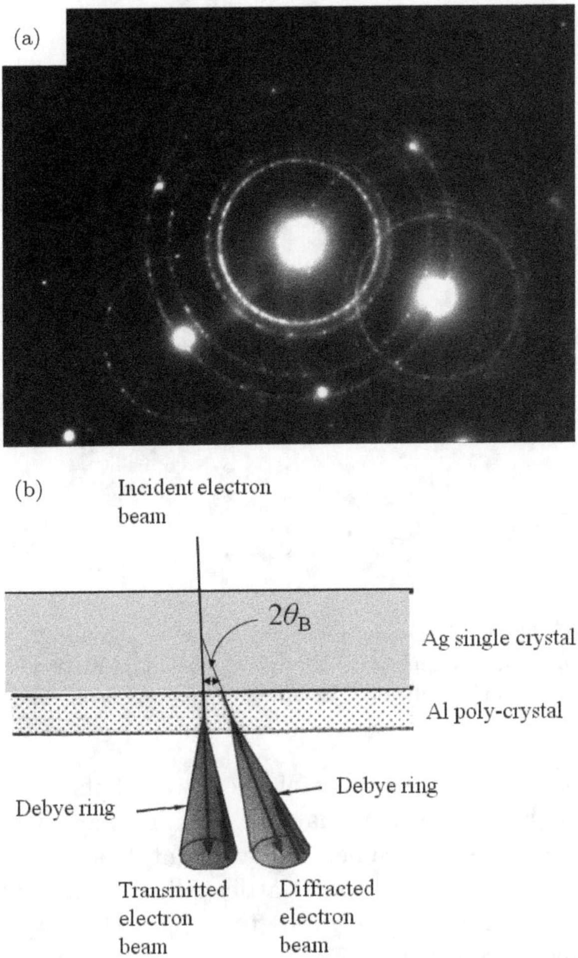

Fig. 5.9: An example of double diffraction. (a) Bragg-diffracted beams in single crystalline Ag behave as if the incident beam in Al poly-crystal. As a result, Debye rings of Al are formed with their centers at diffraction spots of Ag. (b) Poly-crystalline Al is deposited on the surface of Ag single crystal.

Laue zone

So far, the Ewald *sphere* is approximated by a *flat plane*. However, the Ewald sphere is a sphere and, in some cases, this must be taken into consideration.

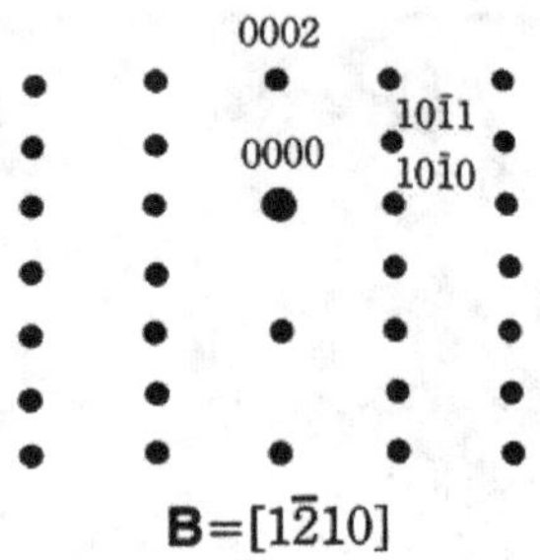

(a)

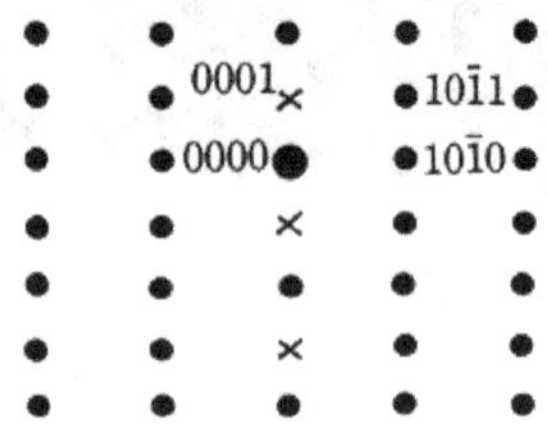

(b)

Fig. 5.10: Forbidden reflections appear as a result of multiple reflections. In HCP, 0001 spot is not to appear (a). However, when a diffraction spot, say, $10\bar{1}0$ reflection behaves as if the incident beam, 0001 spot appears.

Figure 5.11 shows geometry of the Ewald sphere and a reciprocal lattice. Here, the fact that the Ewald sphere is a sphere is a little bit emphasized. When the incident beam **B** is perfectly vertical to a sheet of reciprocal lattice, diffraction spots are perfectly symmetrical with respect to the origin (Fig. 5.11(a)). At a position remote from the origin, the Ewald sphere intersects spots of a reciprocal lattice sheet which does not lie on the origin. The diffraction spots are co-centric and the circle containing the origin ($\mathbf{g}^* = 000$) is referred to as the 0th order Laue zone (ZOLZ), followed by the first order Laue zone (FOLZ), and the second order Laue zone (SOLZ), etc.

When **B** is tilted a little bit, the center of the 0th order Laue zone (ZOLZ) deviates from the origin of the reciprocal lattice as shown in Fig. 5.11(b). On further tilting, only part of the 0th order Laue zone (ZOLZ) intersects the Ewald sphere (Fig. 5.11(c)).

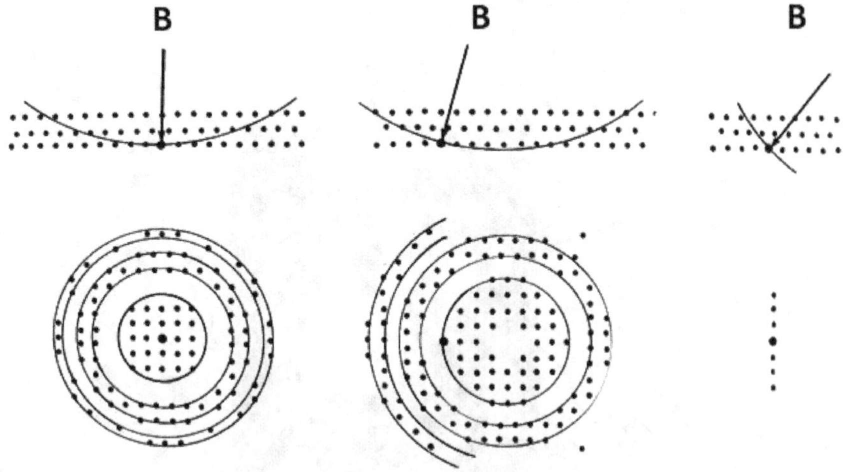

(a) (b) (c)

Fig. 5.11: Relationship between the direction of the incident beam **B** and distribution of the intensity of diffraction spots. (a) When **B** is parallel to a symmetry axis of crystal (symmetrical incidence), the distribution of the intensity of diffraction spots is symmetrical with respect to the transmitted beam (direct spot). When **B** deviates slightly from the symmetrical incidence, the position of diffraction spots does not change significantly, but the distribution of intensity changes (b). (c) **B** deviates from the symmetrical incidence extremely.

5.4 Kikuchi Pattern

Features of Kikuchi pattern

As discussed in the foregoing section, when a crystal tilts slightly, intensity of individual diffraction spots changes. However, the positions of spots remain almost unchanged. In other words, the direction of incident beam (**B**) determined from a diffraction pattern is not accurate. On the other hand, a Kikuchi pattern, a kind of diffraction patterns, changes its position depending sensitively on a slight change in the orientation of crystal. Therefore, by using a Kikuchi pattern a crystal orientation can be determined precisely.

The Kikuchi pattern appears when the sample is fairly thick. Figure 5.12(a) shows a typical Kikuchi pattern, which consists of pairs of parallel white and black lines. The Kikuchi pattern is formed

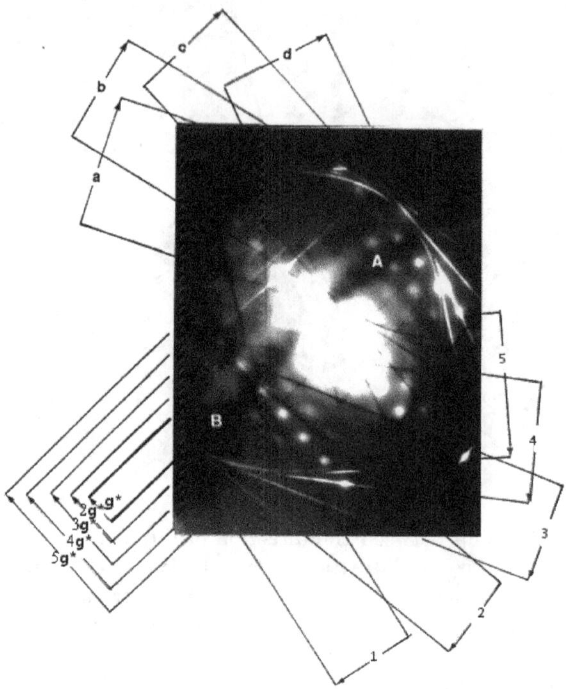

(a)

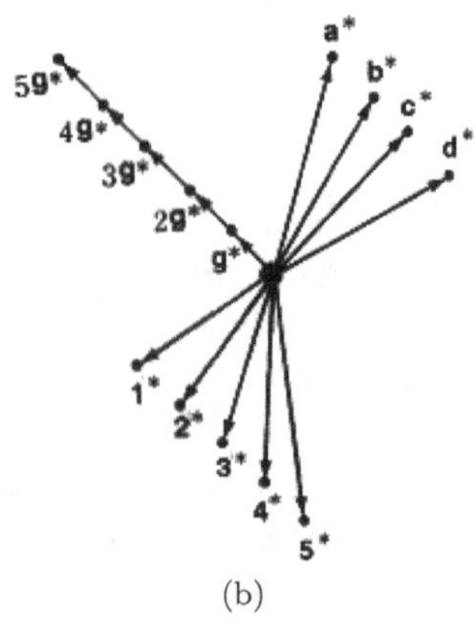

(b)

Fig. 5.12: Indexing Kikuchi lines. From Kikuchi lines of $\mathbf{1}^*\sim\mathbf{5}^*$, $\mathbf{a}^*\sim\mathbf{d}^*$, $\mathbf{g}^*\sim5\mathbf{g}^*$ shown in (a) the diffraction pattern shown in (b) is obtained. Specimen is Cu (FCC).

when those incident electrons which are scattered inelastically[6] within a crystal and lose kinetic energy slightly are Bragg-diffracted again.

Now, let us suppose that an incident electron beam is scattered inelastically at P, propagating to all directions. Among them, those electrons scattered in a direction PA are Bragg-diffracted by (*hkl*) plane and those scattered to a direction PB are Bragg-diffracted by $(\bar{h}\bar{k}\bar{l})$ plane (that is, the back side of (*hkl*) plane). PA is inclined at an angle α with respect to the direction of incident beam OP, and PB is inclined at an angle β. Now, let us assume that $\alpha > \beta$. Intensity of an inelastically scattered electron decreases with increasing the scattering angle: Intensity of those electrons scattered in direction PA is larger than that of electrons scattered in direction PB. In Fig. 5.13(a), PA is printed thicker than PB to show that the intensity of PA is larger than that of PB.

Electrons scattered inelastically to direction PA are Bragg-diffracted by (*hkl*) plane to direction AB′. Similarly electrons scattered inelastically to the direction of PB are Bragg-diffracted by $(\bar{h}\bar{k}\bar{l})$ plane to direction BA′. In other words, electrons are swapped between α- and β-sides. Since the inelastically scattered intensity is larger for α than β, the net balance of electrons is deficient in α-side. In other words, the intensity of electrons in α-side, i.e., travelling along OA′, is weaker than the background, and that of electrons in β-side, i.e., travelling along OB′ is stronger than the background. As a result, a pair of white and black lines appears.

Electrons which are Bragg-diffracted by (*hkl*) and $(\bar{h}\bar{k}\bar{l})$ planes propagate along a cone with a semi-angle of (90-$\theta°$) (Fig. 5.13(b)). In electron diffractions θ is quite small, so that section of the cones with the film can be approximated by two straight lines. These two straight lines are referred to as Kikuchi lines. Kikuchi lines consist of a pair of black and white lines, each of which corresponds to (*hkl*) and $(\bar{h}\bar{k}\bar{l})$ planes, respectively.

[6]Scattering by which an electron does not lose its kinetic energy is referred to as elastic scattering. When an electron loses its kinetic energy by scattering, the scattering is referred to as inelastic scattering.

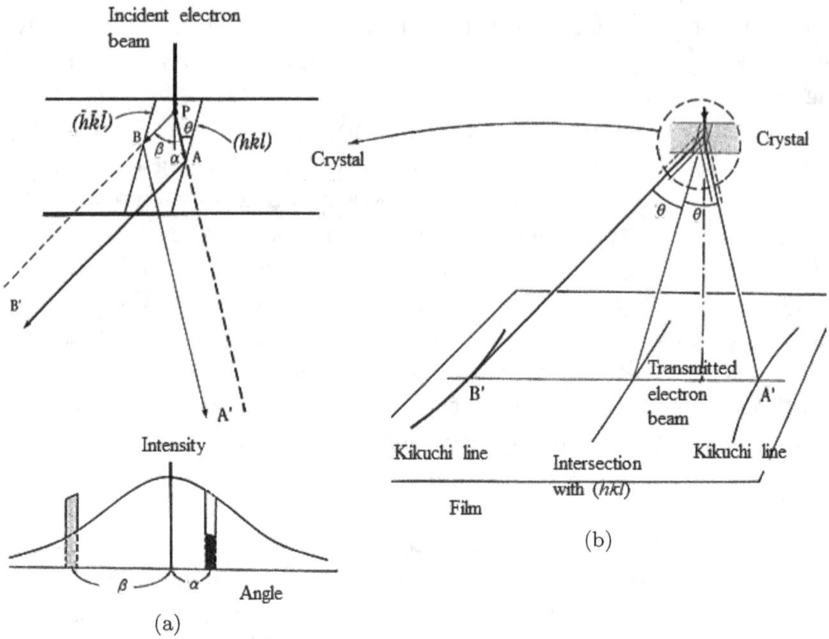

Fig. 5.13: Mechanism by which Kikuchi lines are formed.

Theorem 5.2:

(1) The distance between the black-and-white pair is $|\mathbf{g}^*_{hkl}|$.
(2) Mid-line of the black-and-white pair can be regarded as a projection of (hkl) plane in real space.

Exercise 5.5: Prove Theorem 5.2.

(*Hint*) See Fig. 5.4.

Indexing Kikuchi pattern

Let us index a Kikuchi pattern shown in Fig. 5.12(a). According to Theorem 5.2, the direction of $\mathbf{g}^*_{hkl}$ is perpendicular to the corresponding Kikuchi line and its magnitude $|\mathbf{g}^*_{hkl}|$ is equal to the distance of the black-and-white pair of the Kikuchi line. Thus, the reciprocal lattice can be constructed from a Kikuchi pattern as follows:

Point 5.2.

(1) Draw perpendiculars to $\mathbf{1}^*\sim\mathbf{5}^*$, $\mathbf{a}^*\sim\mathbf{d}^*$ and $\mathbf{g}^*\sim5\mathbf{g}^*$ (Fig. 5.12(a)). Here, $\mathbf{g}^*$, $2\mathbf{g}^*$, $3\mathbf{g}^*$, $4\mathbf{g}^*$, $5\mathbf{g}^*$ are Kikuchi lines of first, second, third, fourth and fifth orders.
(2) Displace these perpendiculars to the origin of the reciprocal lattice (Fig. 5.12(b)).
(3) The end of the perpendicular corresponds to the reciprocal point of (*hkl*) plane, that is, $\mathbf{g}^*_{hkl}$ which produces the Kikuchi lines.
(4) Index these reciprocal points using Eq. (5.3).

Exercise 5.6: Index Kikuchi lines shown in Fig. 5.12(b). The Kikuchi lines cross at A and B. Determine orientations of A and B and the angle between A and B. Here, $\mathbf{g}^*$ is *002*.

Relationship between Kikuchi pattern and diffraction pattern

Let us assume that a thin crystal the orientation of which is shown in Fig. 5.3 is gradually curved, opening upwards (Fig. 5.4). When the incident electron beam **B** is perfectly parallel to the plane (in this case (*100*) plane) shown in Fig. 5.14(a), the reciprocal points are outside the Ewald sphere and Bragg's condition is satisfied neither for $\mathbf{g}^*_{100}$ nor for $-\mathbf{g}^*_{100}$ perfectly. This situation is defined as $\mathbf{s}^*_g < 0$. At the right side of the crystal, where the crystal is bent anticlockwise, the reciprocal lattice rotates anticlockwise accordingly and eventually the Ewald sphere intersects the reciprocal point $\mathbf{g}^*_{100}$. In this case $\mathbf{g}^*_{100}$ satisfies the Bragg's condition perfectly and therefore $\mathbf{s}^*_g = 0$, that is, $\mathbf{g}^*_{100}$ is excited.

On the other hand, at the left side where the crystal is bent clockwise, the Ewald sphere intersects a reciprocal lattice point of $\mathbf{g}^*_{100}$. In other words, $\mathbf{s}^*_{100} = 0$. On further bending clockwise, $\mathbf{s}^*_{100} > 0$. In summary,

Definition 5.1:
When a reciprocal point is inside the Ewald sphere, $\mathbf{s}^*_g > 0$.
When a reciprocal point is outside the Ewald sphere, $\mathbf{s}^*_g < 0$.

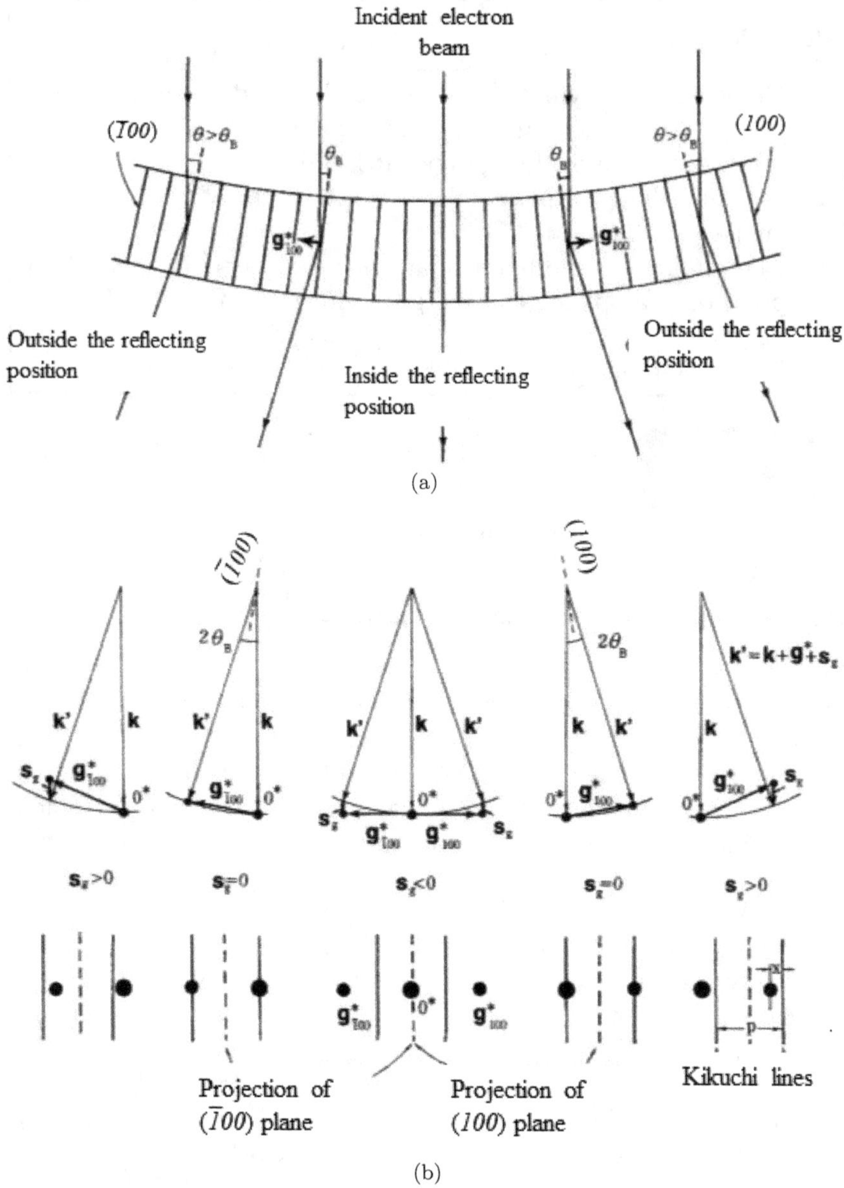

Fig. 5.14: Relationship among a curved crystal, Ewald sphere, reciprocal points, deviation parameters $\mathbf{s}_{\mathrm{g}}^{*}$, diffraction spots $\mathbf{g}_{hkl}^{*}$ and Kikuchi lines. See also Note in page 158.

Exercise 5.7: Discuss on a situation where a foil specimen is curved, opening downwards, which corresponds to the reverse to Fig. 5.14.

Let us go back to the case where the incident electron beam is parallel to (*100*) plane. In this case, the (*100*) plane passes through the direct spot (the origin) and the pair of Kikuchi lines pass midway between the direct spot (the origin) and $\mathbf{g}^*_{100}$ and $\mathbf{g}^*_{\bar{1}00}$. When $\mathbf{s}^*_{100} = 0$, the Kikuchi pair pass through the origin and right above the diffraction spot of $\mathbf{g}^*_{100}$. In summary,

$$|\mathbf{s}^*_{\mathrm{g}}| = \frac{\lambda \mathbf{x}}{(\mathrm{d}^2_{hkl}\mathrm{p})} \tag{5.6}$$

Exercise 5.8: Derive Eq. (5.6) (see Fig. 5.14(b)).

Figure 5.15 shows a concrete example how to adjust Bragg's condition.

Kikuchi map

A Kikuchi pattern provides the exact orientation of a crystal, so that it is convenient to tilt the specimen while watching the Kikuchi pattern. The Kikuchi pattern can be compared to a road map or a car navigation on driving a car. For this reason, a Kikuchi pattern is often referred to as a Kikuchi map.

Figure 5.16(a) illustrates a schematic Kikuchi map of BCC crystal with $\mathbf{B} = [001]$. Pair of indices along a Kikuchi line stand for the indices of the Kikuchi line. Encircled indices stand for the zone axes. Figure 5.16(b) shows a 001 standard projection. It is to be noted that positions of the zone axes on the Kikuchi map (in reciprocal space) are just reverse to those on the standard projection (in real space). This can be accounted for by considering the geometry shown in Fig. 5.16(c).

Appendices K.1–K.4 illustrate Kikuchi maps for FCC (for $\mathbf{B} = [111]$), BCC (for $\mathbf{B} = [001]$, FCC (for $\mathbf{B} = [011]$) and HCP (for $\mathbf{B} = [1\bar{2}11]$ or equivalently $[0\bar{1}0]$). In K.1, is shown also the corresponding Thompson tetrahedra (see Section 3.4.1).

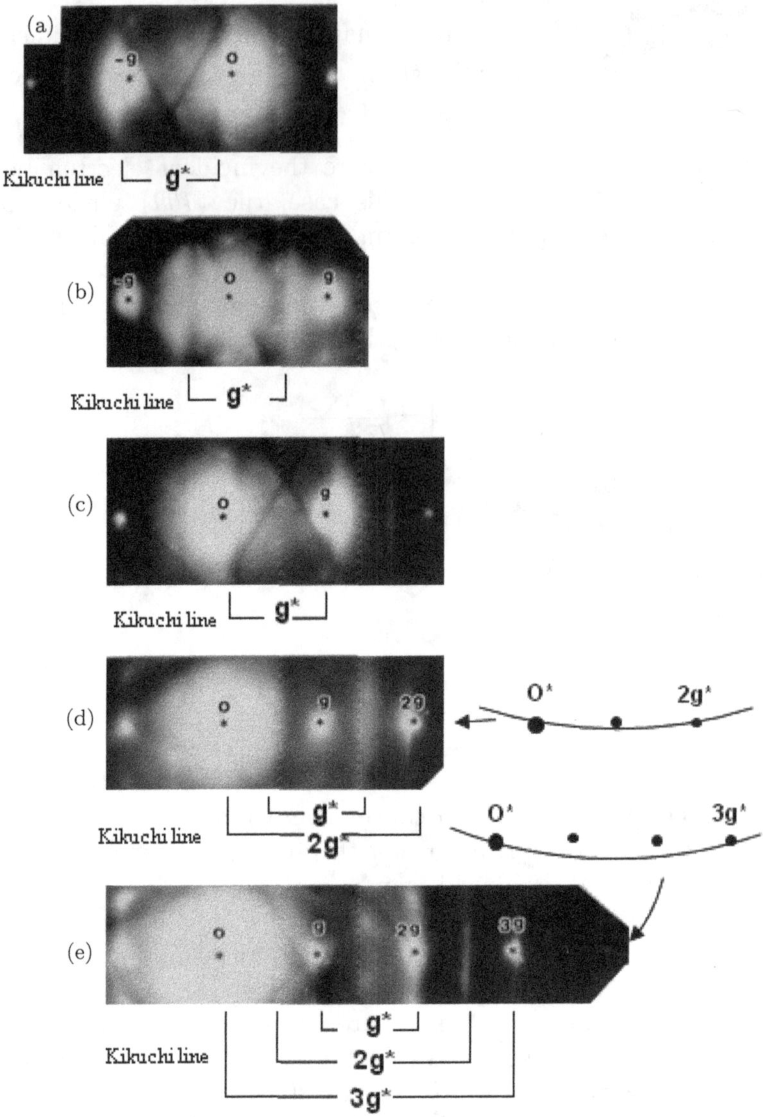

Fig. 5.15: How to adjust the Bragg's condition. Vertical lines drawn below each photograph stand for Kikuchi lines (pairs). In (a), $\mathbf{g}^*_{hkl}$ satisfies the Bragg's condition. In (b), $\mathbf{s}^*_g < 0$ for both $\mathbf{g}^*$ and $\mathbf{g}^*$ (symmetrical incidence). In (c), $\mathbf{g}^*$ satisfies the Bragg's condition. In (d) $2\mathbf{g}^*$ satisfies the Bragg's condition, and in (e) $3\mathbf{g}^*$ satisfies the Bragg's condition. Correspondingly, in (c), the first-order Kikuchi line passes on the diffraction spot $\mathbf{g}^*$. Similarly, in (d) second Kikuchi line and in (e) third Kikuchi line pass on $2\mathbf{g}^*$ and $3\mathbf{g}^*$, respectively.

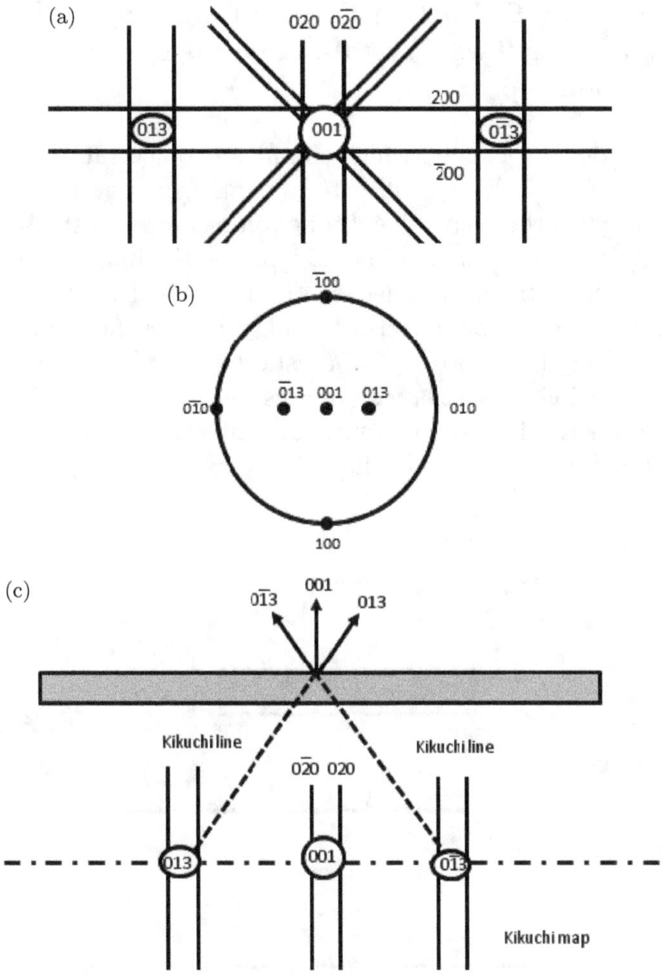

Fig. 5.16: Figures associated with each of Kikuchi lines (e.g., 200, $\bar{2}00$) stand for indices of the Kikuchi lines. Encircled indices stand for zone axes. Note that the positions of zone axes in the Kikuchi map are inversed to those shown in a standard projection in (b). (c) Geometry.

Kikuchi patterns (or lines) are produced by inelastic scattering, so that they are observed in a thick specimen. In other words, they do not appear in a very thin foil specimen. When using a wedge-shaped specimen, it is convenient to adjust Bragg's condition at a fairly thick area and then move to a thinner area.

5.5 Analyzing Diffraction Pattern from a Specimen Containing Two Crystals: Analysis of Orientation Relationship

Let us consider a specimen which contains two crystals ⓐ and ⓑ as shown in Fig. 5.17. Bragg-diffracted beams from crystal ⓐ and ⓑ converse to form the respective diffraction patterns on the back-focal plane and then form images of crystal and on the image plane. Thus, the diffraction pattern obtained on the back-focal plane is a superimposition of diffraction patterns from crystals ⓐ and ⓑ: it provides no information on the position of crystal ⓐ and ⓑ, however, it provides information of orientations of crystals ⓐ and ⓑ. (By contrast, the image on the image-plane provides information on positions of crystals ⓐ and ⓑ.) In what follows some examples are shown.

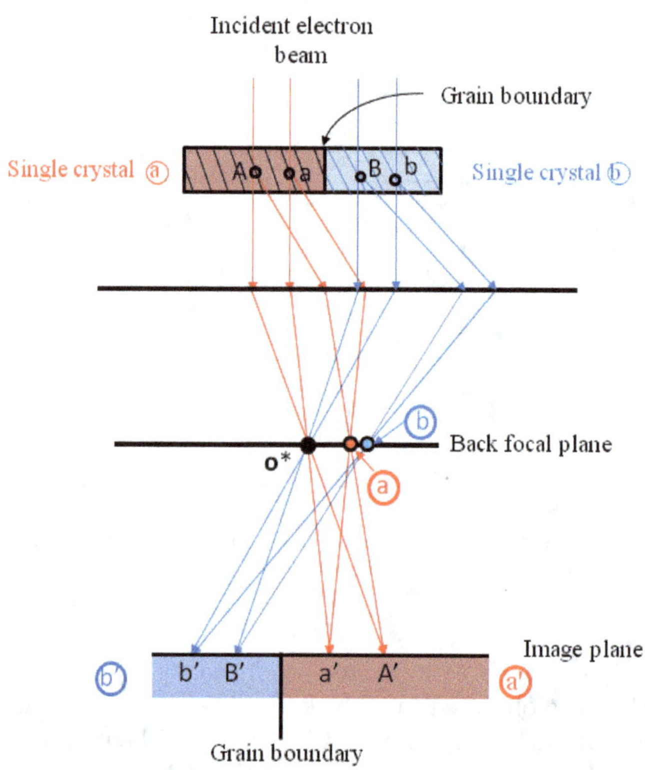

Fig. 5.17: Paths of electron beam from two single crystals ⓐ and ⓑ.

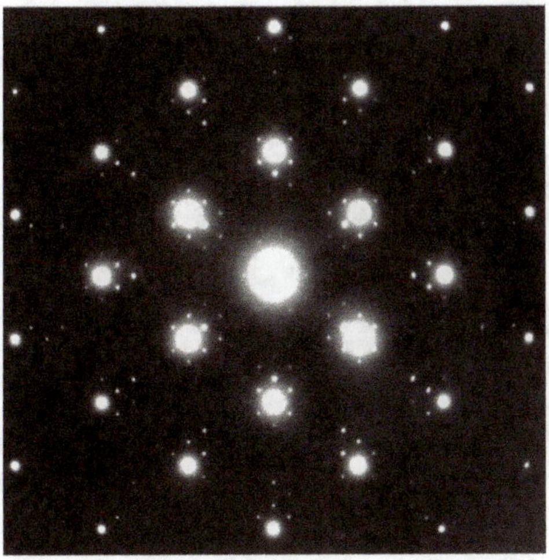

(a)

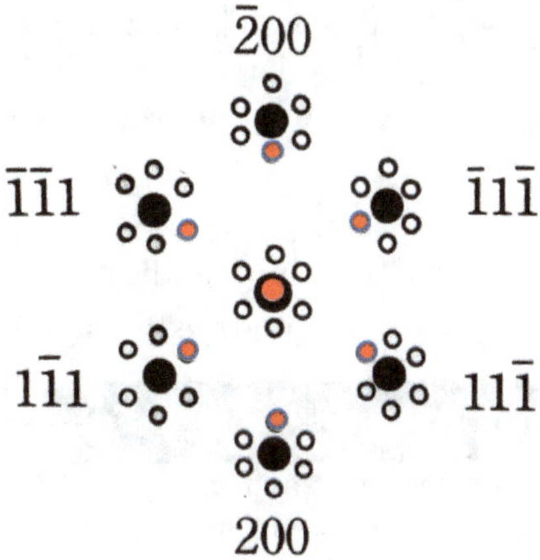

(b)

Fig. 5.18: (a) Diffraction pattern of Pb particles embedded in an Al matrix. **B**//[011]. (b) Key diagram. ● indicates diffraction spots of Al. • indicates those of Pb. ○ are forbidden reflections resulting from multiple reflection (by courtesy of Dr. E. Johnson).

Example 1 (Fine particles of Pb embedded in an Al matrix).

Figure 5.18(a) shows a diffraction pattern obtained on Pb fine particles embedded in an Al matrix. Both Al and Pb crystallize in FCC, the lattice constants of which are 0.40490 nm (for Al) and 0.49495 nm

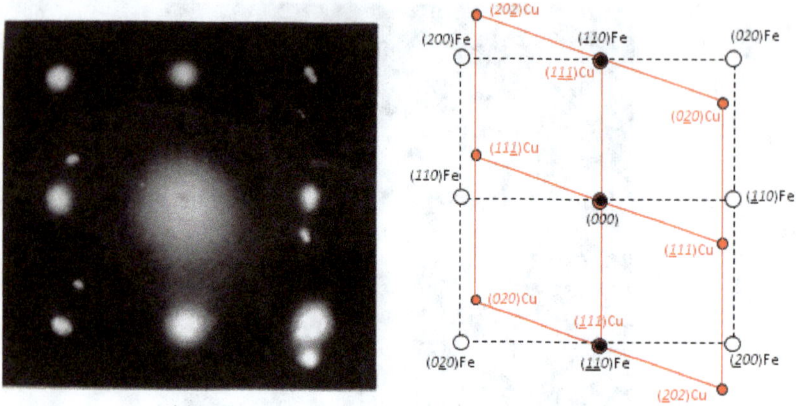

(a) (b)

Fig. 5.19: (a) Diffraction pattern of α-Fe containing precipitates of Cu. (b) Key diagram. Orientation relationship between Cu and α-Fe can be determined.

(for Pb). Figure 5.18(b) illustrates an indexed key-diagram. Since the lattice constant of Pb is slightly larger than that of Al, diffraction spots of Pb (indicated by red circles) appear inside those of Al (see Eq. (5.2)).

Exercise 5.9: Confirm that spots indicated by open circles in Fig. 5.18(b) are forbidden spots produced by a double diffraction.

Example 2 (Precipitates of Cu in α-Fe). Figure 5.19(a) shows a diffraction pattern taken from Cu (FCC: a = 0.3615 nm) precipitates in α-Fe (BCC: a = 0.2866 nm) and Fig. 5.19(b) illustrates the key diagram. Applying Eq. (5.4) to the diffraction spots from α-Fe, the direction of the incident electron beam **B** for α-Fe is determined to be [001]. Similarly, **B** for Cu is [101]. Spot *110* from α-Fe is coincident with spot $1\bar{1}\bar{1}$ from Cu. From these, the following relationship exists between α-Fe and Cu.

$$[001]_{\mathrm{Fe}}//[101]_{\mathrm{Cu}} \quad (110)_{\mathrm{Fe}}//(1\bar{1}\bar{1})_{\mathrm{Cu}} \tag{5.7}$$[7]

[7]In this particular case, spot *110* from α-Fe happens to be coincident with spot $1\bar{1}\bar{1}$ from Cu. However, this is not the necessary condition. Sufficient condition is that *110* and $1\bar{1}\bar{1}$ are parallel.

The relationship like this, which correlates orientations of two crystals, is referred to as an orientation relationship.[8]

Exercise 5.10: In the example shown in Fig. 5.19, the following relationship also holds. Show this.

$$(\bar{1}10)_{\mathrm{Fe}}//(\bar{1}2\bar{1})_{\mathrm{Cu}} \tag{5.7$'$}$$

Example 3. When a precipitate is thin, the reciprocal point elongates in a direction perpendicular to the thinnest direction (see Section 4.5). Therefore, diffraction spots from a thin precipitate are elongated. Such elongated diffraction spots are referred to as streaks. Figures 5.20(a) and 5.20(b) show a TEM micrograph together with a diffraction pattern taken at an initial stage of precipitation where planar Co precipitates from Co_3Ti matrix. Lattice constants of Co_3Ti and Co are very close to each other: Co_3Ti crystallizes in $L1_2$ structure (see Fig. 1.18 in Section 1.4.2), while Co crystallizes in FCC.[9] Therefore, in a diffraction pattern of Co_3Ti with $\mathbf{B} = [001]$ superlattice spots appear as shown in Fig. 5.20(c) (see Fig. J.1).

On the other hand, wide faces of the planar precipitate of Co are $\{100\}$, their normal directions being aligned along [100] and [010] when $\mathbf{B} = [001]$, as shown by A and B in Fig. 5.20(a). Thus, diffraction spots from precipitates A and B are elongated in the direction of [100] and [010], as shown in Figs. 5.20(d) and 5.20(e), respectively. The actual SAD pattern is taken on an area containing both A and B. Therefore, the resultant SAD are superimposition of Figs. 5.20(c)–5.20(e), as shown in Fig. 5.20(f).

Example 4 (Twin). A twin consists of two crystals facing each other with mirror symmetry with respect to a crystallographic plane. For instance, in FCC structure, $[1\bar{1}0]$ direction on (111) plane faces in reverse directions with each other. In Frank's notation (See 3.4.4),

---

[8] Among orientation relationships between BCC and FCC, those expressed by Eq. (5.7) or (5.7′) are referred to as Nishiyama's relationship (N relationship). Besides N relationship, Kurdjumov-Sacks (KS) relationship is well known. KS relationship is expressed by $(011)_{\mathrm{BCC}}//(111)_{\mathrm{FCC}}$, $(11\bar{1})_{\mathrm{BCC}}//(10\bar{1})_{\mathrm{FCC}}$, $(\bar{2}1\bar{1})_{\mathrm{BCC}}//(\bar{1}2\bar{1})_{\mathrm{FCC}}$ (5.8)

[9] Co crystallizes in FCC at high temperature and in HCP at low temperature.

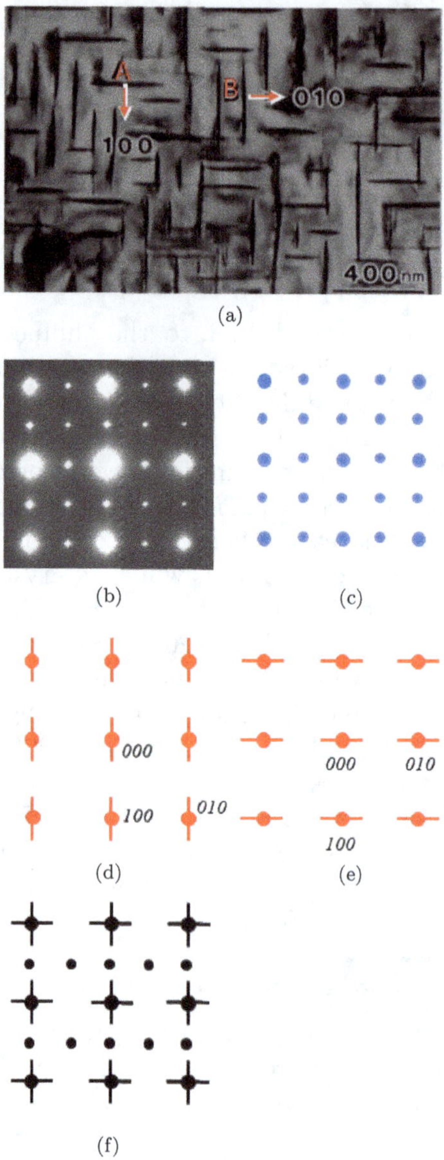

Fig. 5.20: (a) Planar precipitates of Co in Co_3Ti. (b) Diffraction pattern. Fundamental spots from Co are elongated along the normal directions of planar precipitates A and B shown in (a). (c) Illustrates a schematic diffraction pattern of the matrix Co_3Ti. Smaller spots stand for superlattice spots. (d) and (e) Illustrate schematic diffraction patterns of the Co precipitates, showing streaks along the normal directions of A and B. (f) Illustrates superimposition of (d) and (e) and corresponds to the actual diffraction pattern (b).

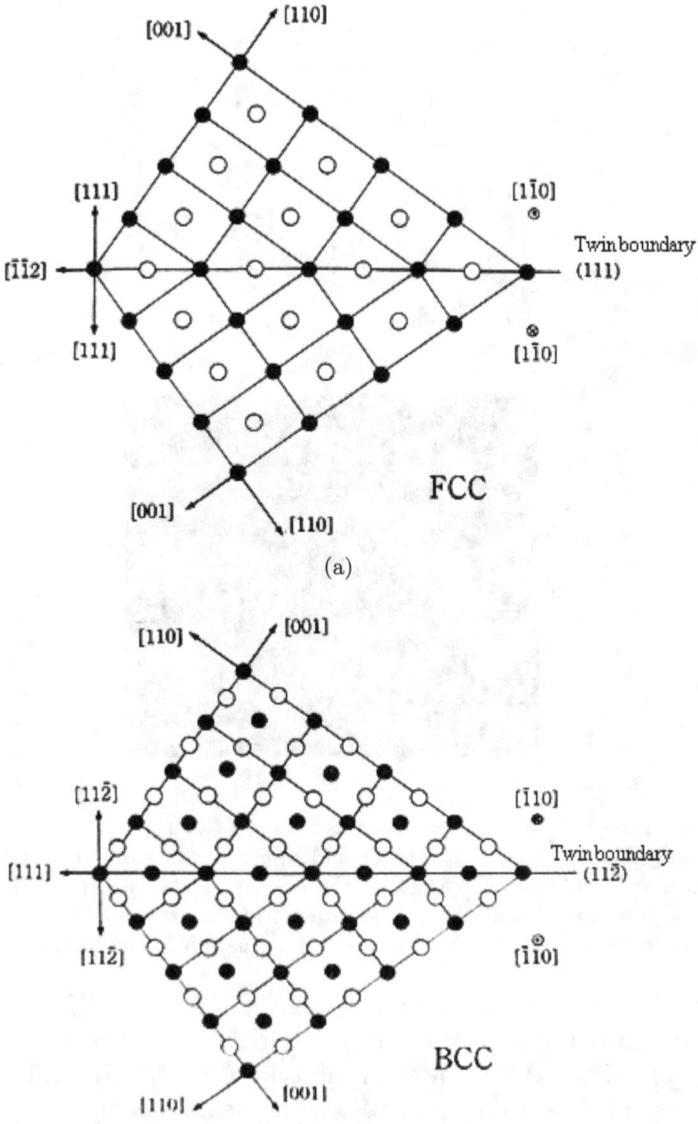

Fig. 5.21: Twin. (a) FCC. (b) BCC. Upper and lower crystals are in mirror symmetry with respect to the twin plane.

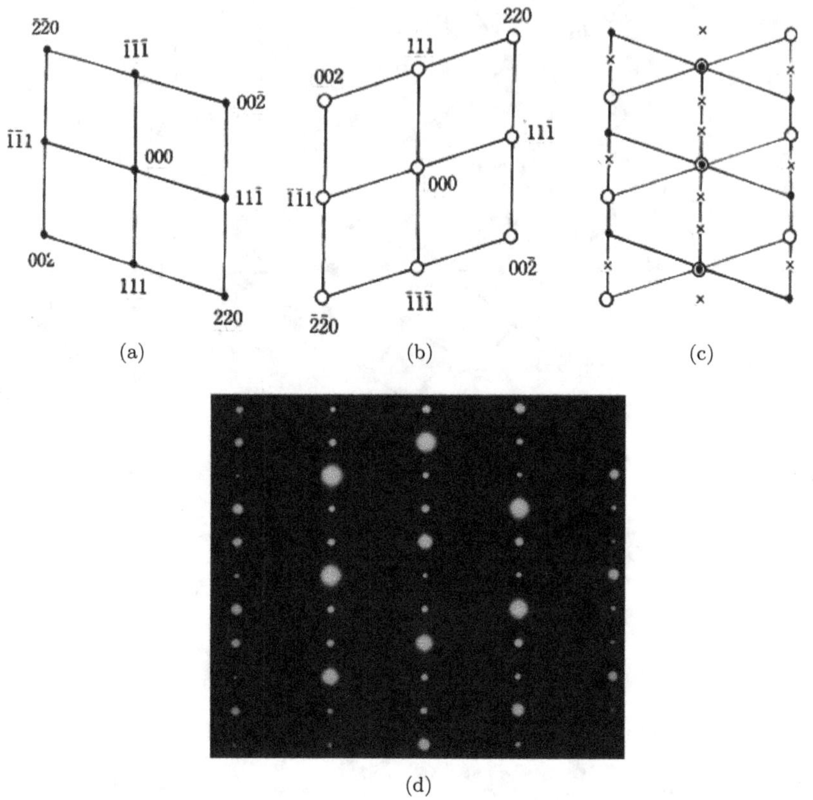

Fig. 5.22: (a)–(d) Diffraction pattern of a twin in FCC. (a) and (b) Correspond to the diffraction pattern of the lower and upper crystals in (a), respectively. (c) Illustrates a diffraction pattern of the overall twin, which corresponds to superimposition of (a) and (b). × indicates forbidden reflections produced by superimposing (a) and (b). (d) An example of actual diffraction pattern.

a twin is described by either △△△△▽▽▽▽ or ▽▽▽▽△△△△. There are two types of stacking fault in FCC, either an intrinsic type (△△△▽△△△) or an extrinsic type (△△△▽▽△△△). They can be regarded as a very thin twin with one atomic layer (intrinsic) or two atomic layers (extrinsic).[10] In BCC, $[\bar{1}10]$ direction on (*112*) planes points in reverse directions with each other (Fig. 5.21(b)).

[10]In a notation introduced by Saka (Classical Theory of Crystal Dislocations — From Iron to Gallium Nitride), △△ and ▽▽ are denoted by □ and ■, respectively, △▽ and ▽△ are denoted by ⬡ and ⬢, respectively. With this notation an intrinsic stacking fault (△ △ △ ▽ △ △ △) is denoted by □□⬡⬢□□ and an

Now let us suppose that an incident electron beam (direction = **B**) is perpendicular to Fig. 5.21(a). The diffraction pattern from the lower crystal of the twin is shown in Fig. 5.22(a) and that from the upper half is shown in Fig. 5.22(b). Then, the diffraction pattern from the twin (Fig. 5.22(c)) is a superimposition of these two diffraction patterns (Figs. 5.22(a) and 5.22(b)). Spots denoted by × are forbidden spots which are produced by double reflection. Figure 5.22(d) shows an example.

Example 5 (Diffraction pattern from a sample containing many twins and/or stacking faults).

Let us consider that many twins and/or stacking faults are present in a sample. When many thin crystals, whether twins or the SF, are present in a sample, the diffraction spots from these thin crystals are elongated in the direction of their perpendiculars and accompanied with streaks (Fig. 5.23).

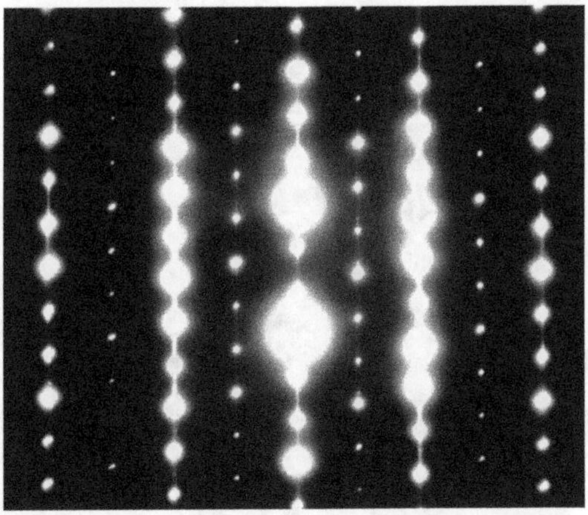

Fig. 5.23: Diffraction pattern of a crystal containing many twins and stacking faults. Specimen is a shape memory alloy of Cu–Zn–Al.

---

extrinsic one (△ △ △ ▽ ▽ △ △△) by □□⬡■⬢□□. In other words, an intrinsic stacking fault contains two atomic layers of HCP and an extrinsic stacking fault contains two atomic layers of HCP plus a twin with one atomic thickness. See also Section 3.4.4.

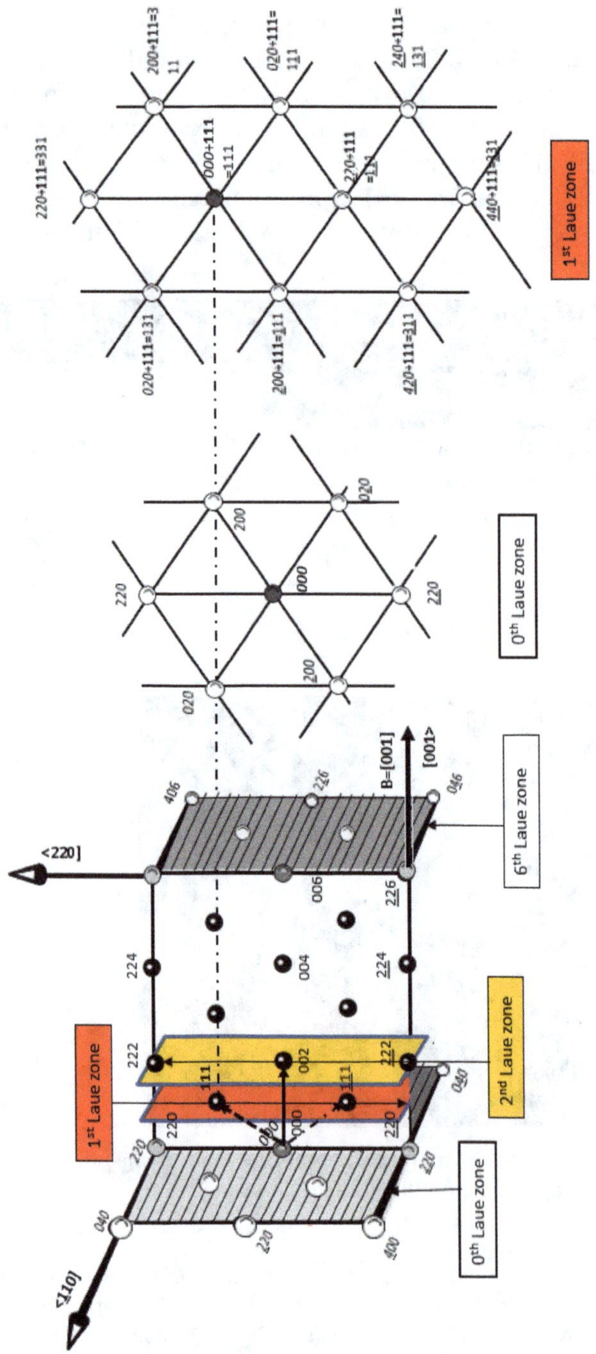

Fig. 5.24: Let the direction of the incident electron beam be $\mathbf{B} = [\mathrm{u\,v\,w}]$ and the diffraction spot be hkl. For the 0th Laue zone $\mathrm{u}h + \mathrm{v}k + \mathrm{w}l = 0$ holds. For the first Laue zone $\mathrm{u}h + \mathrm{v}k + wl = 1$ holds. Similarly, for the Nth Laue zone, $\mathrm{u}h + \mathrm{v}k + \mathrm{w}l = \mathrm{N}$ holds. Underlined index indicates minus, i.e., $\underline{h} = -h = \bar{h}$.

5.6 Indexing Laue Zones

Indexing a Laue zone is nothing but indexing diffraction spots three-dimensionally. Figure 5.24(a) illustrates the reciprocal lattice of FCC three-dimensionally for **B** = [001]. Here, an area shaded heavily is located below ($\bar{1}10$) and that shaded weakly is above the ($\bar{1}10$) plane. Figure 5.24(b) illustrates 0th order Laue zone (ZOLZ). Indexing of second-order Laue zone (SOLZ) is easily performed by displacing ZOLZ to *002*. Indexing first-order Laue zone (FOLZ) appears a little bit complex, but is performed by displacing ZOLZ in the direction indicated by a broken line (Fig. 5.24(c)).

Exercise 5.11: In indexing FOLZ, displacing ZOLZ in a direction indicated by a dashed line, i.e., to $\bar{1}\bar{1}1$ leads to the same result as Fig. 5.24(c). Confirm this.

Exercise 5.12: For **B** = [110] in FCC crystal, when N is odd in uh + vk + wl = N, no spots appear. Discuss reasons for this.

Chapter 6

TEM Image of a Perfect Crystal

6.1 Diffraction Contrast and Bright-Field and Dark-Field Images

As is described in the foregoing section (Section 5.2), a diffraction pattern is obtained when a film is placed at the back-focal plane of the objective lens and an image (inversed) is obtained when a film is placed at the image-plane. In practice, by adjusting the magnitude of an intermediate lens and a projection lens placed below the objective lens the magnification of images (SAD and image) is adjusted. However, quantity and quality of information which SAD and image contain are primarily determined by the performance of the objective lens.

Now, let us consider mechanisms by which contrast is formed in TEM images of crystal. Figures 6.1 is essentially a repeat of Fig. 5.17, but an objective aperture is inserted at the back-focal plane of objective lens. In Fig. 6.1(a), the aperture is inserted in such a way that the diffracted beam (indicated by broken lines) is stopped, allowing the transmitted beam alone (indicated by solid lines) to travel downward.

Now, let us make two assumptions as follows:

Definition 6.1:
(1) Only the transmitted beam and one diffracted beam are considered (referred to as two-beam approximation). (2) The diffracted intensity is so weak that it is not diffracted again (kinematical theory).

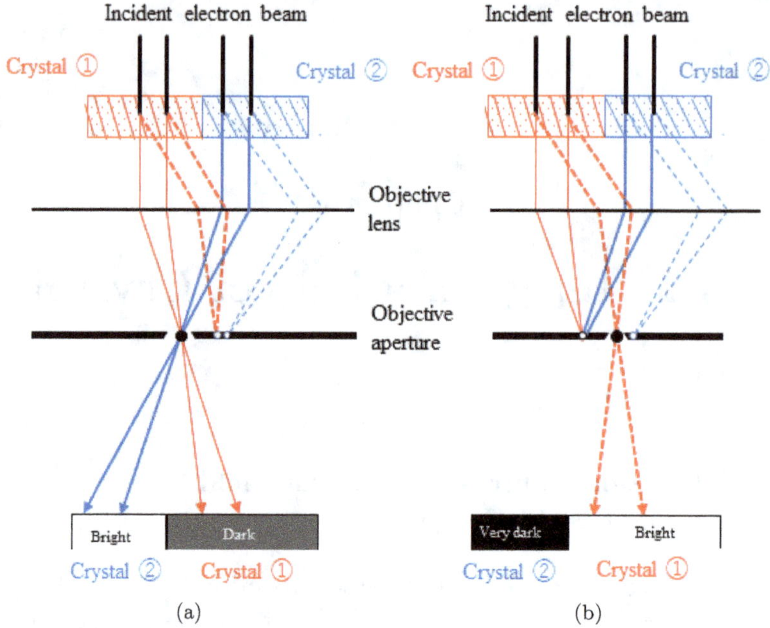

Fig. 6.1: Ray paths of the diffracted and transmitted beams from a specimen containing two crystals ① and ②. (a) Bright-field imaging. (b) Dark-field imaging.

First, let us indicate the diffracted and transmitted amplitudes by ϕ_g and ϕ_0, respectively. Then, assuming the incident intensity being 1, the following relationships hold.

$$\phi_0^2 + \phi_g^2 = 1$$
$$\phi_0^2 \gg \phi_g^2 \tag{6.1}$$

Now let us assume that crystal ① satisfies the Bragg's condition perfectly, while crystal ② does not. In other words, ϕ_g^2 is larger for crystal ① than for ② crystal. Therefore, the transmitted intensity passing through the objective aperture $\phi_0^2(=1-\phi_g^2)$ is much weaker for crystal ① than for crystal ②. As a result, crystal ① appears dark and crystal ② appears bright on a positive film.[1]

However, when the objective aperture is displaced and set to the diffracted beam from crystal ①, then the transmitted beam and the

[1]Throughout this book all the images are positive images printed with an emulsion upwards. Thus, an area illuminated strongly by electron beam appears bright. This situation is the same as a conventional optical photograph.

diffracted beam from crystal ② are stopped completely. As a result, crystal ① appears bright and crystal ② appears dark. This mode of imaging where the objective aperture allows the diffracted beam to pass and stop the transmitted beam is referred to as dark-field (DF) imaging mode. On the other hand, when the transmitted beam passes the objective aperture and the transmitted beams are stopped (corresponding to Fig. 6.1(a)), it is referred to as bright-field (BF) imaging mode.

Contrasts described above are formed depending on whether a crystal satisfies the Bragg's condition or not, and for this reason are referred to as diffraction contrast.

Figure 6.2(a) shows a BF image of Al poly-crystal. Grain ①, which appears dark, satisfies the Bragg's condition, and the

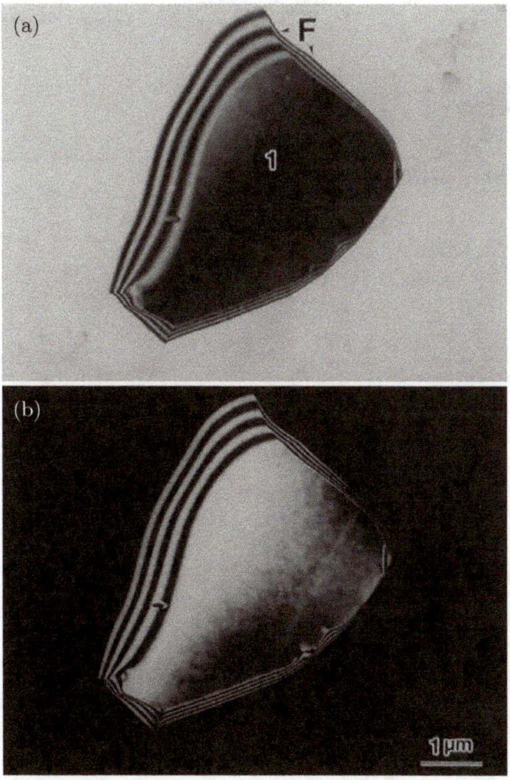

Fig. 6.2: (a) Bright-field (BF) image and (b) dark-field (DF) image of an Al poly-crystal. Crystal ① appears dark in BF and bright in DF. F indicates thickness contours at a grain boundary.

surrounding grains do not. Figure 6.2(b) is a DF image where a spot from grain ① is used. Now, grain ① appears bright and the surrounding grains appear dark.

The concrete procedures are as follows. In the BF imaging mode, first observe the diffraction pattern, then displace the objective aperture to the transmitted spot (direct spot), then come back to the imaging mode (Fig. 6.3(a)). In the DF imaging mode, the objective

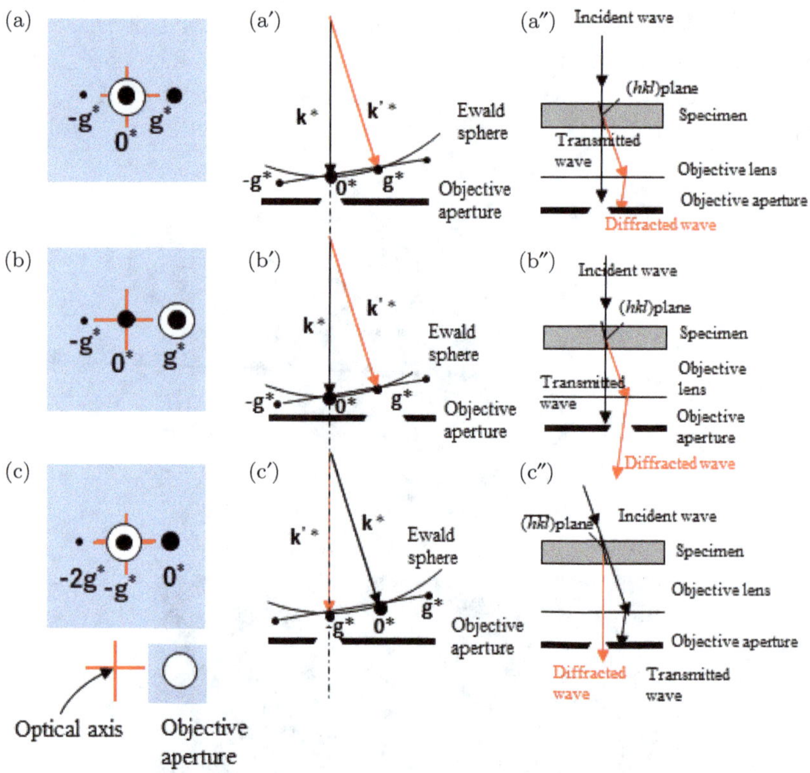

Fig. 6.3: (a) How to set an objective aperture to image in BF mode. (b) A simple method to image in DF mode: An objective aperture is displaced to a diffracted beam. This is an easy method, but since the diffracted beam is tilted from the optical axis, the image is blurred. (c) How to take a sharp DF image. The incident beam is tilted in such a way that the diffracted beam is along the optical axis, then an objective aperture is set to the diffracted beam. (a′), (b′) and (c′) show relationship between reciprocal lattice and the Ewald sphere. (a″), (b″) and (c″) show ray paths of electron beam in real space. In the DF imaged in method (c), the sign of the excited diffraction vector $\mathbf{g}^*$ is reversed with respect to BF.

aperture is displaced to the diffraction spot with which you want to take a DF picture (Fig. 6.3(b)). In this case, however, the diffracted beam is slightly tilted from the optical axis of the objective lens (Fig. 6.3(b″)), resulting in blurring of the DF image obtained. In order to prevent this, the incident electron beam is tilted by $2\theta_B$ with a deflection coil before it enters the sample, so that the diffracted beam is now along the optical axis of the objective lens (Fig. 6.3(c″)).

Concrete procedures are as follows. The diffraction spot with which a DF image is to be formed is adjusted to the optical axis of the objective lens, then set the objective aperture to the diffraction spot (Fig. 6.3(c)). As can be seen from relation between the Ewald sphere and the reciprocal lattice shown in Figs. 6.3(a′) and 6.3(c′), a diffraction spot $-\mathbf{g}^*$ is excited instead of $\mathbf{g}^*$ which is excited in the BF mode. If you want to excite $\mathbf{g}^*$, you have to tilt the sample again to adjust the Bragg's condition.

6.2 Selected Area Diffraction (SAD)

What happens in Fig. 6.1 if an aperture is inserted at the image-plane instead of the back-focal plane? Figure 6.4 illustrates ray paths for

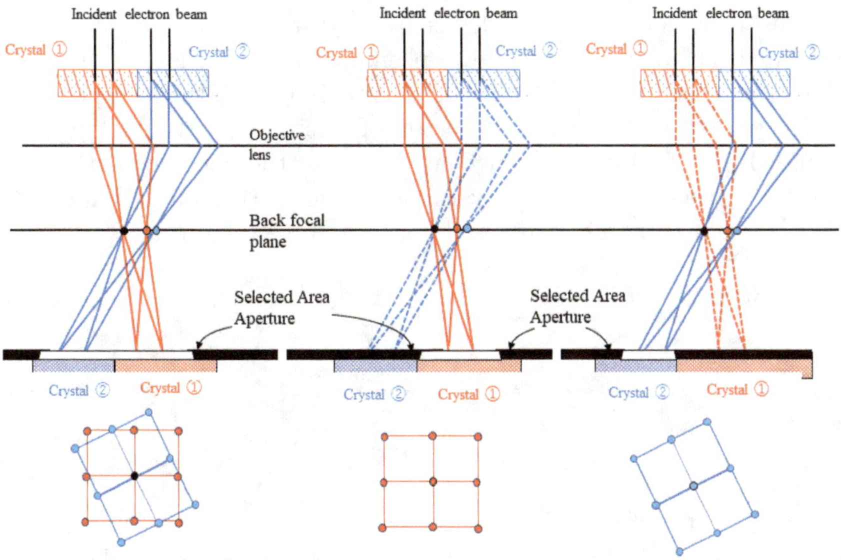

Fig. 6.4: Selected area diffraction.

this case. In the image-plane, an inversed image of a sample, i.e., images of grains ① and ② are formed. If, at this stage, the imaging mode is switched from IMAGE mode to DIFF mode, or in other words, the intermediate lens is focused to the back-focal plane, a diffraction pattern where diffraction patterns from grains ① and ② are superimposed is obtained (Fig. 6.4(a)). If the aperture is set to the image of grain ①, the electron beam from grain ② is stopped, and only the electron beam from grain ① travels downward (Fig. 6.4(b)). Practically, a diffraction pattern is obtained by adjusting the focus of an intermediate lens placed just below the objective lens, and a diffraction pattern from grain ① alone is obtained. In other words, only grain ① is viewed through the aperture. Naturally, if the aperture is set to grain ②, a diffraction pattern from grain ② alone is obtained (Fig. 6.4(c)).

In this way, a diffraction pattern from any selected area is obtained, so that this is referred to as a SAD and the aperture is referred to as a selected area aperture. SAD and DF imaging are inseparably combined.

6.3 Thickness Contour and Bend Contour

Let us consider a perfect defect-free crystal. A whole crystal is regarded as being composed of ensemble of unit cells. Let us consider this ensemble as stacking of n thin slabs with thickness of unit cell (i.e., 1**c**, where **c** is the vector of a unit cell) parallel to x–y plane (Fig. 6.5). An electron beam entering the entrance surface of crystal travels through the crystal as if along a column with a diameter of approximately 1nm and reaches the exit surface (left side

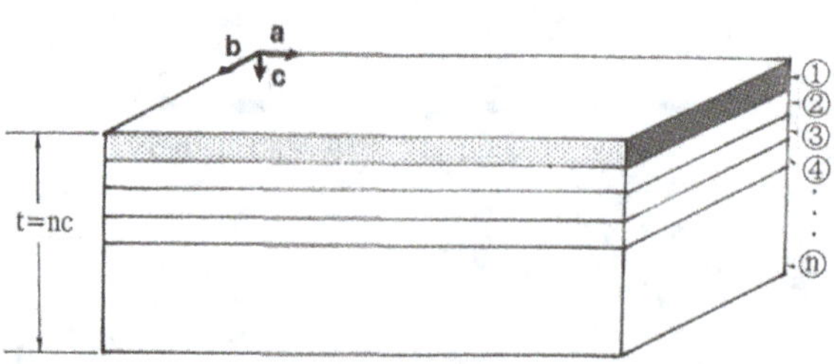

Fig. 6.5: Model of a foil specimen for TEM. A total of n slabs with thickness c (c is the thickness of unit cell) ①, ②, ③, . . . ,ⓝ are stacked.

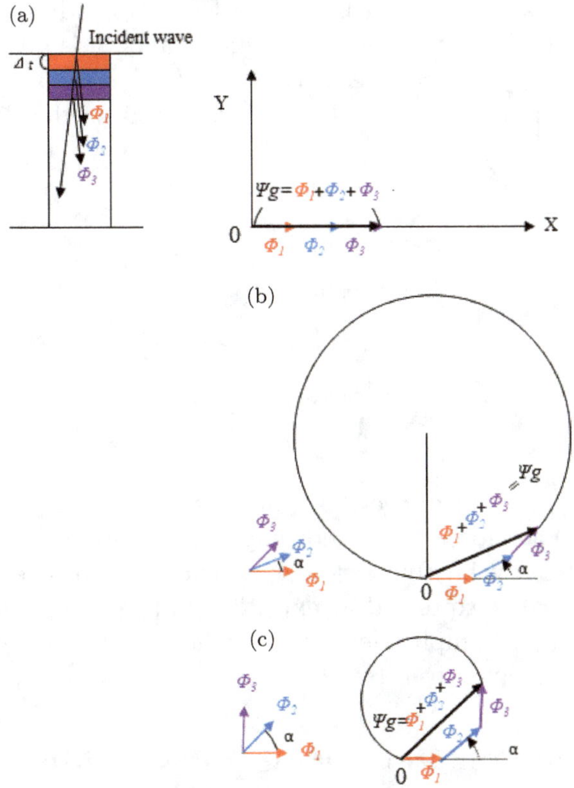

Fig. 6.6: Amplitude-phase diagrams for a foil specimen shown in Fig. 6.5. In (a) the Bragg's condition is satisfied perfectly, while in (b) and (c) the Bragg's condition is not satisfied perfectly.

of Fig. 6.6(a)). This is referred to as column approximation. On traveling through the crystal, the incident electron beam is Bragg-diffracted by each of the unit cells, i.e., scattering centers, which make up the crystal. Let the diffracted beam from the first top unit cell ① be named ϕ_1 and that from the second unit cell ② ϕ_2 and so on.

Now let us suppose that the crystal under question satisfies the Bragg's condition perfectly. In this case, $\phi_1, \phi_2, \phi_3, \ldots$ are perfectly in phase. Therefore, in the amplitude-phase diagram, $\phi_1, \phi_2, \phi_3, \ldots$ point perfectly the same direction. The diffracted amplitude at the exit surface is given by $\phi_1 \times n(n \times |\mathbf{c}| = \mathrm{t}$; t = thickness of crystal), and is proportional to thickness t.

Next, let us consider a case where Bragg's condition is slightly deviated. The phase difference between ϕ_1 and ϕ_2 is given by a difference in the ray path between ① and ② (=**r**) multiplied by $2\pi\mathbf{K}^*$, as is described in Section 4.3. If the deviation from the Bragg's condition is defined by Eq. (4.17), the phase difference between ϕ_1 and $\phi_2(=\alpha)$ is given by

$$\begin{aligned}\alpha &= -2\pi\mathbf{K}^* \cdot \mathbf{r} \\ &= -2\pi(\mathbf{g}^*_{hkl} + \mathbf{s}^*_g) \cdot \mathbf{r} \\ &= -2\pi(\mathbf{g}^*_{hkl} \cdot \mathbf{r} + \mathbf{s}^*_g \cdot \mathbf{r}) \\ &= -2\pi\mathbf{s}^*_g \cdot \mathbf{r} \\ &= -2\pi\mathbf{s}^*_g \cdot \mathbf{c} \end{aligned} \tag{6.2}$$

Here, $\mathbf{g}^*_{hkl} \cdot \mathbf{r}$ is an integer, as can be seen from Eqs. (2.4), (4.7) and (4.12). Therefore, it does not affect the phase difference. When ϕ_1 is expressed by $1 + 0i(=\exp(i0))$, ϕ_2 is expressed by $\exp(i\alpha)$.

Similarly, ϕ_3 is expressed by $\exp(2i\alpha)$ and ϕ_4 by $\exp(3i\alpha)$ and so on. The diffracted amplitude from the whole crystal is a summation of $\phi_1, \phi_2, \phi_3, \ldots$ and given by

$$\phi_\mathrm{g} = \phi_1 + \phi_2 + \phi_3 + \cdots + \phi_\mathrm{n} = \sum_{\mathrm{n}=1}^{\mathrm{n}} \exp\{(n-1)\,\alpha i\} \tag{6.3}$$

When $\alpha \neq 0$, the amplitude-phase diagram becomes a circle as shown in Fig. 6.6(b). ϕ_g increases at first gradually and after reaching a maximum at the diameter of the circle, it decreases and eventually becomes zero. Then, it increases again. This process is repeated. With increasing $\mathbf{s}^*_\mathrm{g}$ the diameter of the circle of an amplitude-phase diagram decreases (Fig. 6.6(c)). This can be compared to driving a car along a circle with a steering wheel rotated. The rotation is compared to α. If the rotation (α) is large, the circle along which a car drives becomes smaller.

To summarize, with increasing the thickness of a crystal, the diffracted amplitude ϕ_g repeats up-and-down periodically.

This can be understood if vectors representing diffracted beam $\phi_1, \phi_2, \phi_3, \ldots$ are drawn from the origin of a complex plane (Fig. 6.7(a)). Now let us assume that, when diffracted beams from first to nth unit cells are superimposed, ϕ_g happens to be zero.

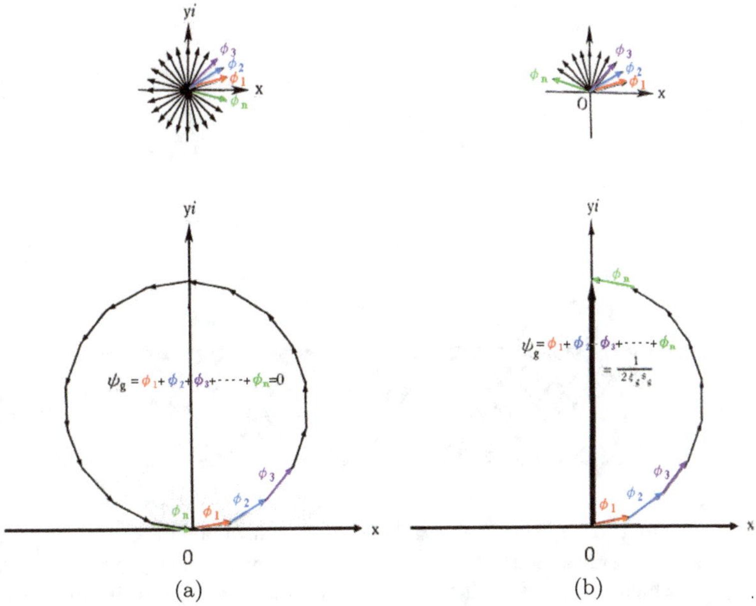

Fig. 6.7: Explanation of Fig. 6.6. In (a), $\phi_1, \phi_2, \phi_3, \ldots, \phi_n$ point all directions, so that $\psi_g = \phi_1 + \phi_2 + \phi_3 + \cdots + \phi_n = 0$, while in (b), all of $\phi_1, \phi_2, \phi_3, \ldots, \phi_n$ direct upward, so that $\psi_g = \phi_1 + \phi_2 + \phi_3 + \cdots + \phi_n$ is a maximum.

That is, a car travels just one complete circuit. At this point, $\phi_1, \phi_2, \phi_3, \ldots, \phi_n$ point all directions, and hence its summation is zero. By contrast, when ϕ_g is a maximum (a car is just at half-round of its circuit), all of $\phi_1, \phi_2, \phi_3, \ldots, \phi_n$ are in the first and second orthants, so that they point upwards. Thus, its summation is a maximum (Fig. 6.7(b)).

Superimposing sinusoidal curves helps an intuitional understanding. A superimposed amplitude $\phi_1 + \phi_2$ is larger than each of ϕ_1 and ϕ_2 and its phase is just in between those of ϕ_1 and ϕ_2 (Fig. 6.8(a)). However, superimposing sinusoidal curves, the phase of which ranges from 0 to 2π with slightly different phase, results in the zero-amplitude (Fig. 6.8(b)).

Exercise 6.1. So far, diffracted amplitudes from $\phi_1, \phi_2, \phi_3, \ldots$ are assumed to be 1 for the sake of simplicity. Calculate the radius of an amplitude-phase diagram when the amplitudes of $\phi_1, \phi_2, \phi_3, \ldots$ are A.

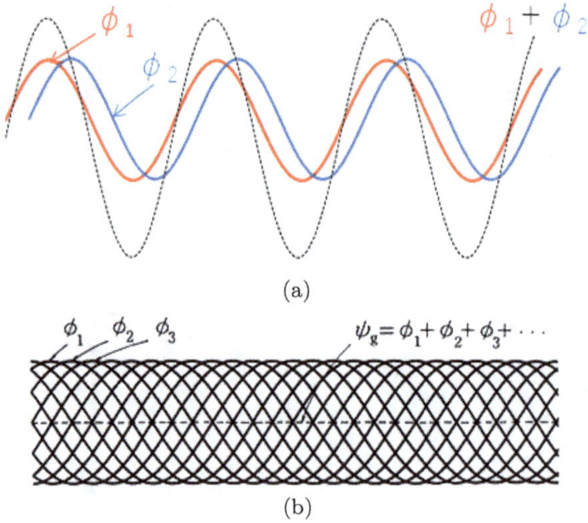

Fig. 6.8: Explanation of Figs. 6.7 and 6.8. When sinusoidal curves ϕ_1 and ϕ_2 are superimposed, the synthesized curve is $\phi_1 + \phi_2$. However, when $\phi_1, \phi_2, \phi_3, \ldots, \phi_n$ the phase of which ranges from 0 to 2π are superimposed, the resultant synthesized wave becomes flat. See also Fig. 4.6.

Mathematically the aforementioned procedure is expressed as follows. Letting the incident intensity be 1, the phase of a diffracted wave when the incident wave travels z is given by

$$\phi_{\mathrm{g}} = \frac{\pi i}{\xi_g} \int_0^{\mathrm{t}} \exp\left(-2\pi i s_g z\right) \mathrm{dz} \tag{6.4}$$

where ξ_g is referred to as an extinction distance for the diffracting vector $\mathbf{g}^*_{hkl}$ and given by

$$\xi_g = \frac{\pi \mathrm{V}_c \cos\theta}{\lambda \mathrm{F}(\theta)} \tag{6.5}$$

where V_c is volume of a unit cell and $\mathrm{F}(\theta)$ is structure factor (see Eq. (4.15)).

Integrating Eq. 6.4 by assuming that s_g does not depend on z, we obtain

$$\phi_g = \frac{\pi i}{\xi_g} \frac{\sin \pi \mathrm{t} \mathrm{s}_g}{\pi \mathrm{s}_g} \exp\left\{-i\pi \mathrm{t} \mathrm{s}_g\right\} \tag{6.6}$$

and the intensity is given by

$$I_g = \phi_g^2 = \left(\frac{\pi}{\xi_g}\right)^2 \left(\frac{\sin \pi t s_g}{\pi s_g}\right)^2 \tag{6.7}$$

Equation 6.7 is an approximation to the Laue function (Eq. (4.18)) by approximating sin $\pi s_z \sim \pi s_z$. ϕ_g^2 depends not only on the outer shape of a sample through t but also $\mathbf{s}_g^*$. Effect of the outer shape is described in detail in Section 4.6.

Exercise 6.2. Show that if Eq. (6.6) is plotted in a complex plane, it becomes a circle with a radius of $1/(2\,\xi_g s_g)$. This is in keeping with result of Exercise 6.1.

Figures 6.9 and 6.10 illustrate the diffracted and transmitted intensities as functions of t and $\mathbf{s}_g^*$. The followings are noted.

Theorem 6.1:
(1) The diffracted intensity changes periodically with crystal thickness, the period of which is $1/s_g$.[2] (2) With increasing the deviation from the Bragg's condition ($\mathbf{s}_g^*$), the diffracted intensity decreases and its period becomes shorter.

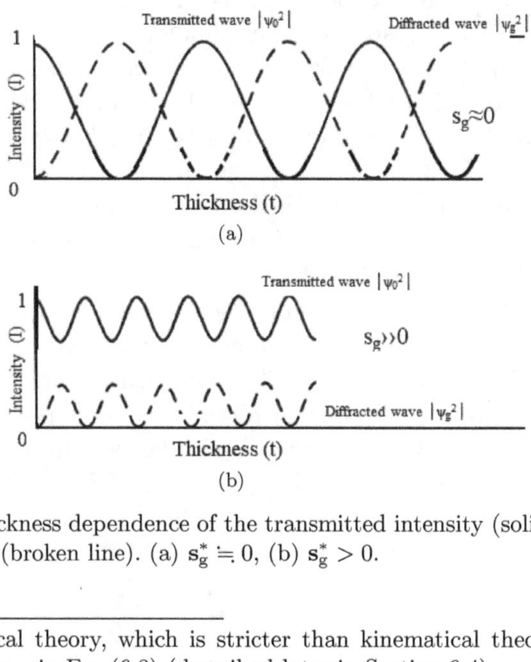

Fig. 6.9: Thickness dependence of the transmitted intensity (solid line) and the diffracted one (broken line). (a) $\mathbf{s}_g^* \fallingdotseq 0$, (b) $\mathbf{s}_g^* > 0$.

[2]In dynamical theory, which is stricter than kinematical theory, s_g is to be replaced with s_{eff} in Eq. (6.8) (described later in Section 6.4)

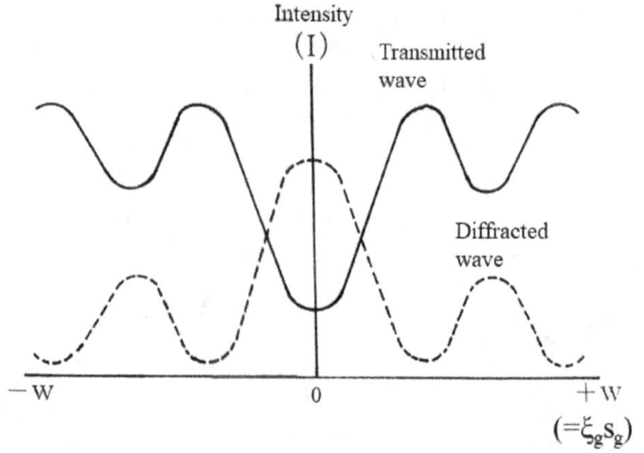

Fig. 6.10: Profile of bend contours. Transmitted wave (solid line) and diffracted wave (broken line). $w = s_g^* \xi_g$ is dimensionless parameter of $\mathbf{s}_g^*$ (see Eq. (6.9)).

Exercise 6.3. Explain Theorem 6.1 by Eq. (6.1).

When the sample is wedge-shaped, thickness t changes from place to place. When $\mathbf{s}_g^*$ is constant, fringes appear at positions where the thickness is equal. Figures 6.11(a) and 6.11(b) show an example where Fig. 6.11(a) shows a BF image and Fig. 6.11(b) a DF image ($\mathbf{s}_g^* \approx 0$). Since the fringes appear at positions where the thickness is equal, this fringe is referred to as a thickness contour. Figures 6.12(a)–6.12(c) show effect of $\mathbf{s}_g^*$ on thickness fringes, where $\mathbf{s}_g^* \approx 0$ in Fig. 6.12(a). With increasing $\mathbf{s}_g^*$, as in Figs. 6.12(b) and 6.12(c), the spacing of the thickness contours decreases.

Let us suppose that a sample contains two grains ① and ② with a tilted boundary, as shown in Fig. 6.13(a). Now let us suppose that crystal ① satisfies the Bragg's condition, while crystal ② does not. An electron beam which was diffracted by crystal ① travels through crystal ② without any interfere: For the beam diffracted by crystal ①, crystal ② behaves as if it didn't exist, as shown in Fig. 6.13(b). As a result, fringes which are quite similar to thickness contours appear at a boundary between crystal ① and ②, as F indicated in Fig. 6.2.

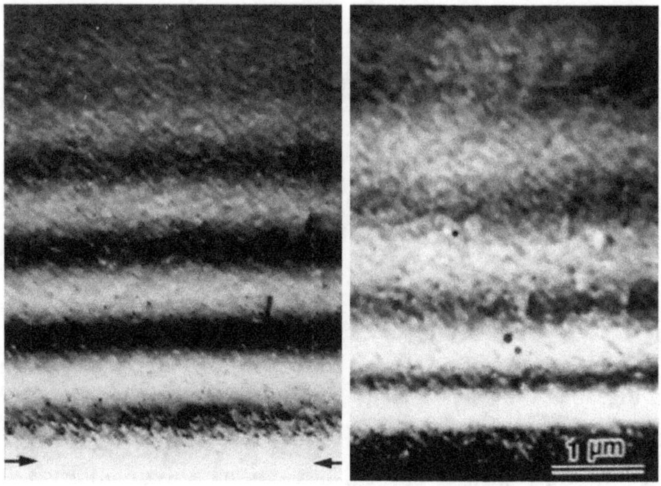

(a) (b)

Fig. 6.11: Examples of thickness contours. (a) BF, (b) DF.

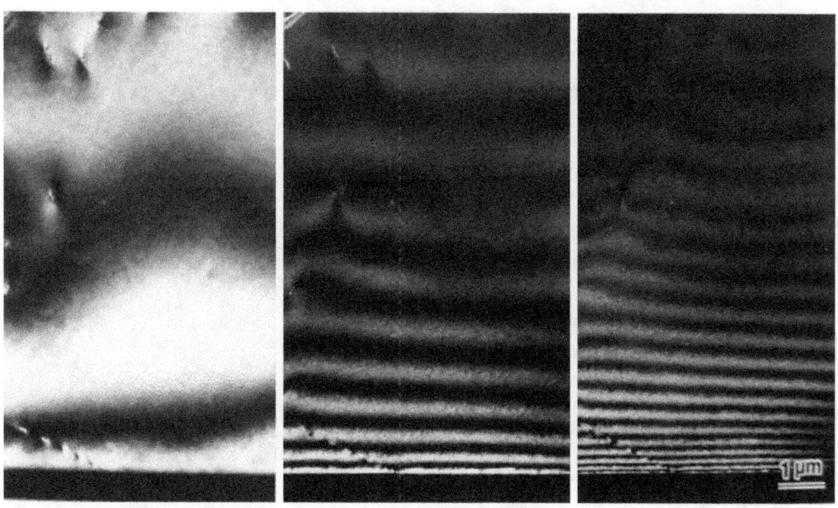

(a) (b) (c)

Fig. 6.12: Effect of $\mathbf{s}_g^*$ on the spacing of thickness contours. The value of $\mathbf{s}_g^*$ increases in the order of (a)→(b)→(c) with the spacing decreasing.

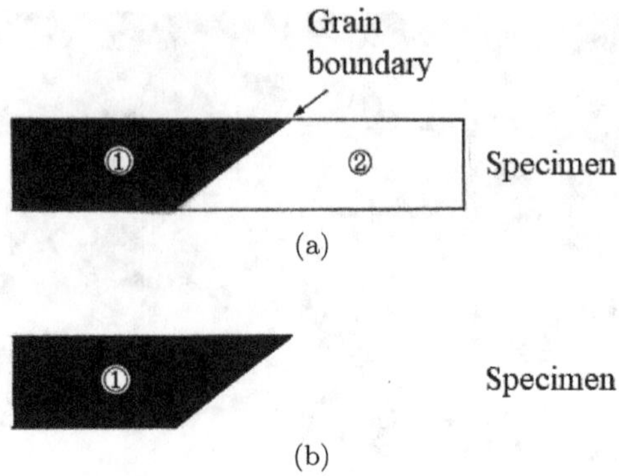

Fig. 6.13: Thickness contours at a tilted grain boundary.

When a wedge-shaped crystal is bent as shown in Fig. 6.14, the so-called bamboo-shoot pattern[3] appears as shown in Fig. 6.15(a). If a SAD is taken at one of the bamboo-shoots, say along A–A′, $\mathbf{g}^* = \bar{2}20$ is excited and satisfies the Bragg's condition (Fig. 6.15(b′)). Correspondingly, if a DF image is taken using a diffraction spot $\mathbf{g}^* = \bar{2}20$, A–A′ appears bright (Fig. 6.15(c)). In a DF image using a spot $\mathbf{g}^* = 2\bar{2}0$, B–B′ appears bright. Contrasts such as A–A′ and B–B′ appear since the Bragg's conditions are satisfied locally as a result of bending of a crystal, so that they are referred to as bend contours. Profiles of contrasts along $t'' \cdots t'$ and $t' \cdots t$, where the thickness is constant, correspond to Fig. 6.10. At the deepest position of the bamboo-shoot, the period of thickness contours is largest, implying that Bragg's condition is satisfied perfectly. In other words, in an

[3]  Bamboo shoot.

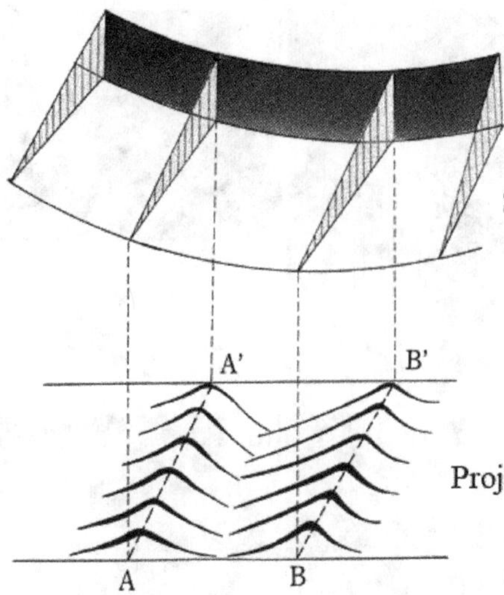

Fig. 6.14: Bamboo-shoot pattern in a curved specimen with a wedge-shaped cross-section.

area where a Bamboo-shoot pattern is observed the sample (crystal) is bent.

When a crystal is bent in a complex way, the bend contours become complex as well, as shown in Fig. 6.16(a). The SAD taken from such an area containing complex bend contour shows an apparently high symmetry with respect to the direct spot as shown in Fig. 6.16(e); however, this does not imply that the whole crystal is orientated along a high-symmetry axis. For instance, Fig. 6.16(a) is composed of three sets of bend contours shown in Figs. 6.16(b)–6.16(d). Here, Fig. 6.16(b) is a superimposition of two DF images each of which was taken using $\mathbf{g}^* = +0\bar{2}2$ and $\mathbf{g}^* = -0\bar{2}2$ (corresponding to Fig. 6.15(b) and 6.15(c)), respectively. Similarly, Figs. 6.16(c) and 6.16(d) are superimpositions of DF images taken using $\mathbf{g}^* = \pm 2\bar{2}0$ and $\pm\bar{2}02$, respectively.

Point 6.1.
In an area where bend contours are bent in a complex manner, diffracting conditions change severely from one area to another, so that such an area is not appropriate for TEM observation.

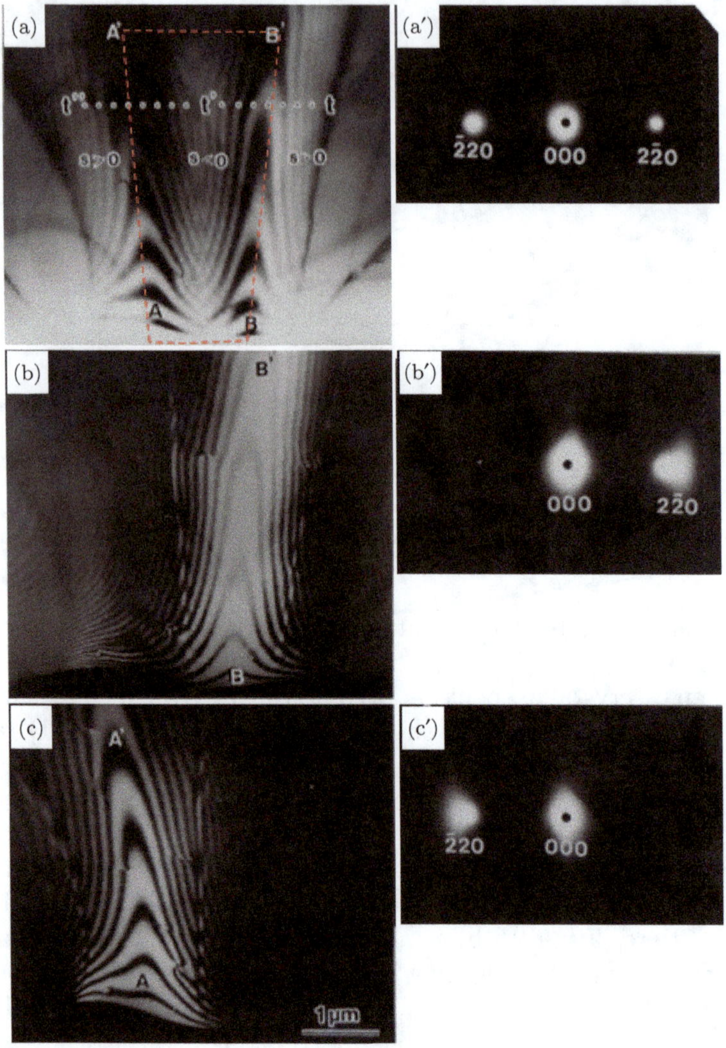

Fig. 6.15: Bamboo-shoot patterns. (a) BF, (b) (c) DF. In (a) two shoots A–A′ and B–B′ are observed. (a′) shows a diffraction pattern from square bounded by A–A′–B′–B (broken line in red). (b′) and (c′) show diffraction patterns from B–B′ and A–A′, respectively. (b) and (c) show DF images using strong diffraction spots shown in (b′) and (c′), respectively.

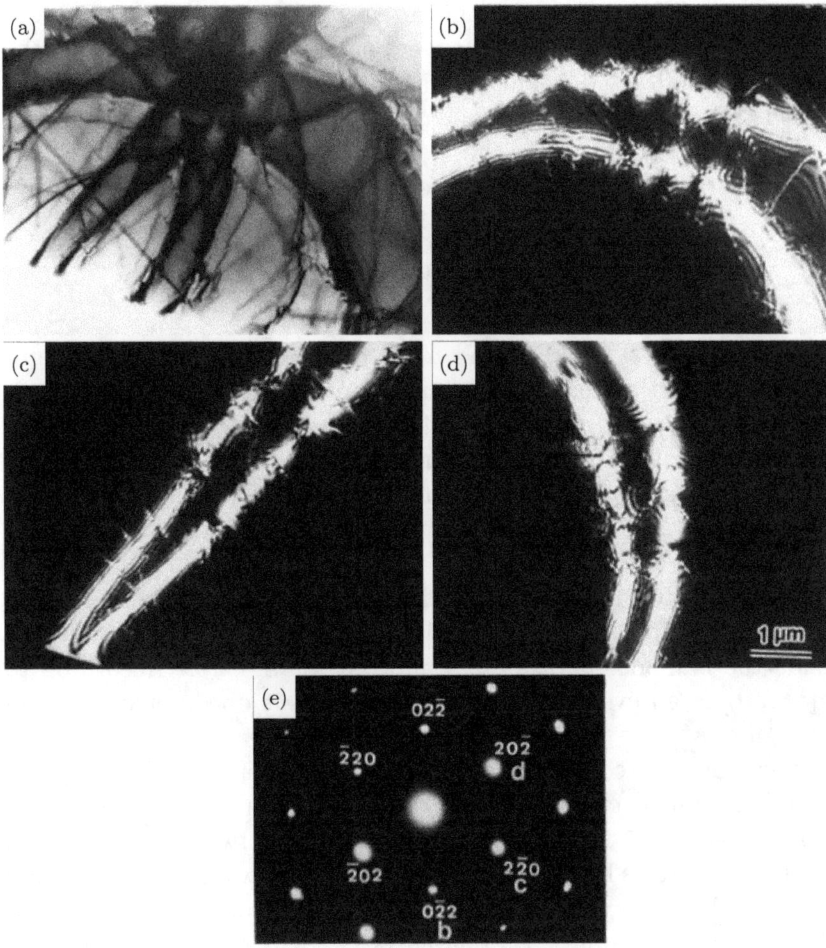

Fig. 6.16: (a) BF image of a complexly bent specimen and its diffraction pattern (e). In (b) DF images taken using $\mathbf{g}^* = \pm 0\bar{2}2$ are superimposed. Same procedures are applied in (c) for $\mathbf{g}^* = \pm 2\bar{2}0$ and in (d) for $\mathbf{g}^* = \pm 20\bar{2}$. (e) shows a diffraction pattern with a high symmetry, but Bragg's condition is satisfied only locally.

6.4 Dynamical Theory of Diffraction

When the Bragg's condition is satisfied ($\alpha = 0$), with increasing the thickness of specimen ϕ_g increases and eventually exceeds the incident amplitude. This cannot occur physically. This inconsistency stems from the breakdown of kinematical theory, which tells that the once-Bragg-diffracted beam is never Bragg-diffracted again. In order to overcome this difficulty, it must be taken into account that the once-Bragg-diffracted beam is Bragg-diffracted again. Such a treatment is referred to as a dynamical theory. This book does not treat the dynamical theory, but the essence of dynamical theory is that the diffracted intensity changes periodically with a period given by

$$\frac{1}{\mathrm{s_{eff}}} = \xi_g^{\mathrm{w}} = \left\{ \mathrm{s}_g^2 + \left(\frac{1}{\xi_g}\right)^2 \right\}^{-1/2} = \xi_g\left(1+\mathrm{w}^2\right)^{-1/2} \tag{6.8}$$

where $\mathrm{w} = \mathrm{s}_g \xi_g$. s_{eff} is referred to as an effective deviation parameter, and w is the dimensionless parameter of s_g. When $\mathrm{s}_g \gg 1/\xi_g$ in Eq. 6.8, s_{eff} approaches s_g. Also, when the Bragg's condition is satisfied ($\mathrm{s}_g = 0$), $1/\mathrm{s_{eff}}$ becomes maximum and $1/\mathrm{s_{eff}} = \xi_g$, and the period of thickness fringes becomes ξ_g as well. For this reason, ξ_g is referred to as an extinction distance. ξ_g^w is referred to as an effective extinction distance.

In Fig. 6.15(a), the region inside the bamboo-shoot A–A′–B′–B (bounded by broken line in red) is darker than outside. In a region inside the bamboo-shoot $\mathbf{s}_\mathrm{g}^* < 0$ and outside $\mathbf{s}_\mathrm{g}^* > 0$.[4] According to Eq. 6.7 I_g is an even function of $\mathbf{s}_\mathrm{g}^*$, so that it does not depend on the sign of $\mathbf{s}_\mathrm{g}^*$.

This can be accounted for by considering effects of absorption of electrons in a crystal. Electron beam which has a single wave number vector $\mathbf{k}^*$ in vacuum is separated into two waves $\psi_{(1)}$ and $\psi_{(2)}$ (referred to as Bloch waves) which have two wave number vectors $\mathbf{k}^*_{(1)}$ and $\mathbf{k}^*_{(2)}$ in a crystal, as shown in Fig. 6.17. A wave number vector $\mathbf{k}^*$ and kinetic energy E are related by

$$k^2h^2/2m = E$$

where m is mass of an electron and h is the Planck's constant. According to energy conservation law, for $\psi_{(1)}$ which has the smaller

[4]In "Electron microscopy of thin crystals" by Hirsch *et al.*, which is called *Yellow Bible*, (see Bibliography) the inside of a bamboo-shoot is referred to as *inside the reflecting position, and the outside is referred to as outside the reflecting position. In in the reflecting position* $\mathbf{g}^*_{hkl}$ is outside an Ewald sphere and in *outside the reflecting position* $\mathbf{g}^*_{hkl}$ is inside an Ewald sphere. Do not confuse yourself (see Fig. 5.14).

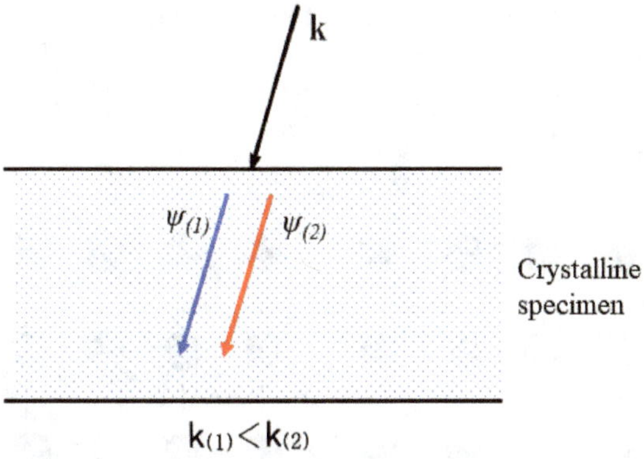

Fig. 6.17: Bloch waves in a crystal.

$\mathbf{k}^*_{(1)}$ a possibility to stay in a higher potential-energy site is high. By contrast $\psi_{(2)}$ which has the larger $\mathbf{k}^*_{(2)}$ a possibility to stay at a lower potential site is high. A potential energy is lower where atoms are aligned (referred to as atomic columns) and higher in between atomic columns. In other words, $\psi_{(1)}$ travels between atomic columns, while $\psi_{(2)}$ travels over atomic columns (Fig. 6.18(a)). Naturally, $\psi_{(2)}$ is more likely scattered by atoms. The scattered electrons are stopped by an objective aperture and do not travel downwards. In other words, $\psi_{(1)}$ is apparently absorbed in a crystal.

This phenomenon can be compared to a car which drives on a high way in the suburb (in vacuum) coming across a junction near a town where the road splits into two. One is a by-pass and the car continues driving comfortably, while the other is an old road where conjunction occurs.

For $\mathbf{s}^*_{\mathrm{g}} > 0$ the amplitude of $\psi_{(1)}$ is high, while for $\mathbf{s}^*_{\mathrm{g}} < 0$ the amplitude of $\psi_{(2)}$ is high. In other words, inside a bamboo-shoot where $\mathbf{s}^*_{\mathrm{g}} < 0, \psi_{(2)}$ is excited more than $\psi_{(1)}$, resulting in a higher absorption.

Intensity profiles of bend contours in dynamical theory are illustrated in Fig. 6.19. In BF the intensity is stronger when $\mathbf{s}^*_{\mathrm{g}} > 0$ and the image is brighter. When $\mathbf{s}^*_{\mathrm{g}} < 0$, the intensity is weaker and the image is darker. The profile of a BF image in Fig. 6.19 corresponds to $\mathrm{t}' \cdots \mathrm{t}$ in Fig. 6.15(a), with t-side corresponding to $\mathrm{w} > 0$ and t′-side corresponding to $\mathrm{w} < 0$. The profile is asymmetrical in intensity with respect to the sign of $\mathbf{s}^*_{\mathrm{g}}$. By contrast, the intensity profile of a DF

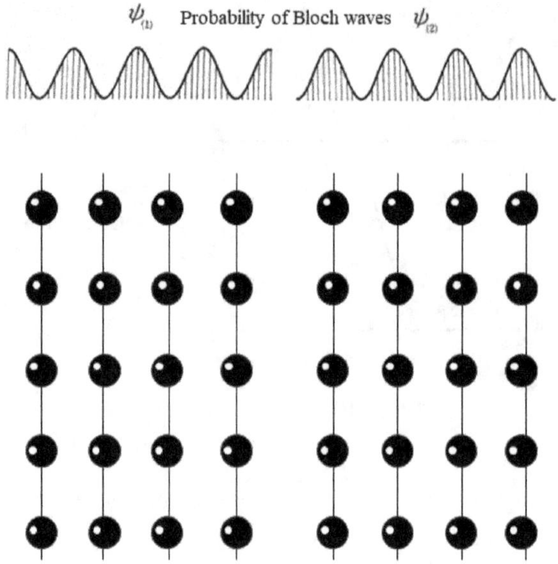

Fig. 6.18: Bloch waves in a crystal. $\psi_{(2)}$ is scattered by atoms more severely than $\psi_{(1)}$, resulting in more severe absorption (redrawn from Hashimoto *et al.* (1962)).

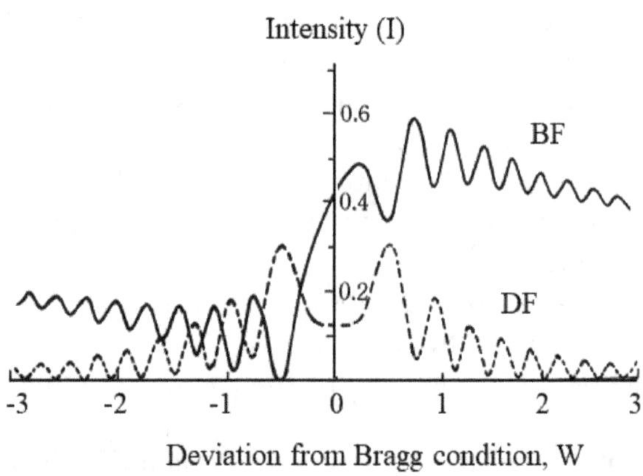

Fig. 6.19: Profile of bend contours with absorption being taken into account. Note that the profile in BF is asymmetrical with respect to the sign of $w = s_g \xi_g$ (see also Fig. 6.11) (redrawn from Hashimoto *et al.* (1962)).

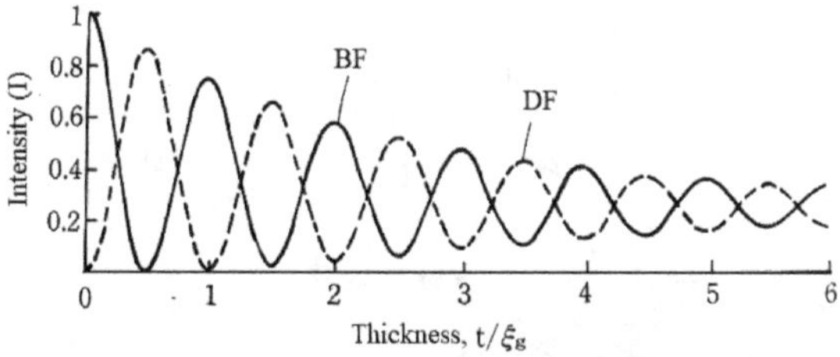

Fig. 6.20: Profile of thickness contours with absorption being taken into account. The intensity decreases with thickness (see also Fig. 6.10). t: Specimen thickness, ξ_g: Extinction distance.

image is symmetrical with respect to the sign of $\mathbf{s}_g^*$ (Figs. 6.15(b), 6.15(c) and 6.19). An intensity profile along $t'' \cdots t'$ in BF shown in Fig. 6.15(a) is coincident to that inversed with respect to $\mathbf{s}_g^*$.

Point 6.2.
It is recommended to observe in $\mathbf{s}_g^* > 0$, that is, *outside the reflecting position*, i.e., "outside the bamboo-shoots".

Figures 6.11 and 6.12 show that the contrasts of the thickness contours diminish with increasing the thickness of specimen. Theoretical profiles (calculated by kinematical theory) shown in Fig. 6.9 shows that the intensity of thickness contour does not depend on thickness. This difference between theory and experiment is also accounted for by considering absorption of electrons inside the specimen in a dynamical theory. Figure 6.20 shows the profile of thickness contours calculated by a dynamical theory: it decreases with increasing thickness in agreement with experimental results shown in Figs. 6.11 and 6.12.

6.5 CAT Method to Estimate Sharpness of an Interface in an Artificial Super-Lattice

In an artificial super-lattice where more than one layers are deposited artificially, sharpness of the interfaces is an important issue. Each of the deposited layers has different chemical compositions, so that their extinction distances differ as well. In other words, the period of the thickness contours differs. A composition analysis by thickness-fringes (CAT) method utilizes this phenomenon to evaluate the

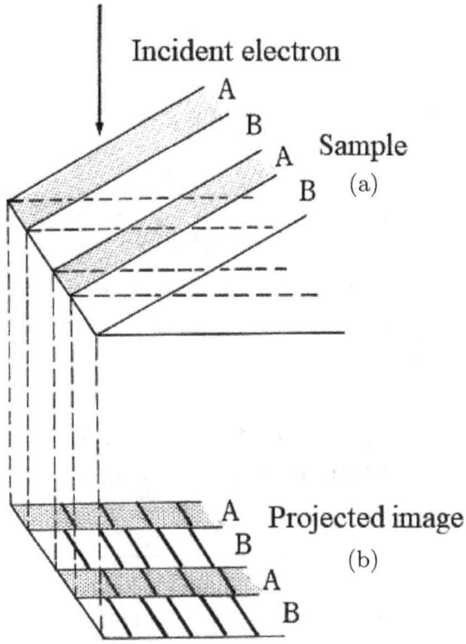

Fig. 6.21: Principle of CAT method. In an artificial super-lattice consisting of two materials A and B, extinction distances (ξ_g) of A and B are different so that from positions of thickness contours, the positions of A/B interfaces can be determined.

sharpness of an interface in an artificial super-lattice. First an interface between layers is set parallel to the direction of the incident electron beam **B**, as shown in Fig. 6.21(a). The thickness fringes stagger at the interface as shown in Fig. 6.21(b). Figures 6.22(a) and 6.22(b) show CAT images of artificial interfaces in $GaAs/Al_{0.3}Ga_{0.7}As$. In Fig. 6.22(a), the interface is sharp, while in Fig. 6.22(b) the interface is less sharp.

6.6 Supplement

Diffraction contrast

Figure 6.23 illustrates schematically how a crystal appears when it is imaged in $\mathbf{g}_1^*$, $\mathbf{g}_2^*$ and $\mathbf{g}_3^*$. In the corresponding SAD's, diffraction spots of $\mathbf{g}_1^*$, $\mathbf{g}_2^*$ and $\mathbf{g}_3^*$ are extremely strong ($\mathbf{g}_1^*$, $\mathbf{g}_2^*$ and $\mathbf{g}_3^*$ are excited in two-beam conditions). Next, let us consider physical meaning of

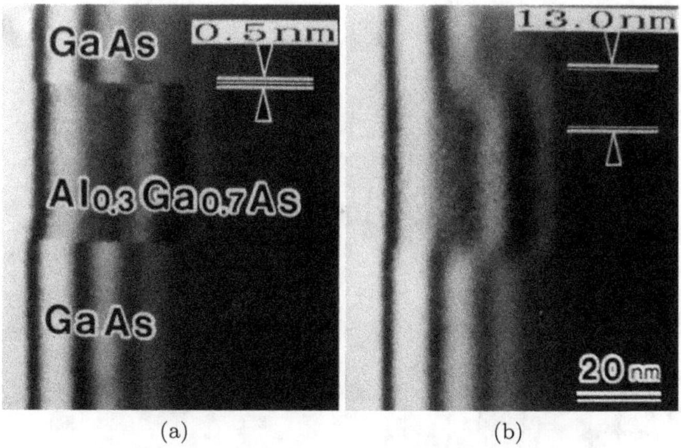

Fig. 6.22: Interfaces in an GaAs/$Al_{0.3}Ga_{0.7}$ As artificial super-lattice as observed by CAT method. The interfaces in (a) are sharper than those in (b) (by courtesy of Dr. F. Nagata).

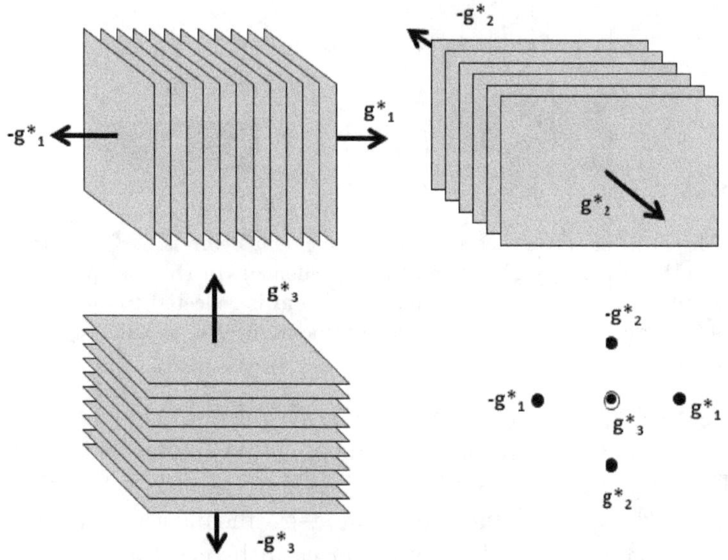

Fig. 6.23: Physical significance of the two-beam approximation. A crystal is assumed as stacking of lattice planes perpendicular to $\mathbf{g}^*$. Other lattice planes are ignored.

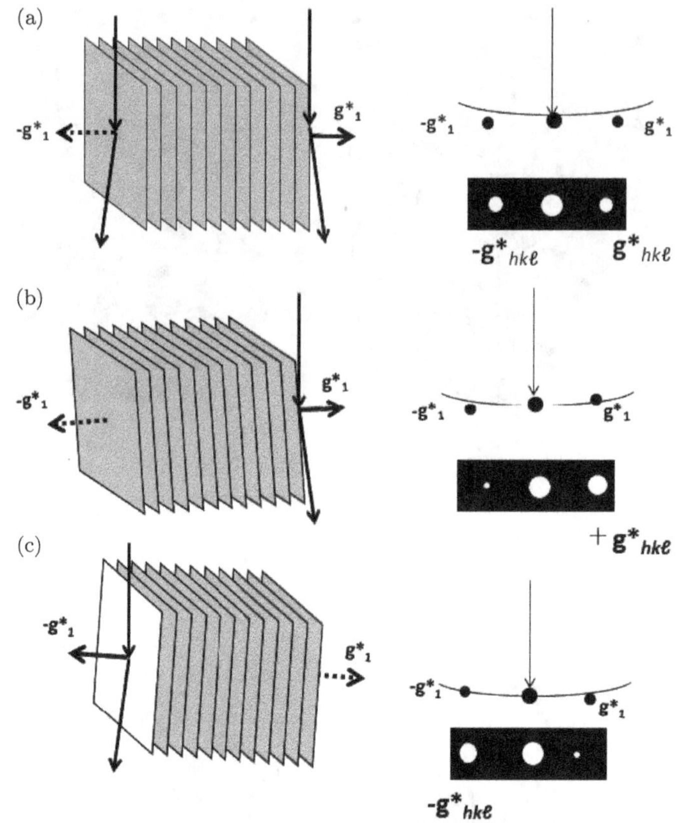

Fig. 6.24: (a) illustrates a case where the incident beam is perfectly perpendicular to both $+\mathbf{g}_1^*$ and $-\mathbf{g}_1^*$ (symmetrical incidence). In (b) the specimen is tilt slightly anticlockwise, so that the incident beam is reflected toward $+\mathbf{g}_1^*$. (c) is the case reverse to (b) and the incident beam is reflected toward $-\mathbf{g}_1^*$.

imaging in $+\mathbf{g}^*$ and $-\mathbf{g}^*$ (Fig. 6.24). Locating the origin at the center of the crystal, let us call the right side plus (+) and the left side minus (−). Exciting $+\mathbf{g}^*$ means that the incident beam is reflected to the right. On the other hands, exciting $-\mathbf{g}^*$ means that the incident beam is reflected to the left. In either case the crystal must be tilted slightly. In other words, a (*hkl*) plane is a both-sided mirror. After getting this knowledge let us consider diffraction contrasts of lattice defects.

What is significance of two-beam approximation in diffraction crystallography? Observing a crystal under two-beam condition means nothing but observing the crystal with a crystallographic plane (*hkl*) being Bragg-reflected. In other words, the crystal under observation is assumed to be a stacking of (*hkl*) planes alone, so that no information other than (*hkl*) plane is obtained.

Now among three-dimensionally arranged crystallographic planes, let us picture the crystal with a set of planes perpendicular to one direction (say, x-axis). Figure 6.23(a) illustrates the picture. If sets of planes perpendicular to y- and z-axes are used, the crystal is pictured as shown in Figs. 6.23(b) and 6.23(c). Let us suppose that the incident beam is perfectly parallel to the plane under question (perpendicular to x-axis, $\pm(100)$) (Fig. 6.24(a)). In this case, neither $+\mathbf{g}^*$ nor $-\mathbf{g}^*$ satisfies the Bragg's condition perfectly, however, they satisfy the Bragg's condition to some extent. As a result, the incident beam is reflected by both (100) and $(\bar{1}00)$ planes equally, resulting in both of $+\mathbf{g}^*$ and $-\mathbf{g}^*$ being excited simultaneously. This situation is referred to as a symmetry incidence, where three beams, that is, ① the incident beam +② diffracted beam $(+\mathbf{g}^*)$ +③ diffracted beam $-(\mathbf{g}^*)$, are excited and the two-beam approximation holds no longer.

Now, if the crystal is tilted anticlockwise slightly, $+\mathbf{g}^*$ satisfies the Bragg's condition perfectly and $-\mathbf{g}^*$ deviates from the Bragg's condition significantly: Two-beam condition, ① the incident beam +② diffracted beam $(+\mathbf{g}^*)$, is satisfied. If the crystal is tilted clockwise, two-beam condition with $-\mathbf{g}^*$ excited is satisfied. It is noted that these correspond to Fig. 5.14.

Important here is that a small objective aperture, by which only one spot is allowed to travel downward while the remaining spots are stopped, is necessary. Modern TEM capable of high-resolution imaging has a large objective aperture as standard: It is cordial to use a small objective aperture when observing diffraction contrast.

With a large objective aperture (as for HREM observation) single crystals ① and ② cannot be distinguished (see Fig. 6.25).

For BF and DF imaging, a small objective aperture must be used.

In a DF imaged using the spot $\mathbf{g}^*$, any region which contributes to imaging (that is, the region reflecting the incident electron beam to the direction of spot $\mathbf{g}^*$) can be located on an enlarged image.

(1) Identifying a region with same crystal structure (identification of phase).

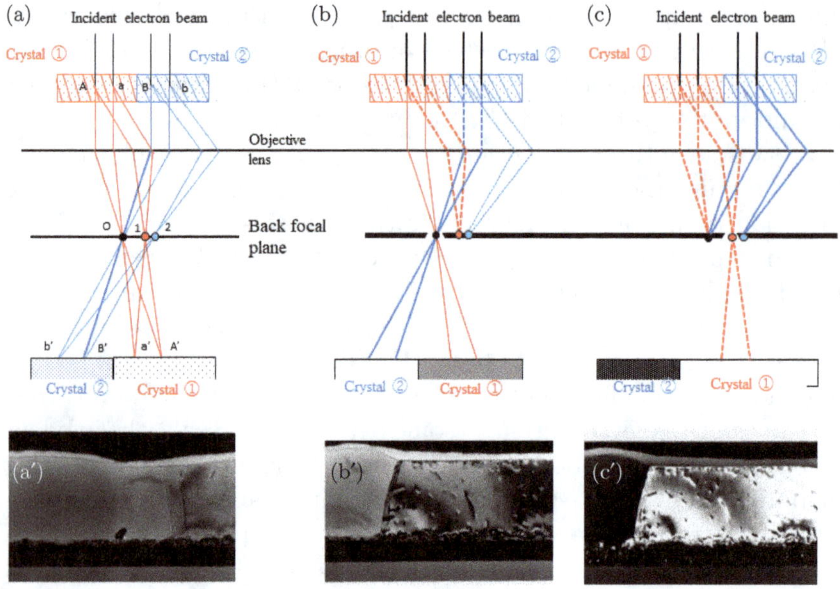

Fig. 6.25: Let us assume that crystal ① satisfies Bragg's condition perfectly and crystal ② deviates from the Bragg's condition. When a large objective aperture is used, both the transmitted and diffracted beam pass through the objective aperture. As a result, no difference in contrast between ① and ② is observed ((a) and (a′)). When a small aperture is set to the diffraction spot of crystal ①, only ① appears very strongly, since the intensity of Bragg-reflected beam from ① is strong ((c) and (c′)). When an objective aperture is set to the transmitted beam (direct spot), the diffracted beam from ① is stopped by the objective aperture. As a result, ① appears dark (b) and (b′).

(2) Identifying grains with same orientation (measurement of grain size).
(3) Detecting a strain within a single crystal.

Dynamical theory

In dynamical theory, let us consider the intensity of incident beam when it just enters the crystal. In this case the transmitted intensity (I_0) is overwhelmingly large as shown in Fig. 6.23(a). However, on propagating downward inside a crystal, the transmitted beam repeats Bragg's reflection with I_0 decreasing gradually. By contrast, the diffracted intensity I_g increases and eventually takes the place of the transmitted beam (Fig. 6.26(c)). Then, the diffracted beam $\mathbf{g}^*$

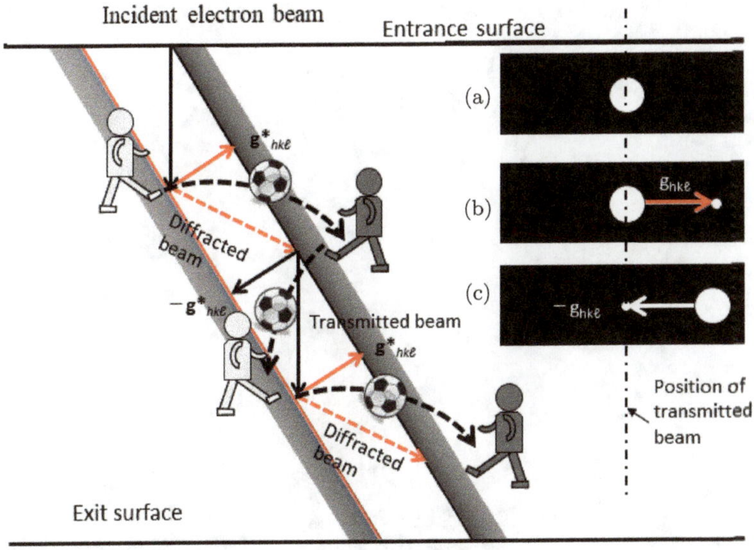

Fig. 6.26: Dynamical behavior of the transmitted and diffracted waves.

behaves as if the incident beam. As a result, the beam $\mathbf{g}^*$ is reflected by $(\bar{h}\bar{k}\bar{l})$ plane, the back side of (hkl) plane: $-\mathbf{g}^*_{hkl}$ is excited. In summary, the transmitted and the diffracted beams propagate downward with exchanging offence-and-defense (Fig.6.26(c)). This can be compared to two players passing a ball (see also Section 6.4).

Chapter 7

Contrast of Planar Defects and Precipitates

7.1 General Theory

If a crystal contains a lattice defect, the nearby region is distorted. Naturally, lattice planes contributing to diffraction are distorted and the diffracted intensity is affected. Let us consider that electrons propagate along a column with diameter of 1 nm as shown in Fig. 7.1. When a crystal contains a lattice defect, the nearby column is distorted slightly. Let the position of the nth unit cell in the column be denoted by

$$\mathbf{r}'_n = \mathbf{r}_n + \mathbf{R}_n \tag{7.1}$$

where $\mathbf{R}_n$ is the displacement from the initial position $\mathbf{r}_n$. The phase difference α (see Eq. (6.2)) is now given by

$$\begin{aligned}\alpha &= -2\pi\mathbf{K}^* \cdot \mathbf{r}'_{\mathrm{n}} = -2\pi\,(\mathbf{g}^*_{hkl} + \mathbf{s}{*}_{\mathrm{g}}) \cdot (\mathbf{r}_{\mathrm{n}}{+}\mathbf{R}_{\mathrm{n}}) \\ &= -2\pi\,(\mathbf{g}^*_{hkl} \cdot \mathbf{r}_{\mathrm{n}} + \mathbf{s}{*}_{\mathrm{g}} \cdot \mathbf{r}_{\mathrm{n}} + \mathbf{g}^*_{hkl} \cdot \mathbf{R}_{\mathrm{n}}) \\ &= -2\pi\,(\mathbf{s}{*}_{\mathrm{g}} \cdot \mathbf{r}_{\mathrm{n}} + \mathbf{g}^*_{hkl} \cdot \mathbf{R}_{\mathrm{n}})\end{aligned} \tag{7.2}$$

where $\mathbf{s}^*_{\mathrm{g}} \cdot \mathbf{R}_n$ is neglected since this is a product of the second-ordered terms. Also, $\mathbf{g}^*_{hkl} \cdot \mathbf{r}_n$ is an integer as can be seen from Eqs. (2.4) and (4.7), so that it does not affect the phase difference. Taking $\mathbf{r}_n$

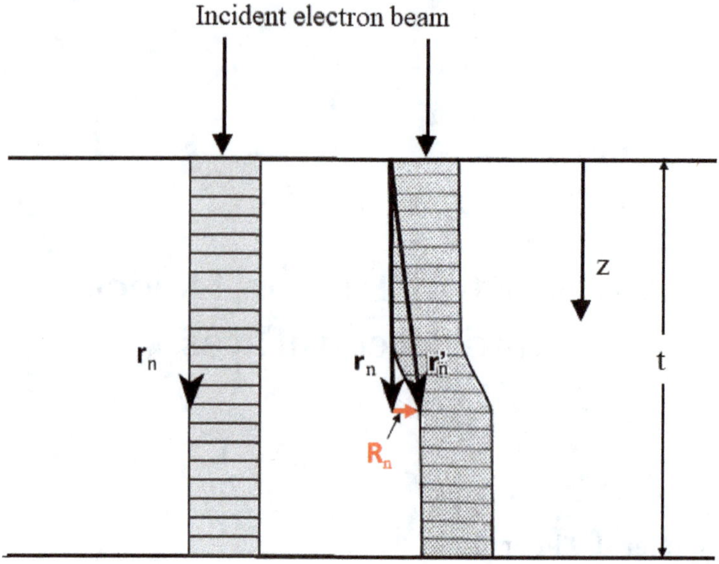

Fig. 7.1: Definition of a displacement vector $\mathbf{R}_n$ in a distorted crystal. $\mathbf{r}_n$ is a vector indicating nth unit cell (from above) in an undistorted crystal.

along the direction of the incident beam, $\mathbf{r}_n$ is coincident with the direction of thickness of the sample, i.e., z-axis.

In summary, the diffracted amplitude given by Eq. (6.4) is modified as

$$\phi_{\mathrm{g}} = \frac{\pi i}{\xi_g} \int_0^{\mathrm{t}} \exp\left[-2\pi i \left\{\mathrm{s_g z} + \mathbf{g}^*_{\mathrm{hkl}} \cdot \mathbf{R}(\mathrm{z})\right\}\right] \mathrm{dz} \tag{7.3}$$

Here, $\mathbf{R}_n$ is a function of z, so that it is written as $\mathbf{R}(\mathrm{z})$ explicitly.

Comparison between Eqs. (6.4) and (7.3) shows that the only difference is that $\mathbf{g}^*_{hkl} \cdot \mathbf{R}(\mathrm{z})$ is added in Eq. (7.3).

Procedures of calculating contrasts of TEM images are (1) calculating $\mathbf{R}_n$ using theory of elasticity and then (2) substituting $\mathbf{R}_n$ into Eq. (7.3). However, a simple inspection of Eq. (7.3) shows the following:

Theorem 7.1:
When

$$\mathbf{g}^*_{hkl} \cdot \mathbf{R}\,(\mathrm{z}) = 0 \tag{7.4}$$

holds, contrast of a defect is invisible.

In what follows, positive images printed with emulsion upward are treated. This corresponds to looking down the fluorescent screen from above.

7.2 Planar Defects

General theory

A planar defect has a simplest displacement and characterized by $(hkl)[\mathrm{h'k'l'}]$. (hkl) is a plane on which the planar defect lies and $[\mathrm{h'k'l'}]$ is a vector $\mathbf{R}_\mathrm{F}$ denoting a displacement between two crystals facing across (hkl) plane.

$$\mathbf{R}_\mathrm{F} = [\mathrm{h'k'l'}] = \mathrm{h'}\mathbf{a} + \mathrm{k'}\mathbf{b} + \mathrm{l'}\mathbf{c} \tag{7.5}$$

Now, let the whole thickness of crystal be t (Fig. 7.2) and assume that a planar defect with $\mathbf{R}_\mathrm{F}$ lies at a depth $\mathrm{z} = t_1$.

$$\mathbf{R}_\mathrm{F} = 0 \quad (0 < z < t_1)$$

$$\mathbf{R}_\mathrm{F} = \mathbf{R}_\mathrm{F} \quad (t_1 < z < t)$$

where $\mathbf{R}_\mathrm{F}$ is defined as follows:

Definition 7.1:
Fixing the upper-half crystal, the lower-half crystal is displaced by $\mathbf{R}_\mathrm{F}$ (Fig. 7.3(a)).

Equation 7.3 is modified as

$$\phi_\mathrm{g} = \frac{\pi i}{\xi_g}\left[\int_0^{\mathrm{t}_1} \exp\{-2\pi i\,(\mathrm{s}_g \mathrm{z})\}\,\mathrm{dz} + \int_{\mathrm{t}_1}^{\mathrm{t}} \exp\{-2\pi i\,(\mathrm{s}_g \mathrm{z}) + \mathbf{g}^* \cdot \mathbf{R}_\mathrm{F}\}\,\mathrm{dz}\right] \tag{7.6}$$

$$= \frac{1}{\xi_g \mathrm{s}_g}\exp(-\pi i \mathrm{s}_g \mathrm{t}_1) \times \{\sin\pi \mathrm{t}_1 \mathrm{s}_g + \exp(-i\alpha)\exp(-\pi i \mathrm{s}_g \mathrm{t})\sin\pi(\mathrm{t}-\mathrm{t}_1)\mathrm{s}_g\} \tag{7.6$'$}$$

$$= \frac{i}{\xi_g \mathrm{s}_g}\exp\left(-\frac{i\alpha}{2}\right)\exp(-\pi i \mathrm{s}_g \mathrm{t}) \times \left[\sin\left(\pi \mathrm{t} \mathrm{s}_g + \frac{\alpha}{2}\right) - \sin\left(\frac{\alpha}{2}\right)\exp\left\{2\pi i \mathrm{s}_g\left(\frac{\mathrm{t}}{2} - \mathrm{t}_1\right)\right\}\right] \tag{7.6$''$}$$

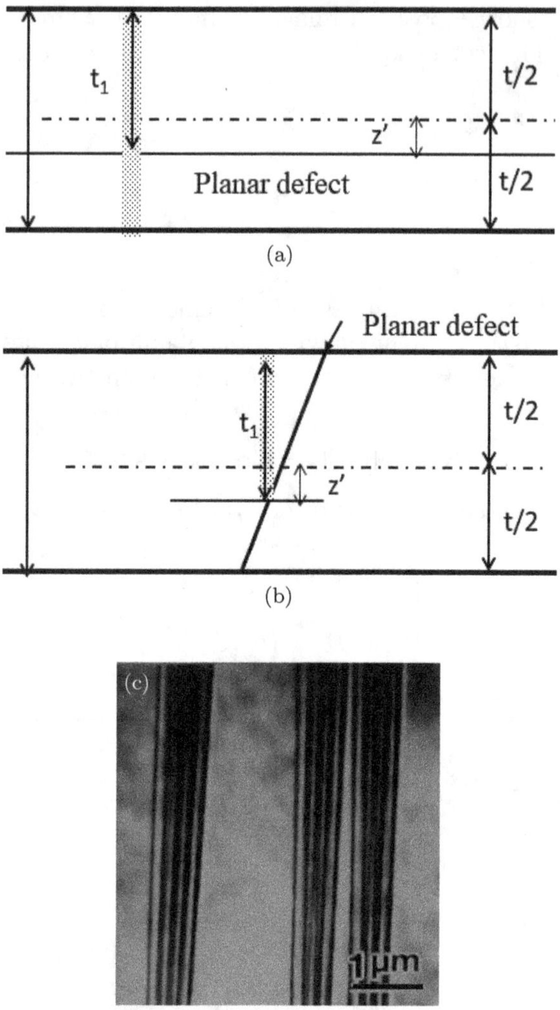

Fig. 7.2: (a) A planar defect is in the plane of foil. (b) An inclined planar defect. (c) TEM of inclined planar defects: Fringes appear parallel to the intersection of planar defects and foil surface.

Here, $\alpha = 2\pi\mathbf{g}^* \cdot \mathbf{R}_F$. The intensity I is given by

$$\mathrm{I} = \phi_g^2 = \frac{1}{(\xi_g s_g)^2}\left\{\sin^2\left(\pi t s_g + \frac{\alpha}{2}\right) + \sin^2\left(\frac{\alpha}{2}\right)\right\}$$
$$-2\sin\left(\frac{\alpha}{2}\right)\sin\left(\pi t s_g + \frac{\alpha}{2}\right)\cos 2\pi s_g z' \qquad (7.7)$$

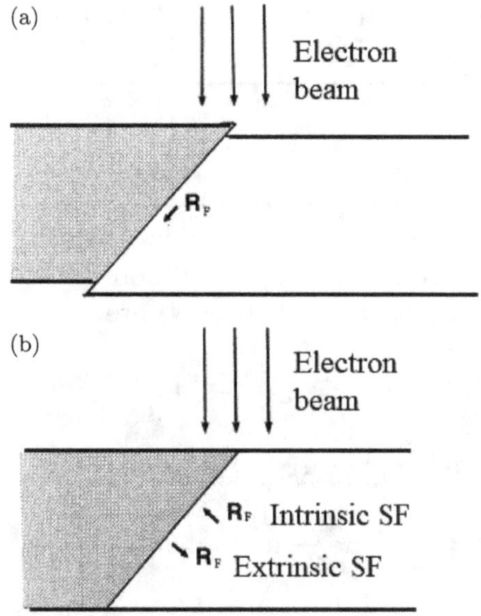

Fig. 7.3: (a) Definition of displacement vector $\mathbf{R}_F$ of a planar defect. The upper crystal is fixed and the lower crystal is displaced by $\mathbf{R}_F$. (b) Definition of a displacement vector $\mathbf{R}_F$ for the intrinsic and extrinsic stacking faults in FCC crystal.

This is a cosine function of z′. Here, $z' = t_1 - t/2$ is a distance from the center of a specimen to the stacking fault (Fig. 7.2(a)). Therefore, when the stacking fault is inclined as shown in Fig. 7.2(b), fringe patterns, which are parallel to the intersection with foil surface, appear (Fig. 7.2(c)).

The stacking faults in FCC crystal

As described in Sections 3.3.2 and 3.4.4, the stacking fault in FCC is characterized by $\{111\}\langle{}^1\!/_3{}^1\!/_3{}^1\!/_3\rangle$: It lies on $\{111\}$ plane and its displacement vector is $\mathbf{R}_F = 1/3\langle 111\rangle$.

That is,

$$\mathbf{R}_F = \frac{1}{3}\mathbf{a} + \frac{1}{3}\mathbf{b} + \frac{1}{3}\mathbf{c} \tag{7.5'}$$

When one $\{111\}$ plane is missing, it is referred to as an intrinsic SF, and when one plane is inserted, it is referred to as an extrinsic SF (see Section 3.4.4). Therefore, from Definition 7.1 it follows:

Definition 7.2:
When $\mathbf{R}_\mathrm{F}$ is upward, the SF is intrinsic.
When $\mathbf{R}_\mathrm{F}$ is downward, the SF is extrinsic.

In FCC, Miller indices of those planes which produce Bragg's reflections are limited to all even or all odd (Theorem 4.2), so that possibilities of $\mathbf{g}^*_{hkl} \cdot \mathbf{R}_\mathrm{F} = \frac{1}{3}(\pm h \pm k \pm l)$ are

$$0, \pm\frac{1}{3}, \pm\frac{2}{3}, \pm 1, \pm\frac{4}{3} \ldots$$

Values of phase angle $\alpha = 2\pi\mathbf{g}^*_{hkl} \cdot \mathbf{R}_\mathrm{F}$ are given by

$$0, \pm\frac{2\pi}{3}, \pm\frac{4\pi}{3}, \pm 2\pi \ldots$$

However, since

$$4\pi/3 = -2\pi/3 + 2\pi, -4\pi/3 = 2\pi/3 - 2\pi, \pm 2\pi = 0 \pm 2\pi$$

these phase angles are equivalent to

$$-2\pi/3, 2\pi/3, 0$$

Thus, phase angles are limited to the following three values, i.e., $-2\pi/3, 2\pi/3, 0$. Features of fringes for the respective value of phase angle are summarized as follows.

Theorem 7.2:

(1) $\mathbf{g}^*_{hkl} \cdot \mathbf{R}_\mathrm{F} = 0$ or integers ($\alpha = 0$ or $2n\pi$) : the SF is invisible.
(2) $\mathbf{g}^*_{hkl} \cdot \mathbf{R}_\mathrm{F} = +1/3$ ($\alpha = +2\pi/3$) : In BF, the contrast is symmetrical and the outermost fringe is bright.
(3) $\mathbf{g}^*_{hkl} \cdot \mathbf{R}_\mathrm{F} = -1/3$ ($\alpha = -2\pi/3$) : In BF, the contrast is symmetrical and the outermost fringe is dark.
(4) $\mathbf{g}^*_{hkl} \cdot \mathbf{R}_\mathrm{F} = \pm 1/3$ ($\alpha = \pm 2\pi/3$) : In BF, the contrast is symmetrical, but in DF it is asymmetrical. The side where BF and DF show similar contrast (both bright or both dark) is the entrance (top) surface of the electrons and the side where BF and DF show different contrast (bright + dark or dark + bright) is the exit (bottom) surface.

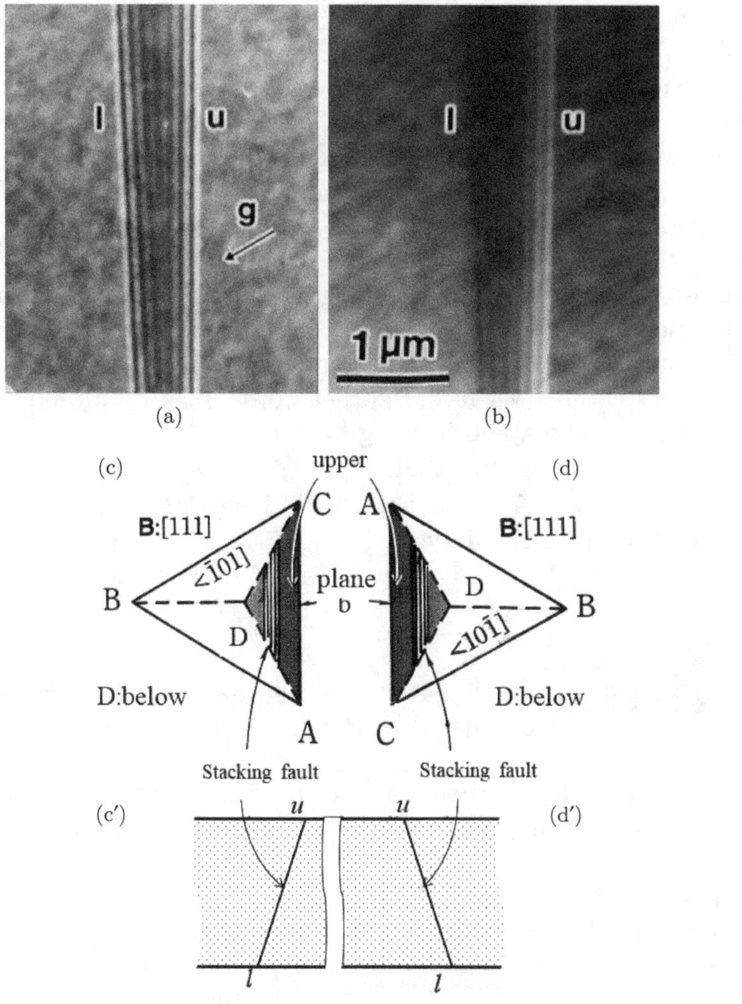

Fig. 7.4: How to determine the nature of stacking faults in FCC crystal. (a) BF. (b) DF. The stacking fault is inclined either like (c)(c′) or (d)(d′).

In order to derive this theorem dynamical theory is necessary, however, in this book the detail is not mentioned.

Figures 7.4(a) and 7.4(b) shows a set of BF and DF images of a SF imaged in $\mathbf{g}^* = \mathit{220}$. Let us take **B** upward (see footnote to Eq. (5.4)). From SAD it is known that **B** is $\langle 111 \rangle$. Now, let **B** be [111]. This corresponds to letting the upper surface of the sample be *d*, i.e., face (*111*) of the Thompson tetrahedron.

When $\mathbf{B} = [111]$, $\mathbf{g}^* = \pm 2\bar{2}0, \pm\bar{2}02, \pm 0\bar{2}2$ are possible candidates out of $\mathbf{g}^* = \{220\}$. Now, let us assume that $\mathbf{g}^* = \pm\bar{2}02$. Referring to the Thompson tetrahedron, $\mathbf{g}^*//\text{BC}$. This means that the SF lies on $(1\bar{1}\bar{1})$ plane (b on the Thompson tetrahedron). In other words, $\mathbf{R}_\text{F}$ is either $-\frac{1}{3}[1\bar{1}\bar{1}]$ or $\frac{1}{3}[1\bar{1}\bar{1}]$.

With regard to inclination of the SF, there are two possibilities, one shown in Fig. 7.4(c′) (i.e., (c)) and the other is Fig. 7.4(d′) (i.e., (d)).[1] In order to determine which is which, let us compare the outermost fringes in BF and DF images. At u both in BF and DF the outermost fringe is bright, while at l the outermost fringe is bright in BF but dark in DF. Thus, Theorem 7.2(4) tells that u is the upper face and l is the lower face. In other words, Figs 7.4(c) and 7.4(c′) is the right geometry, and hence $\mathbf{g}^* = \bar{2}02$.

Next let us assume that $\mathbf{R}_\text{F} = -\frac{1}{3}[\bar{1}11]$, then $\mathbf{g}^*_{hkl} \cdot \mathbf{R}_\text{F} = \frac{4}{3}$, $\alpha = 2\pi\mathbf{g}^*_{hkl} \cdot \mathbf{R}_\text{F} = \frac{8\pi}{3} = 2\pi + 2\pi/3$. In this case the outermost fringe in BF is to be bright. Indeed, in Fig. 7.4(a) the outermost fringe is bright, so that $\mathbf{R}_\text{F} = -\frac{1}{3}\left[\bar{1}11\right]$. Referring to Thompson tetrahedron, $\mathbf{R}_\text{F}$ is upward. Therefore, the SF is intrinsic.

Exercise 7.1. Assuming $\mathbf{B}$=[111] in Fig. 7.4 and perform a similar procedure. Show that the SF is intrinsic.

Extending this method, the nature of the SF can be determined using only a DF image.

Theorem 7.3:
First place the origin of $\mathbf{g}^*_{hkl}$ at the center of fringes of a DF image. For $\mathbf{g}^*_{hkl} = 111, 220$ and 400 (Class B), when the outermost fringe which $\mathbf{g}^*_{hkl}$ points towards is dark, the SF is intrinsic and when bright the SF is extrinsic. For $\mathbf{g}^*_{hkl} = 220, 440$ (Class A) the reverse is true.

Exercise 7.2. Apply this method to Fig. 7.4 and confirm that the result is same.

[1] Determining the sign of $\mathbf{g}^*//202$ is equivalent to determining which out of Figs. 7.4(c) and 7.4(d) is correct.

Point 7.1.

(1) Set $\mathbf{s}_g^* = 0$.
(2) Image in as many different $\mathbf{g}^*$ as possible.
(3) Image both in $+\mathbf{g}^*$ and $-\mathbf{g}^*$.

Now, let us consider the case where two SF's ① and ② overlap closely. Assuming that both ① and ② are intrinsic, then it is denoted by $\Delta\Delta\Delta\ \nabla\ \Delta\Delta \cdots \Delta\Delta\ \nabla\ \Delta\Delta\Delta$ in Frank's notation (see Section 3.4.4), where the first ∇ is SF ① and second ∇ is SF ②. Assuming that ① and ② are very close with each other and the contrast of region between ① and ② can be neglected, the contrast of the overlapped SF's is approximated by $\Delta\Delta\Delta\ \nabla\nabla\ \Delta\Delta\Delta$. This is an extrinsic SF.

This is accounted for as follows in terms of phase difference. Let us suppose phase differences α of both of the intrinsic SF's ① and ② be $\alpha = 2\pi/3$, then α of the overlapped SF is the sum of those of ① and ②, i.e., $\alpha = 4\pi/3$, which is equivalent to $\alpha = -2\pi/3$. Therefore, contrast of the overlapped SF is same as that of an extrinsic SF.

When three SF's are overlapped, $\alpha = 2\pi/3 \times 3 = 2\pi$, so that the contrast disappears (Fig. 7.5(b)). In Frank's notation, it is approximated by $\Delta\Delta\Delta\ \nabla\nabla\nabla\ \Delta\Delta\Delta$. This corresponds to a very thin twin (see Section 5.5). Indeed, such an overlapped SF is observed at an interface of a twin whose thickness decreases (Fig. 7.6).

Anti-phase boundary (APB) in an ordered alloy

As described in Section 3.5, an APB in an ordered alloy is a kind of planar defects and can be characterized by $(hkl)[\mathrm{h'k'l'}]$ where (hkl) is a plane on which the APB lies and $[\mathrm{h'k'l'}]$ is the displacement vector of the two domains across the APB, i.e., $\mathbf{R}_\mathrm{F} = [\mathrm{h'k'l'}]$. $\mathbf{R}_\mathrm{F}$ is a lattice vector when the ordered alloy under question is assumed to have a disordered structure. Therefore, when a fundamental reflection which appears in disordered alloy is excited, $\mathbf{g}^* \cdot \mathbf{R}_\mathrm{F} = 0$ (or integers), so that the APB is invisible. However, when a superlattice reflection is excited, in general $\mathbf{g}^* \cdot \mathbf{R}_\mathrm{F} \neq 0$ (or integers), so that the contrast of APB appears.

Let us consider a CsCl ordered alloy as an example. As mentioned in Section 1.4.2, CsCl structure is based on BCC but with different elements at the center and corners. Therefore, $\mathbf{R}_\mathrm{F} = \frac{1}{2}[111]$. Since

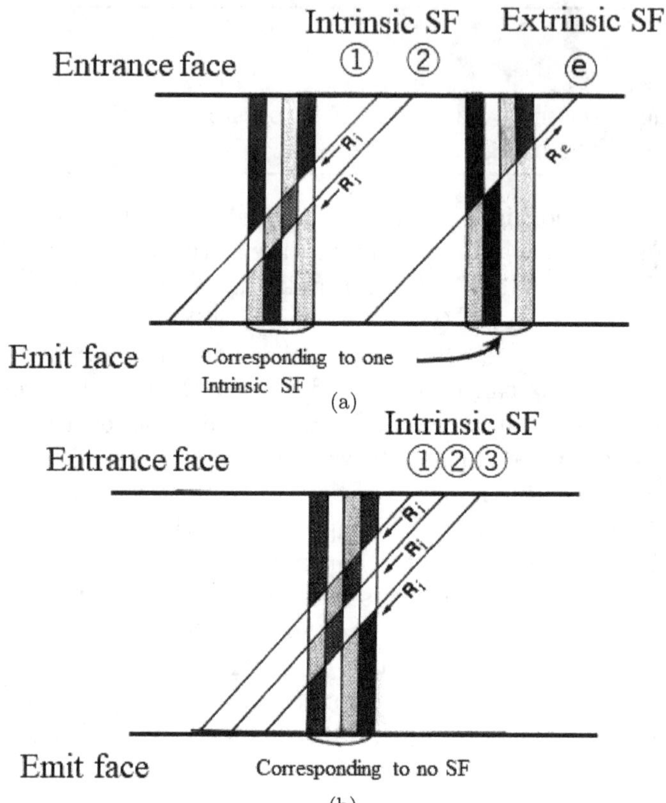

Fig. 7.5: (a) Overlapping two intrinsic stacking faults ① and ② is equivalent to one extrinsic stacking fault ⓔ. (b) Overlapping three intrinsic stacking faults ①, ② and ③ is equivalent to no stacking fault.

a CsCl type ordered alloy is BCC in the disordered state, when a reflection in the disordered state (fundamental reflection), say $\mathbf{g}^* = 200$, is excited, $\mathbf{g}^* \cdot \mathbf{R}_\mathrm{F} = 1$, leading to $\alpha = 2\pi\ \mathbf{g}^* \cdot \mathbf{R}_\mathrm{F} = 2\pi$ and the contrast is invisible (Fig. 7.7(a)).

By contrast, when a forbidden reflection, which appears only in the ordered state, say $\mathbf{g}^* = 100$, is excited, $\mathbf{g}^* \cdot \mathbf{R}_\mathrm{F} = 1/2$, leading to $\alpha = 2\pi\ \mathbf{g}^* \cdot \mathbf{R}_\mathrm{F} = \pi$, and the APB is visible (Fig. 7.7(b)). The contrast which appears when $\alpha = \pi$ is referred to as π-contrast. This corresponds to putting $\alpha = \pi$ in Eqs. (7.5) and (7.6) and when the

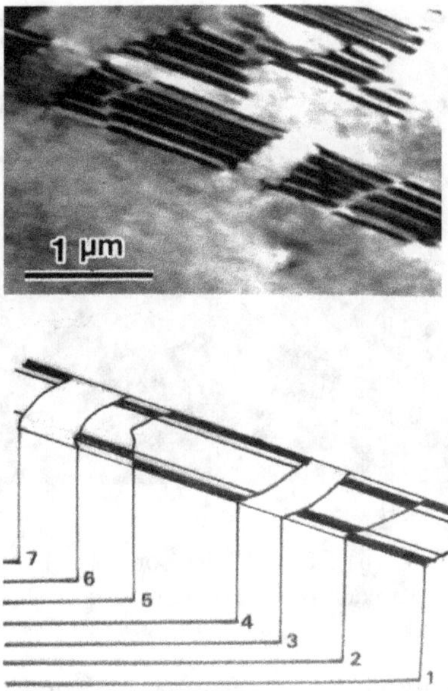

Fig. 7.6: An example where many stacking faults are overlapped. Numbers attached indicate how many stacking faults are overlapped. Between $3n + 1$ and $3n + 2$, the position of fringes staggers. For $3n$ no fringe is observed.

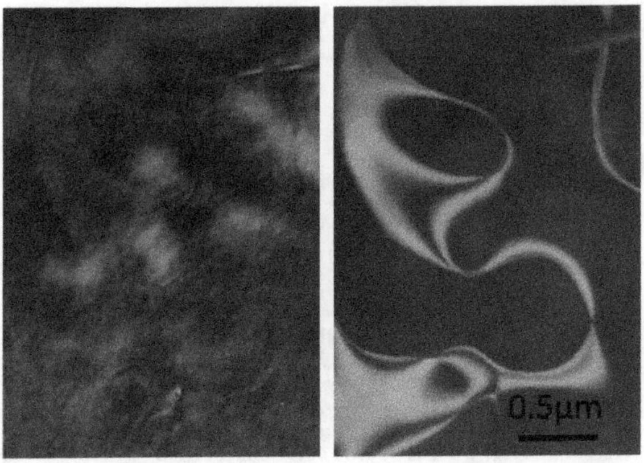

(a) (b)

Fig. 7.7: Anti-phase boundary (APB) in an FeAl (CsCl type) ordered alloy. (a) is DF image taken using a fundamental reflection $\mathbf{g}^* = \mathit{200}$, while (b) is imaged using a superlattice reflection $\mathbf{g}^* = \mathit{100}$ (by courtesy of Dr. Kuroda).

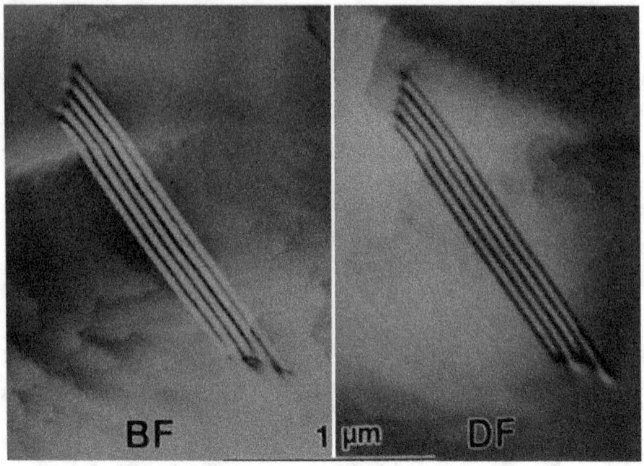

(a) (b)

Fig. 7.8: π-fringes in rutile. (a) BF, (b) DF. Contrasts in BF (a) and DF (b) are exactly complementary with each other (by courtesy of Dr. J. van Landuyt).

APB is inclined, a fringe contrast appears.[2] Such a fringe is referred to as π-fringe. The π-fringes are observed not only in ordered alloys but also minerals. Figures 7.8(a) and 7.8(b) show BF and DF images of π-fringes in a rutile (TiO_2). As can be seen from this,

> **Theorem 7.4:**
> π-fringes both in BF and DF are symmetrical and complementary.

7.3 Precipitates

Spherical precipitates

When a second phase is formed in the matrix, this phenomenon is referred to as precipitation and the second phase is referred to as a precipitate. The structures of interfaces between the matrix and the precipitates are categorized into three as shown in Figs. 7.9(a)–7.9(c).

When lattices of the matrix and a precipitate are distorted only elastically, and a one-to-one correspondence is kept as shown in Fig. 7.9(a), the interface is referred to as a coherent interface. On

[2]Structure factor for superlattice reflection is small, so that the extinction distance is long (see Eq. (6.5) in Exercise 6.2). Thus, the period of the fringe is long as well.

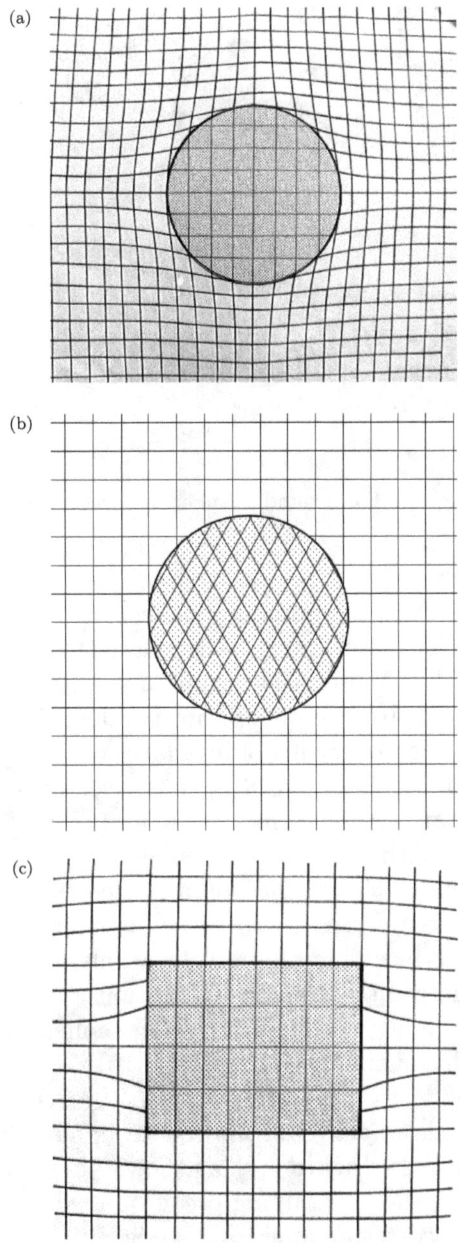

Fig. 7.9: Interfaces between matrices and precipitates. (a) Coherent interface, (b) incoherent interface, (c) partial coherent precipitate with a coherent interface and a semi-coherent interface.

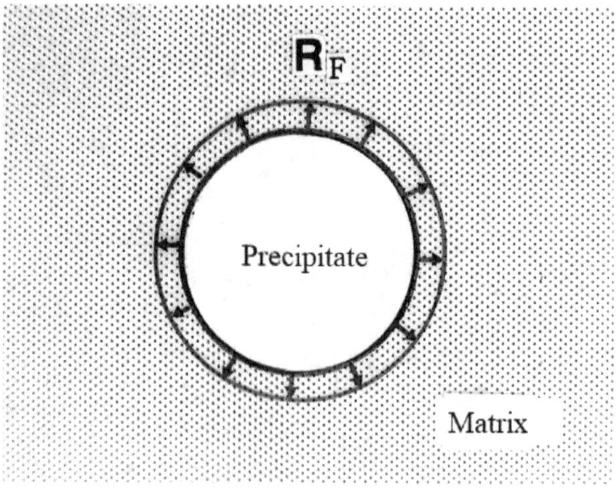

Fig. 7.10: Displacement $\mathbf{R}_F$ around a spherical precipitate, which is radially symmetric.

the other hand, when the correspondence does not exist at all as shown in Fig. 7.9(b), the interface is referred to as an incoherent interface. When all the interfaces around a precipitate are incoherent as shown in Fig. 7.9(b), the precipitate is referred to as an incoherent precipitate. When only part of interfaces covering the precipitate is coherent as shown in Fig. 7.9(c), it is referred to as a partially coherent precipitate. An interface such as the vertical one shown in Fig. 7.9(c) is neither completely coherent nor completely incoherent, with a mismatch being accommodated by dislocations. This type of interface is referred to as a pseudo-coherent interface or a semi-coherent interface, and those dislocations which accommodate mismatch are referred to as misfit dislocations.[3]

Lattices of the matrix around a spherical coherent precipitate are distorted radically symmetrically (Fig. 7.10). As a result, along a direction vertical to $\mathbf{g}^*_{hkl}$, $\mathbf{g}^* \cdot \mathbf{R}_F = 0$, so that the contrast disappears. This produces a pair of strain contrasts along $\mathbf{g}^*$, producing a butterfly-like contrast. Figure 7.11(a) shows an example, which is a BF image of an Al alloy containing precipitates of Al_3Sc precipitates together with the corresponding SAD. In addition to spots from the

[3]Do not confuse partial coherency with semi-coherency (or pseudo-coherency).

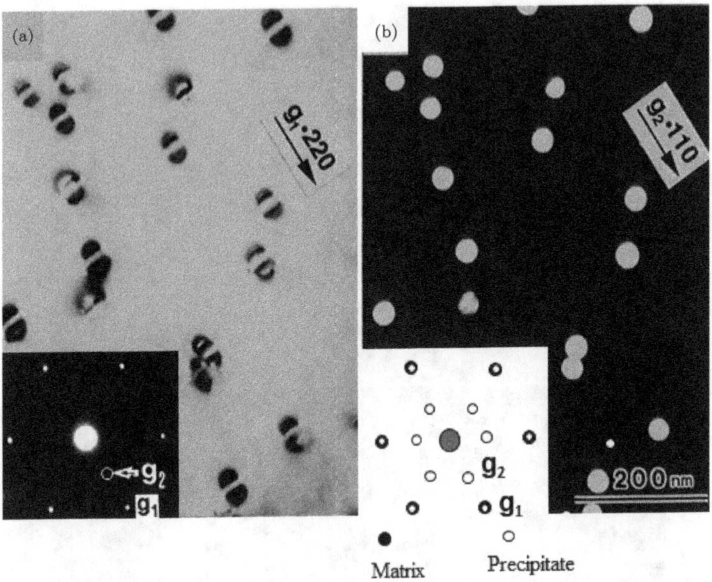

Fig. 7.11: (a) BF image of spherical coherent precipitates with $\mathbf{g}_1^*$ excited. (b) DF image of the same area as (a) but using $\mathbf{g}_2^*$ reflection. Precipitates appear bright (by courtesy of Dr. Y. Miura).

matrix ($\mathbf{g}_1^*$), finer spots from Al_3Sc precipitates ($\mathbf{g}_2^*$) are observed. Here, in the SAD pattern the incident electron beam $\mathbf{B}$ is aligned exactly along [111]. BF image (a) was taken with the crystal slightly tilted in such a way $\mathbf{g}_1^*$ satisfies the Bragg's condition.

Dark field imaging

In observing a precipitate, DF imaging is quite effective. Figure 7.11(b) shows a DF image, which is a counterpart of the BF image shown in Fig. 7.11(a). Here, the DF image was taken using a forbidden fine diffraction spot $\mathbf{g}_2^*$ in the accompanying SAD shown in Fig. 7.11(a). Fine precipitates of Al_3Sc precipitates are observed clearly. Also, it is easy to measure the exact size of the precipitates in the DF image, which is quite difficult in the BF image.

Dark field imaging can be applied to not only crystalline precipitates but also liquid or amorphous precipitates. Figure 7.12 shows an example where indium (In) liquid droplets are present in a crystalline InSb matrix. Figure 7.12(a) shows a BF image and the dark

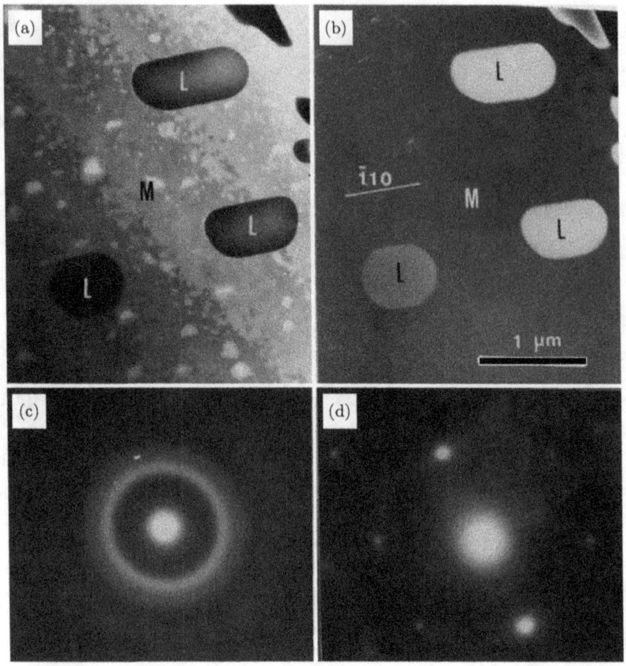

Fig. 7.12: Liquid droplets (L) dispersed in the matrix of InSb crystal (M). (a) BF image. (c) and (d) show selected area diffraction (SAD) patterns taken on L and M, respectively. (b) shows a DF image taken with part of a hallow ring shown in (c).

areas denoted by L are liquid droplets of indium (In), and the area denoted by M is the matrix crystalline InSb. SAD's from L and M are shown in Figs. 7.12(c) and 7.12(d), respectively. In Fig. 7.12(d), ordinary diffraction spots are observed, while in Fig. 7.12(c) halo-rings are observed. Figure 7.12(b) shows the DF image taken using part of the halo-rings, where liquid (L) appears bright.

7.4 Moire Fringes

When two crystals overlap up and down, and both of them satisfy the Bragg's conditions, moire fringe appears at the interface. Moire is a wavy fringe pattern which is observed when two geometrically regular patterns overlap.

Let us consider that two crystals with slightly different orientation and/or lattice spacing are overlapped. A set of parallel fringes that

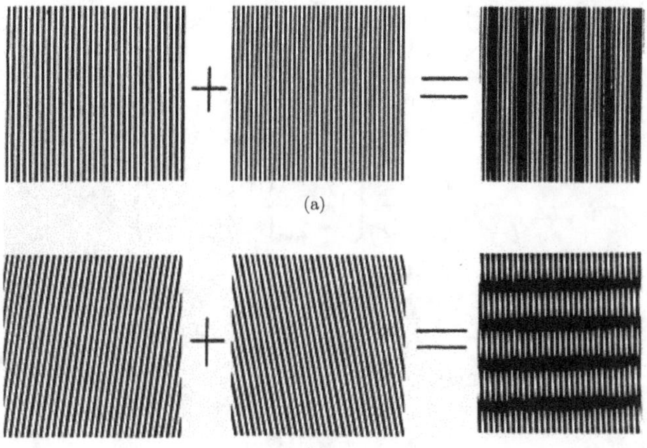

Fig. 7.13: Mechanism by which moire fringes appear. (a) Parallel moire. (b) Rotation moire.

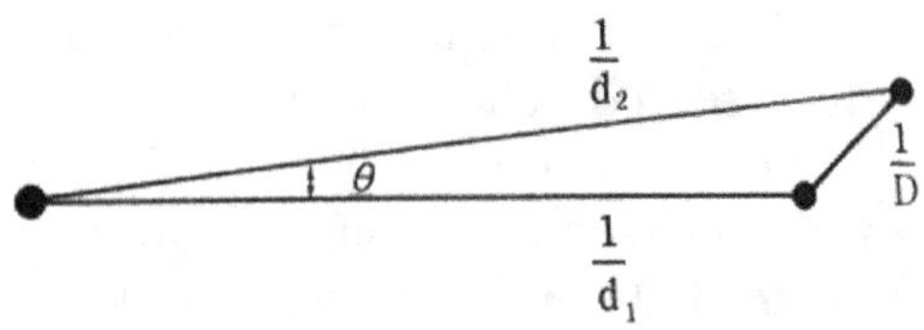

Fig. 7.14: Hint to Exercise 7.3.

appears when two lattices with slightly different spacing are superimposed is referred to as a parallel moiré and that which appears when two lattices with same spacing but with slightly different orientations are superimposed is referred to as a rotation moiré. A spacing of moiré fringes is calculated as follows (Fig. 7.13).

Let us suppose that out of the two overlapping crystals, the upper half satisfies Bragg's condition for $\mathbf{g}_1^*$, and the lower half satisfies $\mathbf{g}_2^*$. When $\mathbf{g}_1^*$ and $\mathbf{g}_2^*$ differ slightly, $\triangle\mathbf{g}^* = |\mathbf{g}_1^* - \mathbf{g}_2^*|$ is a reciprocal lattice corresponding to the spacing of moire D. That is, the moiré fringe is perpendicular to $\triangle\mathbf{g}^*$ and its period is $1/|\triangle\mathbf{g}^*|$.

Exercise 7.3. Show that the spacings of moiré fringe D are given by the following equations.

(*Hint*) See Fig. 7.14.

When $d_1 \neq d_2$ (parallel moiré)

$$D = \frac{d_1 d_2}{\left(d_1^2 + d_2^2 - 2d_1 d_2 \cos\theta\right)^{\frac{1}{2}}} \tag{7.8}$$

$$\sim \frac{d_1 d_2}{d_1 - d_2}\left\{1 - \frac{d_1 d_2}{2(d_1 - d_2)^2}\theta^2\right\}$$

$$\sim \frac{d_1 d_2}{d_1 - d_2}\cos\theta \tag{7.8'}$$

When $d_1 = d_2$ (rotation moiré)

$$D = \frac{d}{\theta} \tag{7.8''}$$

Figures 7.15(a) and 7.15(b) show a BF image of fine Pb precipitates dispersed in an Al matrix. Moire fringes along one direction are observed in Fig. 7.15(a), while in Fig. 7.15(b) moiré fringes along three directions are observed. The SAD is shown in Fig. 7.15(c). As mentioned in example 1 in Section 5.5, both Al and Pb crystallize in FCC, with lattice constant of Pb slightly larger than Al (lattice constant of Al: a = 0.40490 nm, Pb: 0.49495 nm), so that diffraction spots from Pb are inside those from Al.

Moiré fringes observed in Figs. 7.15(a) and 7.15(b) are parallel moiré resulting from the difference in lattice constants between Al and Pb. In Fig. 7.15(b) moiré fringes along three directions appear. The reason for this is that the incident electron beam is exactly along [011] as shown in Fig. 7.15(c), so that all of $\mathbf{g}_1^* = \pm\overline{1}\overline{1}1$, $\mathbf{g}_2^* = \pm\overline{1}1\overline{1}$ and $\mathbf{g}_3^* = \pm 200$ satisfy the Bragg's condition to similar extent. By contrast, in Fig. 7.15(a) only $\mathbf{g}_1^* = \pm\overline{1}\overline{1}1$ satisfies the Bragg's condition. As a result, moiré fringes along one direction perpendicular to $\mathbf{g}_1^* = \pm\overline{1}\overline{1}1$ appear.

7.5 Summary of Contrasts from Planar Defects

Important characteristics for different types of planar crystal defects.

Defect	Types of fringes	Spacing	Visibility	Characteristics	Quantitative information
Stacking fault forms α fringe $\alpha \neq n\pi$	$\alpha = 2\pi \mathbf{g}^* \cdot \mathbf{R}_F$ $\mathbf{R}_F$: displacement vector $\alpha = 0(2n\pi)$ or $\pm(2/3)\pi$, $(4/3)\pi$	$(1/2)\xi_g$ at $s^* = 0$ $\xi_g/(1+w^2)^{1/2}$ at $\mathbf{s}^* \neq 0$	Best at $s^* = 0$ for both BF and DF	BF image symmetric, that is fringes at top and bottom of foil either bright or dark. DF image asymmetric, that is opposite contrast fringes at top and bottom of foil.	Fringes invisible $\mathbf{g}^* \cdot \mathbf{R}_F = 0$ Use to determine direction of $\mathbf{R}_F$. Intrinsic/extrinsic nature determined by a number of techniques for $s^* = 0$ only Simplest is use dark field image, transfer $\mathbf{g}^*$ to positive print, origin of $\mathbf{g}^*$ set at the center of fringes If $\mathbf{g}^*$ points towards bright outer fringe, fault is extrinsic If dark fringe, fault is intrinsic for class $B = \{111\}\{220\}\{400\}$ reflections. For class A $\{200\}\{222\}\{440\}$ reflection reverse is true

(*Continued*)

(Continued)

Defect	Types of fringes	Spacing	Visibility	Characteristics	Quantitative information
Planar precipitate with small **misfit forms** $\boldsymbol{\alpha}$ **fringe** $\boldsymbol{\alpha \neq n\pi}$	displacement vector $\alpha = 0(2n\pi)$ or **fraction**	$(1/2)\xi_g$ at $\mathbf{s}^* = 0$	best at $s^* = 0$ for both BF and DF	BF symmetric, DF asymmetric	Fringes invisible $\mathbf{g}^* \cdot \mathbf{R}_F = 0$. Use to determine direction of $\mathbf{R}_F$. **N.B. only valid if precipitate is very thin and not diffracting**
Antiphase boundary $\alpha = \pi$	$\alpha = 2n\pi\mathbf{g}^*\mathbf{R}_F =$ 0 or π	$(1/2)\xi_g$ at $s^* = 0$	best at $s = 0$ for both BF and DF	BF symmetric, DF symmetric and exactly opposite of BF	Fringes invisible $\mathbf{g}^*\mathbf{R}_F = 0$ (can be used to determine direction of $\mathbf{R}_F$)
Moiré	fringes perpendicular to the operative $\Delta\mathbf{g}^* = \mathbf{g}_1^* - \mathbf{g}_2^*$	$1/\lvert\Delta\mathbf{g}^*\rvert$	$\mathbf{s}_g$ small and positive with strong intensity in diffracted beam in each crystal	Direction and spacing changes with operative reflection	Moiré fringes may be recognized by changing the operative reflection when the direction of the fringes will change

Thickness fringes	Fringes following regions of uniform thickness, that is depth contours at edge of foil or grain boundary fringes when only one grain diffracting	ζ_g at $\mathbf{s}_g^* = 0$ $\zeta_g/(1 + \mathrm{w}^2)^{1/2}$ at $\mathbf{s}_g^* \neq 0$	Best at $\mathrm{s}^* = 0$	Fringes reverse in BF and DF	Used to determine foil thickness $= (n + 1/2)\zeta_g$ black fringes, BF, at $\mathbf{s}_g^* = 0$, where $n =$ number of fringes

Source: Edington (1976).

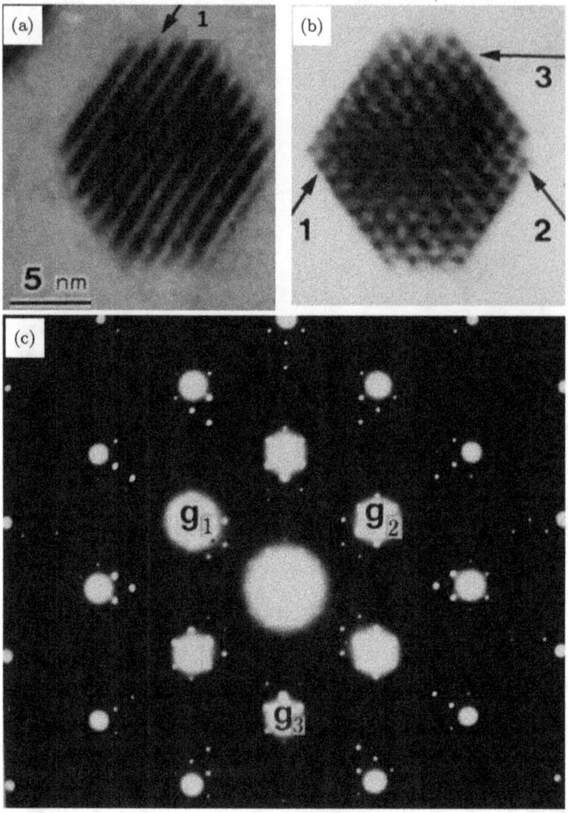

Fig. 7.15: (a) and (b) BF images of a Pb particle dispersed in the matrix of Al. (c) shows the SAD under which (b) was taken. The incident beam is symmetrical with respect to diffracting planes $\mathbf{g}_1^*$, $\mathbf{g}_2^*$ and $\mathbf{g}_3^*$, so that in (b) three moiré fringes corresponding to $\mathbf{g}_1^*$, $\mathbf{g}_2^*$ and $\mathbf{g}_3^*$ appear. (a) was taken with $\mathbf{g}_1^*$ alone excited, so that moiré fringe corresponding $\mathbf{g}_1^*$ alone appears (by courtesy of Dr. E. Johnson).

7.6 Supplement

Some useful examples of DF imaging
(A) Determining crystal orientations.
(A-1) Whisker ①

Figure 7.16 shows a DF image of a bent whisker: It is obvious that the whisker is a single crystal.

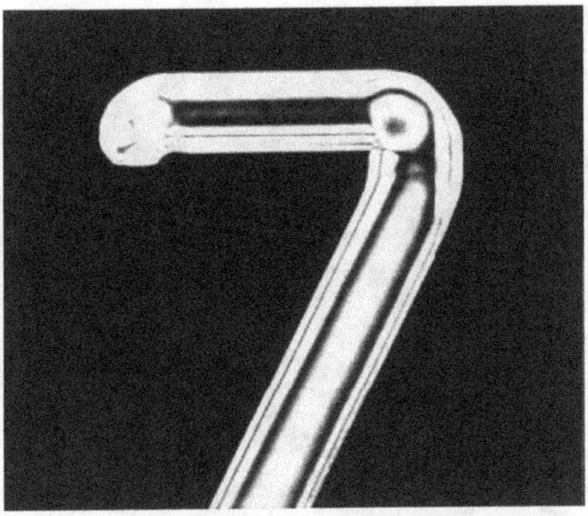

Fig. 7.16: A bent single crystalline whisker of Al.

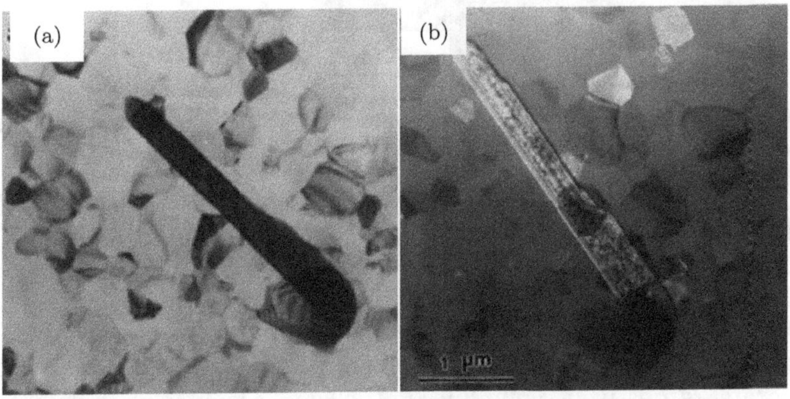

Fig. 7.17: Whisker grown on bottom grain.

(A-2) Whisker ②

Figures 7.17(a) and 7.17(b) show BF and DF images, respectively, of a whisker which nucleates on the surface of Al. In BF in (a), the whisker appears dark just in a similar way as the bottom grain from which the whisker nucleates and this may lead to a conclusion that orientations of the whisker and the bottom grain are the same. However, in DF (Fig. 7.17(b)), the whisker appears bright, while the bottom grain appears dark: They are differently orientated.

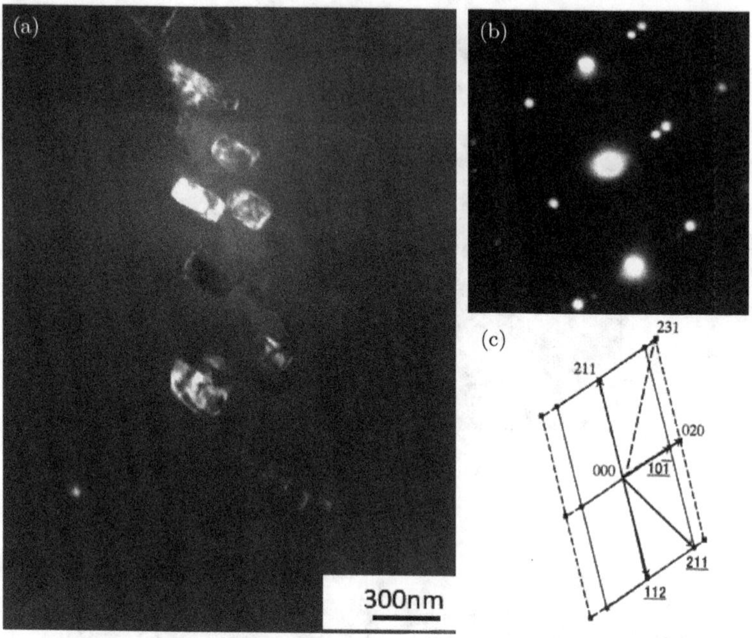

Fig. 7.18: Precipitates of Ag_3Sn in a leadfree solder. Reproduced from Torazawa *et al.* (2000).

(B) Orientation relationship between precipitates and matrix
(B-1) Structure of a solder

Figure 7.18(a) shows a DF image of a leadfree solder containing Ag. Figures 7.18(b) and 7.18(c) show the corresponding SAD and its key diagram, respectively. Solid lines in the key diagram stand for Ag_3Sn: From this, the orientation relationship between Ag_3Sn and the matrix is determined. The DF image shown in Fig. 7.18(a) was taken using a diffraction spot from Ag_3Sn and the size and distribution of Ag_3Sn can be determined.

(B-2) Precipitation hardening 17-7PH steel

Diffraction pattern shown in Fig. 7.19(b) indicates that precipitates are NiAl with an ordered structure (B2) and are coherent with the matrix. From a DF image taken using a forbidden spot (indicated by white arrow in (b)), size and distribution of the precipitates can be obtained (Fig. 7.19(a)).

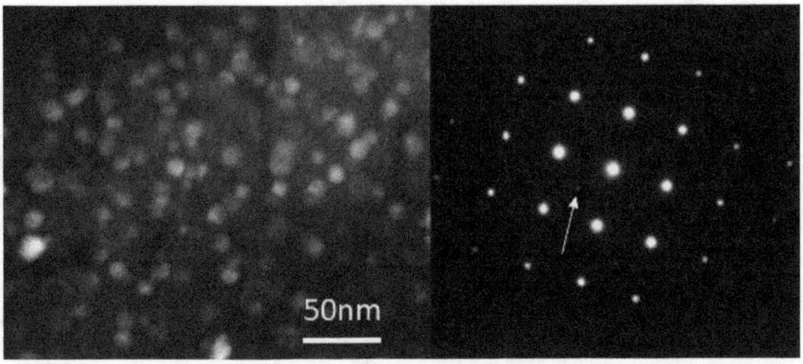

Fig. 7.19: (a) Precipitate of NiAl in 17-7PH stainless steel. (b) SAD. DF image in (a) was taken using a small diffraction spot indicated by an arrow.

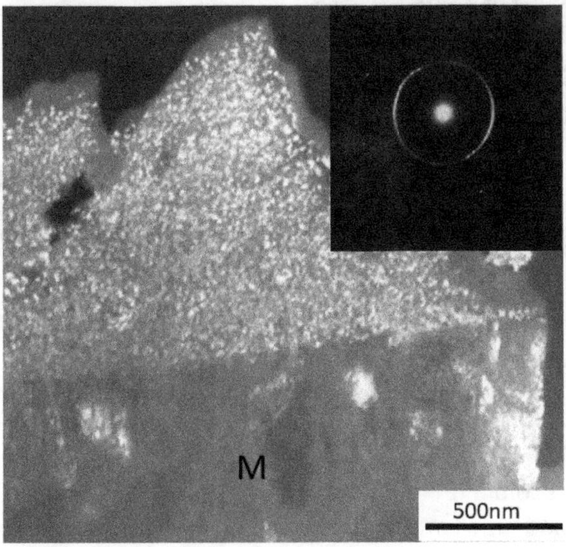

Fig. 7.20: Fine-grained BCC Fe. Inset shows the SAD. Reproduced from Muroga and Saka (1995).

(C) Fine grained structure and amorphous phase
(C-1) Structure in the vicinity of a crack formed by rolling contact fatigue (Fig. 7.20).

From the accompanying SAD in the inset of Fig. 7.20, it is evident that the structure is a polycrystal of α-Fe (BCC) composed of very

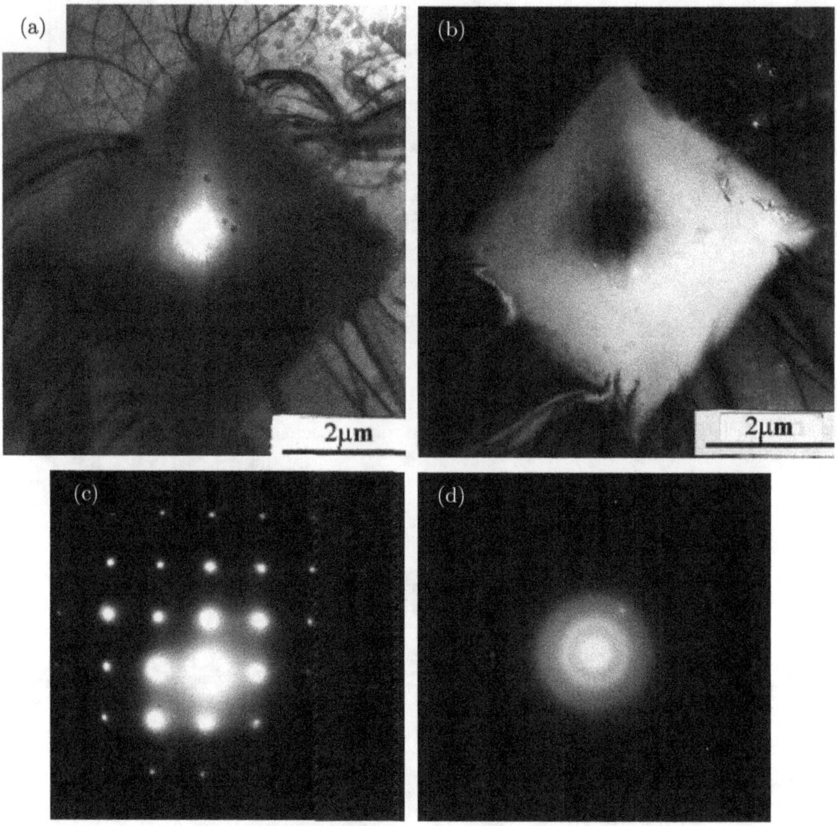

Fig. 7.21: Amorphous phase formed under image of indentation. (a) BF. (b) DF taken with part of halo-ring shown in (d). (c) SAD taken at crystalline Si. (d) SAD taken at amorphous Si.

fine grains. From DF image obtained using part of the Debye ring the grain size is measured. Coarse structure below (indicated by M) is a martensite.

(C-2) Vickers indentation in Si (plan view)

Figures 7.21(a) and 7.21(b) show BF and DF images, respectively, of a Vickers indent formed on Si. Squares in the center show contrast quite different from the surrounding region. Figure 7.21(c) shows SAD from the surrounding region: Diffraction spots from single crystal of Si are evident. By contrast, SAD taken on the square at the center shows halo-rings: This region is an amorphous phase. In the

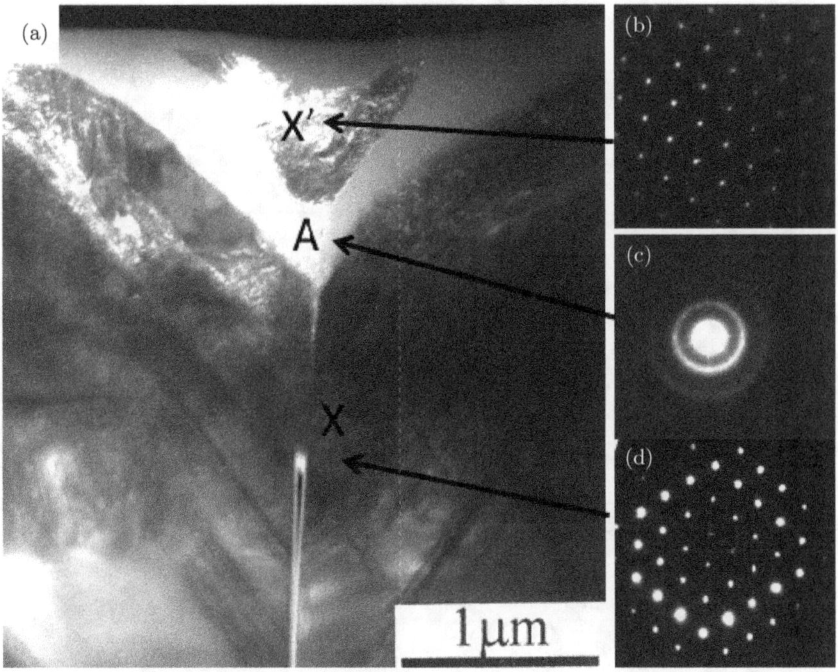

Fig. 7.22: Cross sectional view of indent in Si. Regions indicated by X′ and X in (a) is crystalline as can be seen from SAD's in (b) and (d). Region A is amorphous as can be seen from SAD shown in (c). Reproduced from Suprijadi *et al.* (2002).

DF image taken using part of the halo-rings (shown in Fig. 7.21(d)) the square appears bright, indicating that the square is indeed amorphous (see also Fig. 7.12).

(C-3) Vickers indentation in Si (cross-sectional view)

Figure 7.22 shows a Vickers indent in Si but viewed cross-sectionally. SAD from region A shows halo-rings, indicating that A is amorphous. In a DF obtained using part of the halo-rings A appears bright. SAD's from X and X′ show ordinary diffraction patterns of single crystal of Si. The amorphous phase is formed underneath the indent, however, its geometry is complicated. It is difficult to attribute formation of amorphous phase solely to a static hydraulic pressure caused by indentation.

Chapter 8

Contrasts of Dislocations

8.1 Determining Burgers Vector of Dislocations

Screw dislocation

Let us suppose that a screw dislocation AB lies parallel to the foil surface. Displacement around such a screw dislocation $\mathbf{R}$ is given by

$$\mathbf{R} = \frac{\mathbf{b}}{2\pi}\tan^{-1}\left(\frac{\mathrm{z}-\mathrm{y}}{\mathrm{x}}\right) \tag{8.1}$$

Therefore,

Theorem 8.1:
For a screw dislocation when $\mathbf{g}^* \cdot \mathbf{b} = 0$ and $\mathbf{g}^* \cdot \mathbf{R} = 0$ so that the dislocation is invisible.

Edge dislocation

For an edge dislocation $\mathbf{R}$ is given by Eqs. (3.4) and (3.4′) as follows:

$$\mathbf{R} = \frac{1}{2\pi}\left\{\mathbf{R}_1 + \left(\frac{\mathbf{b}}{\mathrm{b}}\times\varsigma\right)\cdot|\mathbf{R}_2|\right\} \tag{8.2}$$

Here,

$$\mathbf{R}_1 = \frac{\mathbf{b}}{2\pi}\left\{\phi + \frac{\sin 2\phi}{4(1-\nu)}\right\} \tag{8.2′}$$

$$\mathbf{R}_2 = -\frac{\mathbf{b}}{2\pi}\left\{\frac{1-2\nu}{2(1-\nu)}\ln|\mathbf{r}| + \frac{\cos 2\phi}{4(1-\nu)}\right\} \tag{8.2″}$$

Therefore, the following theorem holds.

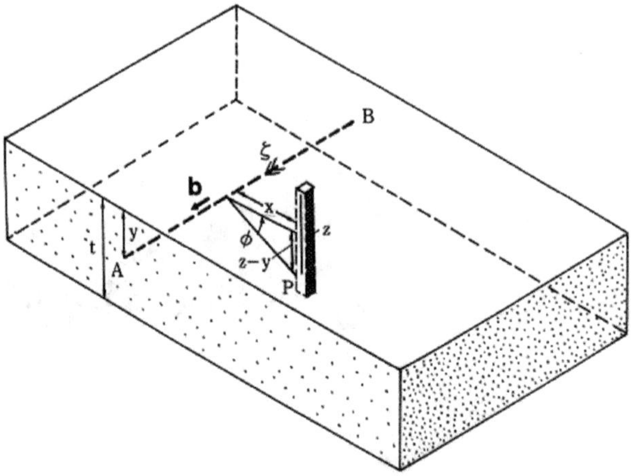

Fig. 8.1: A screw dislocation in a thin crystal with thickness t. Note that the co-ordinates are different from those in Fig. 3.11.

Theorem 8.2:
For an edge dislocation, even when $\mathbf{g}^* \cdot \mathbf{b} = 0$, a weak contrast appears in the case of $\mathbf{g}^* \cdot (\mathbf{b} \times \varsigma) \neq 0$. This weak contrast is referred to as a residual contrast.

This can be accounted for by considering bending of lattices near an edge dislocation. Figure 8.2 illustrates an edge dislocation. As can be seen, a lattice plane ⓐ is not affected at all by the existence of edge dislocation. In other words, an electron beam propagates through the crystal without noticing the existence of the edge dislocation. For plane ⓐ, $\mathbf{g}^* \cdot \mathbf{b} = 0$ and $\mathbf{g}^* \cdot (\mathbf{b} \times \varsigma) = 0$ hold. By contrast, plane ⓑ contains an extra-half plane, so that it is bent severely at both sides of the extra-half plane clockwise and anti-clockwise. Therefore, in the vicinity of dislocation the Bragg's condition changes significantly and strong contrasts appear (see Section 8.2). On the other hand, plane ⓒ, remote from the dislocation, is not affected, but slightly bent in the very vicinity of the dislocation core. This slight bending of lattice plane ⓒ corresponds to $\mathbf{g}^* \cdot (\mathbf{b} \times \varsigma) \neq 0$, producing a residual contrast.

Figure 8.3 shows an example of residual contrasts. The sample is Ag (FCC) and Frank's loop (see Section 3.3.4) is present in the plane of foil of (*111*). $\mathbf{b}$ and $\mathbf{R}_\mathrm{F}$ of this Frank's loop is $\frac{1}{3}[111]$. Therefore, when imaged in any of $\mathbf{g}^* = 0\bar{2}2$ (b), $\mathbf{g}^* = \bar{2}20$ (c) and $\mathbf{g}^* = \bar{2}02$ (d),

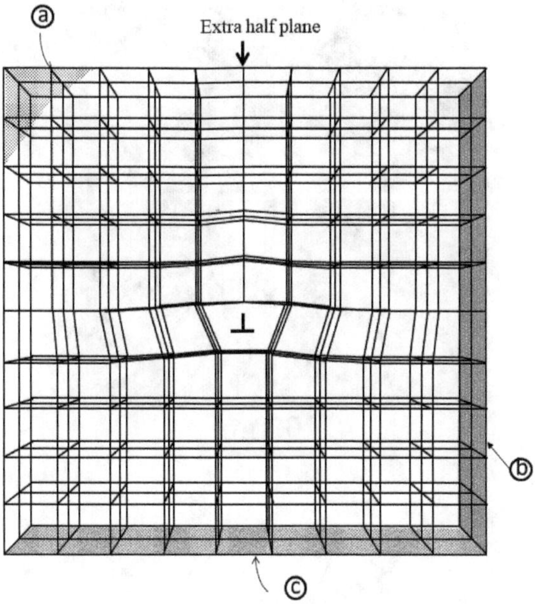

Fig. 8.2: Bending of lattice plane around an edge dislocation, (a) is in the plane of paper.

$\mathbf{g}^* \cdot \mathbf{b} = 0$ and $\mathbf{g}^* \cdot \mathbf{R}_F = 0$. Thus, the SF is invisible. In $\mathbf{g}^* = \bar{1}1\bar{1}$ (a), $\mathbf{g}^* \cdot \mathbf{R}_F = -1/3$, so that contrast appears (Theorem 7.2). However, since the SF is not tilted, fringe contrast does not appear. Now, in either case of (a)–(c), a residual contrast appears except along a direction parallel to $\mathbf{g}^*$. When a dislocation ς is parallel to $\mathbf{g}^*$, $\mathbf{g}^* \cdot (\mathbf{b} \times \varsigma) = 0$ so that the residual contrast disappears (Theorem 8.2).

When an identical dislocation is observed in different $\mathbf{g}^*$, nature of the dislocation can be determined. This type of experiment is referred to as a contrast experiment; an example is shown in Fig. 8.4. Dislocation A shows strong contrast in $\mathbf{g}_1^* = 0\bar{1}1$ (a), while it disappears in $\mathbf{g}_2^* = 211$ (b). Thus, it can be determined that Burgers vector ($\mathbf{b}$) of dislocation A is perpendicular to [211].

Theorem 8.3:
If for two diffracting vectors $\mathbf{g}_1^*$ and $\mathbf{g}_2^*$ a dislocation disappears or shows a residual contrast, the direction of Burgers vector $\mathbf{b}$ can be identified uniquely by

$$\mathbf{b} // \mathbf{g}_1^* \times \mathbf{g}_2^* \tag{8.3}$$

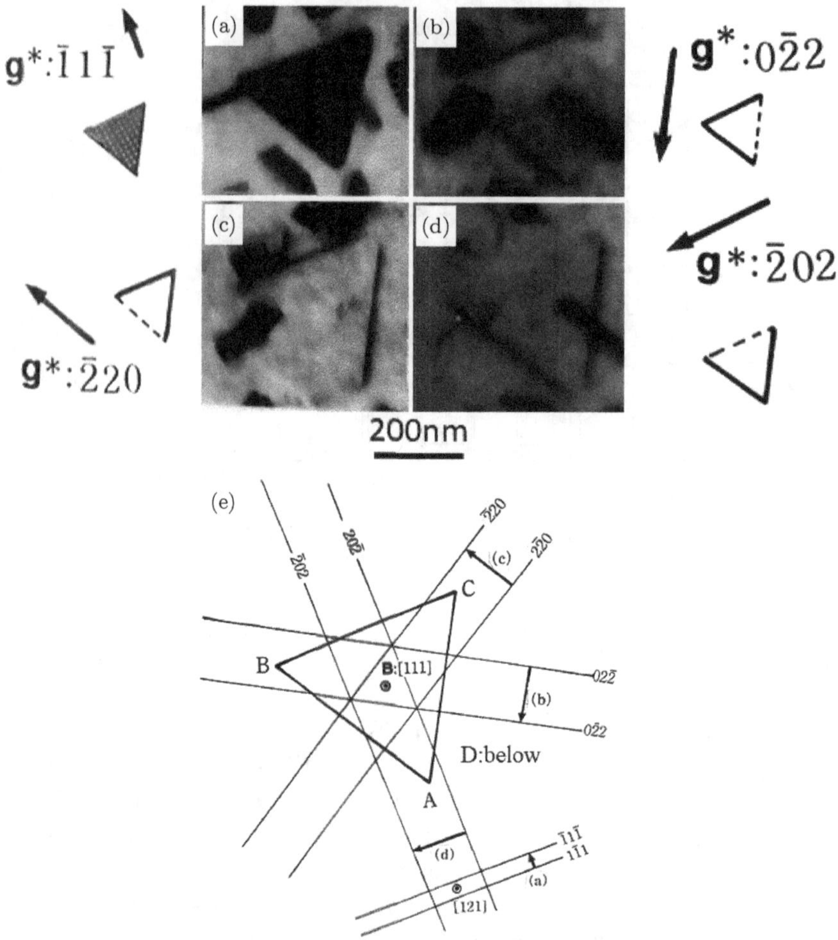

Fig. 8.3: Residual contrasts of a Frank loop containing the SF in Ag. (a) An equilateral triangular Frank loop is in the plane of foil. (e) Illustrates a Kikuchi map and Thompson tetrahedron together with diffraction conditions under which (a)–(d) were imaged.

> **Point 8.1.**
> Practically it is not so easy to find such a pair of $\mathbf{g}^*$. Therefore, information obtained by a contrast experiment is to be supplemented with other information (knowledge) such as crystal structure and slip system etc.

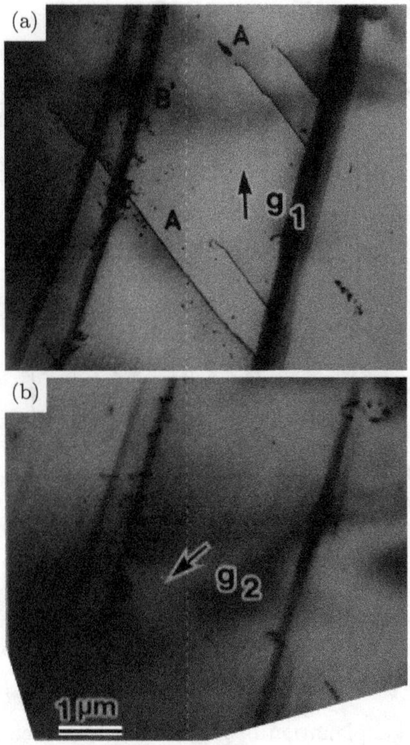

Fig. 8.4: Contrast experiments on dislocations in Mo. In (a) dislocation A is visible in $\mathbf{g}_1^* = 0\bar{1}1$, while in (b) it is invisible in $\mathbf{g}_2^* = 211$.

8.2 Determining the Sign of Burgers Vector

The method described in the foregoing section allows one to determine the direction of **b**, but it does not allow to determine the sign of **b**. In order to determine the sign of **b**, it is necessary to examine an identical dislocation with different signs of $\mathbf{g}^*$ and/or $\mathbf{s}_\mathrm{g}^*$. In order to understand this method, let us consider mechanism by which contrast of a dislocation appears. Let us suppose a crystal containing an edge dislocation as shown in Fig. 8.5. Taking the direction of the dislocation ς as pointing into the plane of paper, FS/RH (perfect) convention is employed. Then, **b** points to the left as shown in Fig 8.5. Let the extra half plane (ehp) of the edge dislocation lie on (hkl) planes, which are bent at the vicinity of the ehp. Let us suppose that an electron beam is incident from upper right with respect to

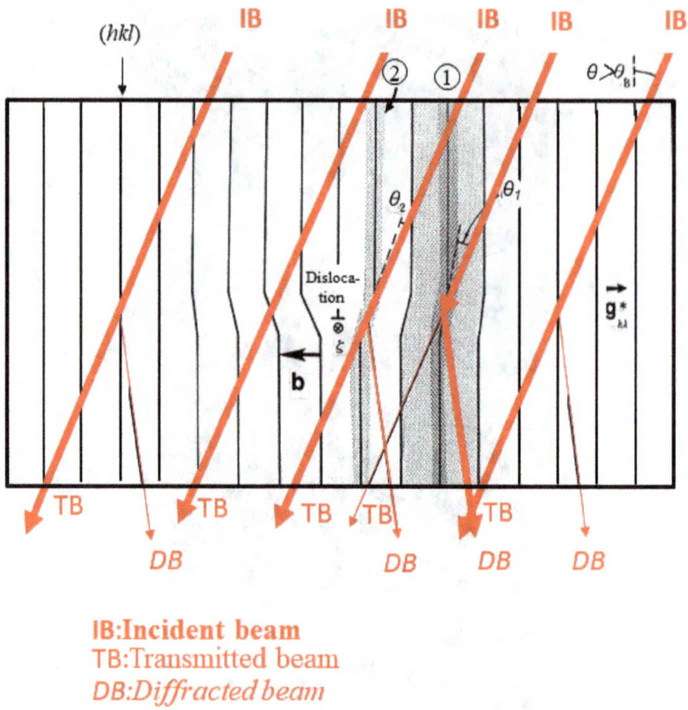

Fig. 8.5: Qualitative explanation of mechanism by which diffraction contrasts appear around a dislocation. Column ① at the right side of a dislocation satisfies Bragg's condition.

(hkl) planes and that an angle of θ which the incident beam makes with (hkl) planes remote from the dislocation core is larger than the Bragg's angle θ_B, that is, $\theta > \theta_B$. Then, according to Theorem 5.1, $s_g^* > 0$.

Under these circumstances let us discuss on the diffracted intensity as a function of the distance from the dislocation core. At positions remote from the dislocation core, $\theta > \theta_B$. That is, the Bragg's condition is not satisfied fully, so that the diffracted intensity is not so strong. This produces the contrast of the background.

On the other hands, in the column ①, which is located to the right of the dislocation core, (hkl) plane is rotated clockwise with respect to the dislocation line ς. This results in decreasing θ locally, leading to $\theta_1 = \theta_B$. That is, the Bragg's condition is satisfied. In column ②, which is closer to the dislocation core, $\theta_2 < \theta_B$ and again the Bragg's condition is not satisfied. Thus, the diffracted intensity is maximum at (or near) column ①. Since a contrast in BF corresponds to the incident intensity minus diffracted intensity, black contrast appears near column ①. In summary,

Theorem 8.4:
Contrast appears at a position where the lattice plane is rotated so as to satisfy the Bragg's condition.

In the case shown in Fig. 8.5, $(\mathbf{g}^* \cdot \mathbf{b})\mathbf{s}_\mathrm{g}^* < 0$ ($\because$ $\mathbf{s}_\mathrm{g}^*>0$, $\mathbf{g}^* \cdot \mathbf{b} = -|\mathbf{g}^*||\mathbf{b}|$). It follows that

Theorem 8.5:
When $(\mathbf{g}^* \cdot \mathbf{b})\mathbf{s}_\mathrm{g}^* < 0$, contrast appears at the right side of the dislocation core. When $(\mathbf{g}^* \cdot \mathbf{b})\mathbf{s}_\mathrm{g}^* > 0$, contrast appears at the left side of the dislocation core.

Exercise 8.1. Consider a case where $\mathbf{s}_\mathrm{g}^* < 0$ and prove Theorems 8.4 and 8.5.

Exercise 8.2. Prove that Theorem 8.5 is true when an extra-half plane is below.

The foregoing discussion is applied to an edge dislocation. However, the same discussion is applied to a screw dislocation.

Figure 8.6 shows a pair of dislocation images taken in $\pm\mathbf{s}_\mathrm{g}^*$ while keeping $\mathbf{s}_\mathrm{g}^* > 0$. Relative positions of images of dislocations ①, ②, ③ and ④ are reverse in (a) and (b).

Point 8.2.
As is obvious from Theorem 8.5, it is necessary to observe an identical dislocation in $\pm\mathbf{g}^*$ or $\pm\mathbf{s}_\mathrm{g}^*$. However, when $\mathbf{s}_\mathrm{g}^* < 0$, the background is darker due to absorption (see Section 6.4). Therefore, it is recommended to observe in $\pm\mathbf{g}^*$ rather than in $\pm\mathbf{s}_\mathrm{g}^*$.

8.3 Determining Natures of Dislocation Loop

The method described in the foregoing section can be applied to determining the natures of dislocation loops, i.e., whether they are formed by precipitation of vacancies (vacancy-type; V-type) or interstitials (interstitial-type; I-type). Figures 8.7(a) and 8.7(b)

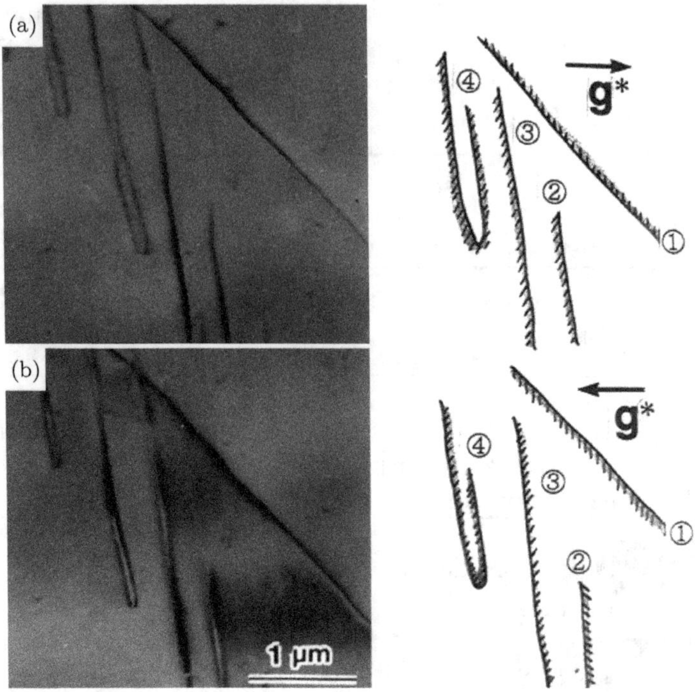

Fig. 8.6: BF images of dislocations imaged in $\mathbf{g}^*$(a) and $-\mathbf{g}^*$(b) while keeping $\mathbf{s}_g^* > 0$. Positions of contrasts with respect to the dislocations are reversed on the reversal of sign of $\mathbf{g}^*$.

illustrate schematically dislocation loop of V-type and I-type, respectively.

Theorem 8.6:
Relationship between the sense of direction of a dislocation line ς and $\mathbf{b}$ is of a right-handed screw for a V-type loop, and of a left-handed screw for an I-type loop (see Exercise 3.4).

Here,

Definition 8.1:
Looking down a dislocation loop, take ς clockwise.

From Theorem 8.4, it follows:

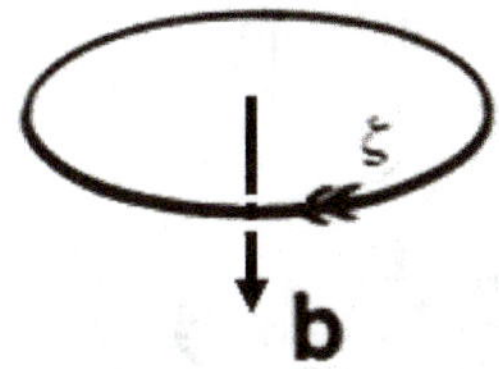

(a) Vacancy-type dislocation loop

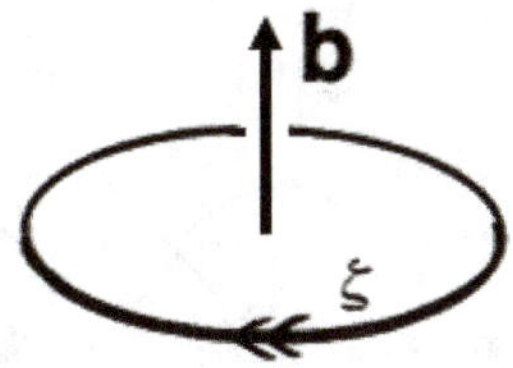

(b) Interstitial-type dislocation loop

Fig. 8.7: Relation between **b** and the dislocation directions ζ for vacancy type (V-type) (a) and interstitial type (I-type) (b) loops. For V-type the relation is of a right-handed screw and for I-type it is of a left-handed screw.

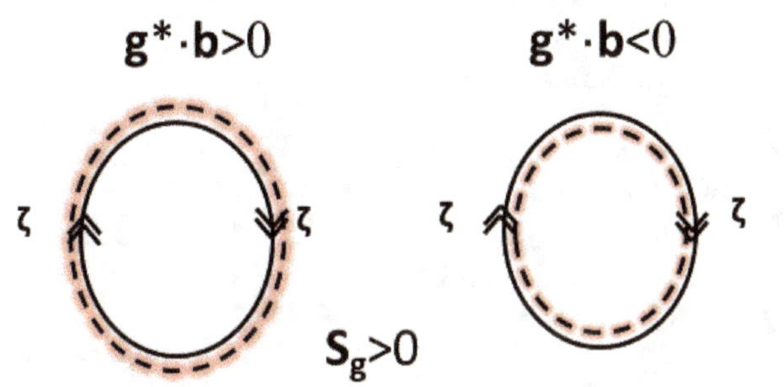

(a) (b)

Fig. 8.8: Theorem 8.7. Solid lines illustrate positions of real dislocations and broken lines (in red) illustrate positions where the images appear. For $\mathbf{s}_{\mathrm{g}}^* > 0$, the image appears outside the dislocation loop when $\mathbf{g}^* \cdot \mathbf{b} > 0$, while it appears inside the dislocation loop when $\mathbf{g}^* \cdot \mathbf{b} < 0$.

Theorem 8.7:
When $\mathbf{s}_{\mathrm{g}}^* > 0$, as can be seen from Figs. 8.8(a) and 8.8(b), if $\mathbf{g}^* \cdot \mathbf{b} > 0$, strong contrast appears at the left side, that is, outside of a dislocation loop. If $\mathbf{g}^* \cdot \mathbf{b} < 0$, strong contrast appears at the right side, that is, inside of a dislocation loop.

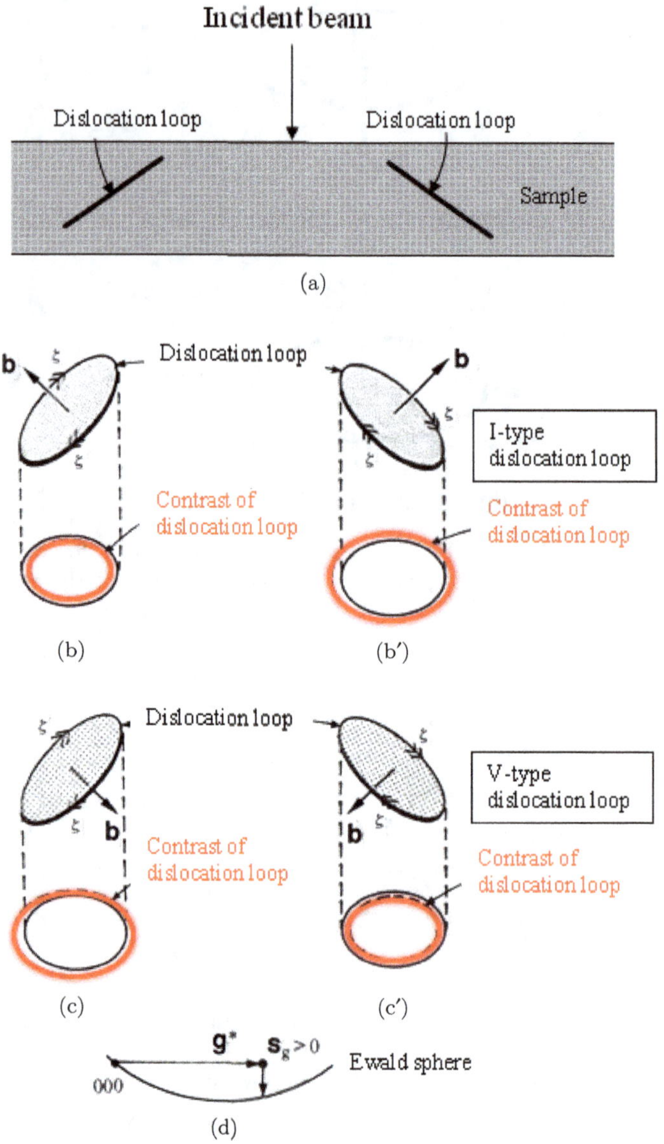

Fig. 8.9: Where dislocation contrasts appear when inclined I-type ((b)(b′)) and V-type ((c)(c′)) dislocation loops are present in a foil specimen? Black lines illustrate positions of real dislocations and red lines illustrate positions of the diffraction contrasts. It is necessary to know the sense of tilting of the dislocation loops to determine whether I-type or V-type.

As is shown in Fig. 8.9, let us suppose that $\mathbf{g}^*$ points to the right and that $\mathbf{s}_g^* > 0$, that is, the reciprocal vector $\mathbf{g}^*$ is inside the Ewald sphere (d) (see also Fig. 4.16). Also let us assume that the dislocation loop is inclined as shown in (a). For an I-type ((b)(b′)), $\mathbf{b}$ is of a left-handed screw with respect to ς. Then, if the loop is inclined clockwise as shown in (b), $\mathbf{g}^*$ and $\mathbf{b}$ make an obtuse angle, so that $\mathbf{g}^* \cdot \mathbf{b} < 0$: Since $\mathbf{s}_g^* > 0$, $(\mathbf{g}^* \cdot \mathbf{b})\mathbf{s}_g^* < 0$. From Theorem 8.5, the image of dislocation is formed on the right side of the actual position of the dislocation, i.e., inside the loop.

When an I-type loop is inclined anti-clockwise (b′), the image is formed outside the loop. For a V-type loop the reverse is true ((c)(c′)). In other words, if the plane on which a loop lies is determined, the nature of the loop, i.e., whether V-type or I-type, can be identified. When the sign of the diffracting vector $\mathbf{g}^*$ is reversed, the results are also reversed.

Exercise 8.3. Figure 8.10 illustrates an inclined loop of I-type. Assuming that solid lines (nearly vertical) stand for (*hkl*) plane, which is the diffracting plane, discuss whether a strong image is formed inside or outside the dislocation loop, based on Theorem 8.4.

Figures 8.11(a)–8.11(d) show an example. These images were taken on a specimen inclined considerably with respect to the incident electron beam $\mathbf{B}$. That is, in (a) $\mathbf{B} = [0\bar{1}1]$, in (b) $\mathbf{B} = [011]$, and in (c)(d) $\mathbf{B} = [001]$. $\mathbf{B}$'s are shown in the accompanying stereographic projection in Fig. 8.12(a) and the geometry of dislocation loop is shown in Fig. 8.12(b). Since in $\mathbf{B} = [0\bar{1}1]$, the loop shows residual contrasts along a direction perpendicular to $\mathbf{g}^* = \bar{4}00$, $\mathbf{b} // \pm[0\bar{1}1]$.

Now, the residual contrast is formed inside in Fig. 8.11(c) ($\mathbf{g}^* = 040$) and outside in (d)($\mathbf{g}^* = 0\bar{4}0$). Since these images are taken in $\mathbf{s}_g^* > 0$, $\mathbf{g}_{040}^* \cdot \mathbf{b} < 0$ and $\mathbf{g}_{0\bar{4}0}^* \cdot \mathbf{b} > 0$. In other words, $\mathbf{b}$ and $\mathbf{g}_{0\bar{4}0}^*$ makes an acute angle, while $\mathbf{b}$ and $\mathbf{g}_{040}^*$ makes an obtuse angle. In summary $\mathbf{b}$ points upward and the relation between ς and $\mathbf{b}$ is of a left screw. From Theorem 8.6, this loop is identified as an I-type loop.

This method is applicable to a Frank's loop which contains the SF inside. Figures 8.13(a)–8.13(d) show an example. Diffracting vectors in which Figs. 8.13(a)–8.13(d) are imaged are shown in Fig. 8.13(e). In Thompson tetrahedron nomenclature, the upper foil surface is *d*

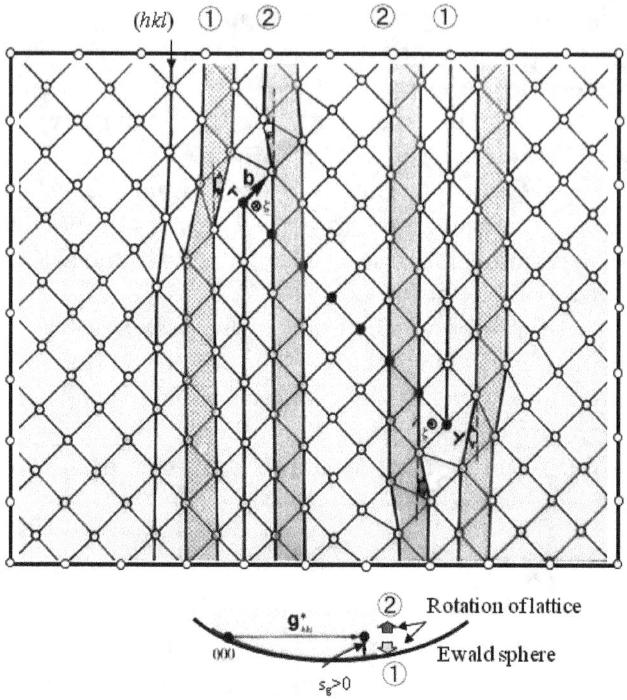

Fig. 8.10: Bending of (*hkl*) planes near an inclined dislocation loop.

and the vertex D is below. Micrograph (a) is imaged in $\mathbf{g}^* = 02\bar{2}$, while (b) is imaged in $\mathbf{g}^* = 0\bar{2}2$. Therefore, the contrast of Frank's loop ⓑ which lies on *b* face (of Thompson tetrahedron) is invisible.

On the other hand, (c) is imaged in $\mathbf{g}^* = \bar{2}20$, while (d) is imaged in $\mathbf{g}^* = 2\bar{2}0$ and Frank loop ⓒ lying on *c* face is invisible. According to Definition 8.1, the sense of the dislocation loop ς is clockwise when viewed from above, so that ς's for dislocations ⓐ, ⓑ and ⓒ are defined as indicated in (f). Furthermore, since **b** lies on *a*, *b* and *c* faces of the Thompson tetrahedron, the respective **b**'s are perpendicular to *a*, *b* and *c* faces.

Now, coming back to (a) and (b), dislocation loop ⓐ shows an inside-contrast, while in (b) it shows an outside-contrast. On the contrary, dislocation loop ⓒ shows an outside-contrast in (a) and an inside-contrast in (b). In summary, **b** of the dislocation loop ⓐ makes an acute angle with $\mathbf{g}^* = 0\bar{2}2$, and **b** of dislocation loop ⓒ makes an acute angle with $\mathbf{g}^* = 02\bar{2}$. Therefore, relationships between ς and

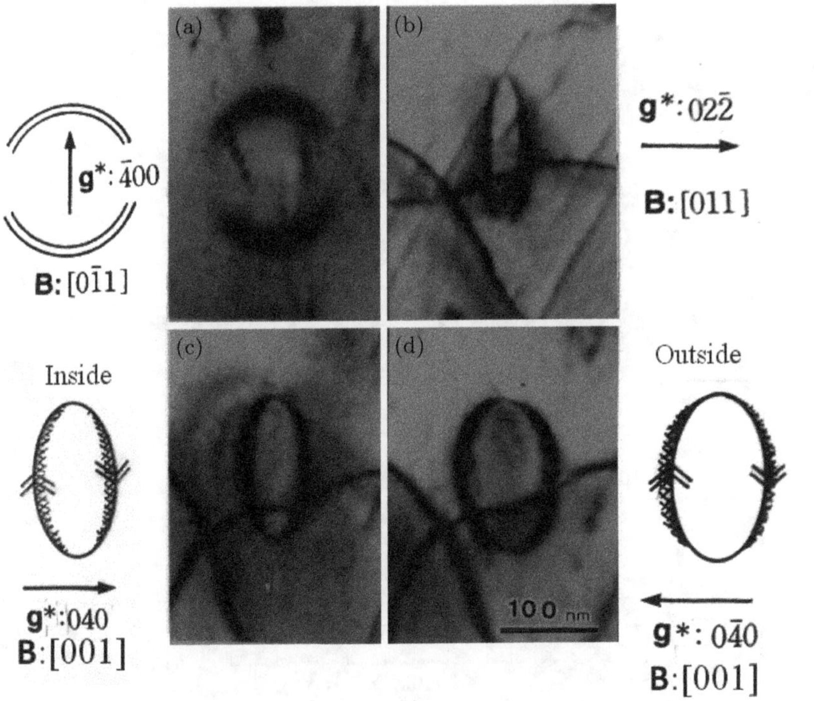

Fig. 8.11: Determining nature of a dislocation loop (reproduced with permission from Nishigaki *et al.* (1991)).

b for ⓐ and ⓒ are of left-handed screw, so that from Theorem 8.6 they are identified as I-type.

Exercise 8.4. Show that both of the dislocation loops ⓐ and ⓑ shown in Figs. 8.6(c) and 8.6(d) are of I-type.

The method described above determines types of a dislocation loop by observing whether a contrast is formed inside or outside of the (real) dislocation loop; this method is referred to as inside/outside method.[1]

---

[1]Tilting angle of a dislocation loop can be determined by a stereographic method (see Section 9.2.1).

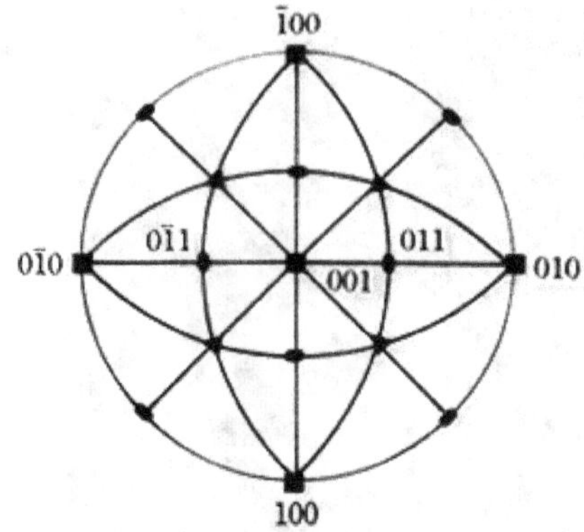

(a)

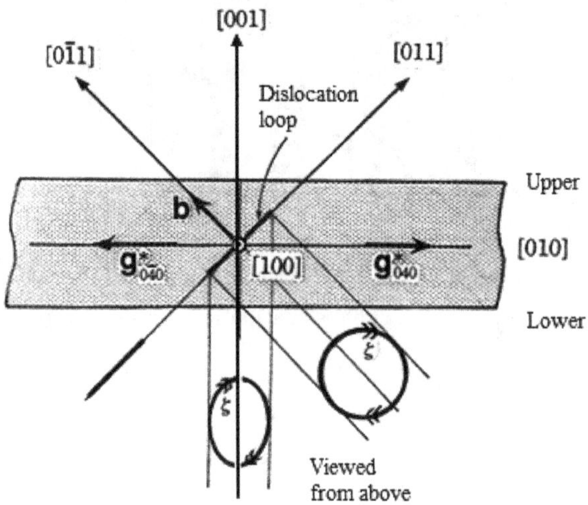

(b)

Fig. 8.12: Explanation of Fig. 8.11.

8.4 Miscellaneous Contrasts

Contrasts of inclined dislocations

When an inclined dislocation pierces a foil specimen, an oscillating contrast such as shown in Figs. 8.14(a) and 8.14(b) appears. However, this does not mean that the actual shape of dislocation is zigzagged. Interference between the transmitted and diffracted beams produces

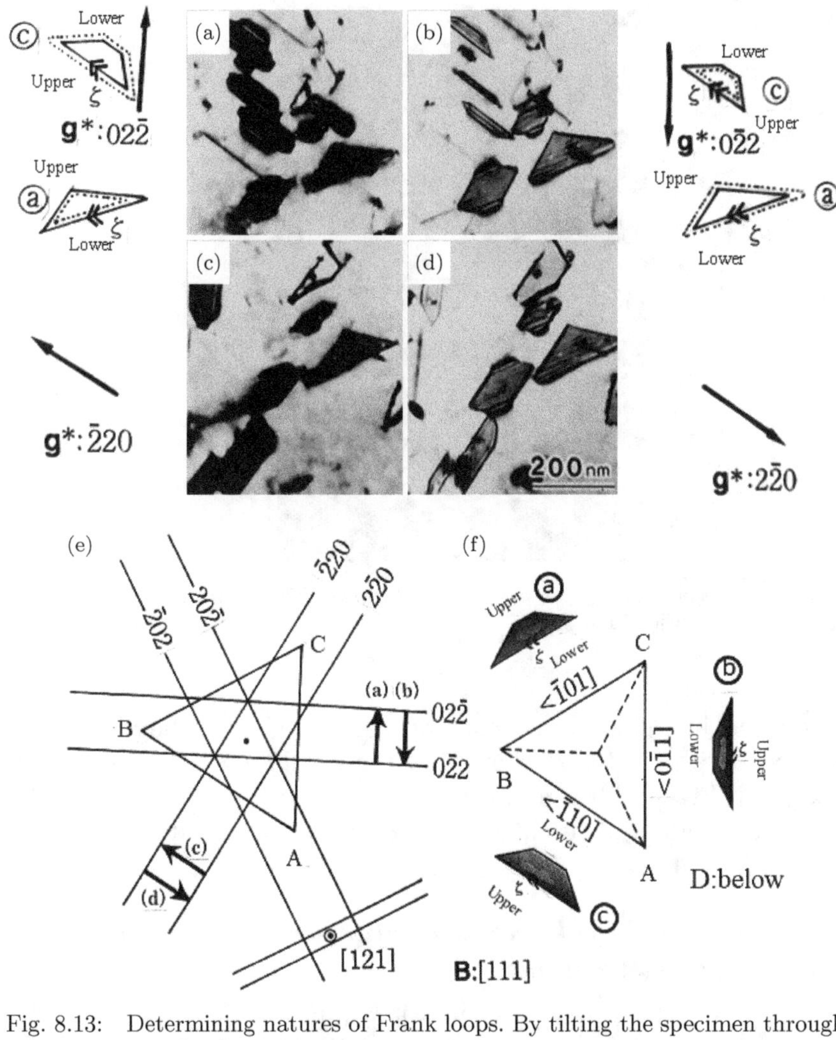

Fig. 8.13: Determining natures of Frank loops. By tilting the specimen through large angles the [121] pole is determined. Referring to Kikuchi map in Fig. K.1, those planes on which dislocation loops ⓐ, ⓑ and ⓒ lie are determined as shown in (f). Dislocation loop ⓐ shows an outside-contrast in $\mathbf{g}^* = 0\bar{2}2$ (b) and $2\bar{2}0$ (d), while it shows an inside-contrast in $\mathbf{g}^* = 02\bar{2}$ (a) and $\bar{2}20$ (c). In other words, in (b) and (d) **b** and $\mathbf{g}^*$ make an acute angle, while in (c) and (d) **b** and $\mathbf{g}^*$ make an obtuse angle.

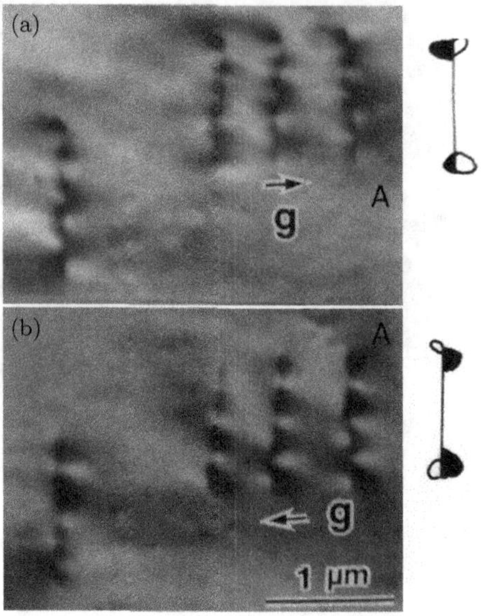

Fig. 8.14: Oscillatory effect at inclined dislocations. Sketches outside micrographs are for dislocation A.

this effect, just like a fringe contrast associated with the SF. Therefore, when $\mathbf{g}^*$ and/or $\mathbf{s}_g^*$ change(s), the oscillation changes as well. Images in Figs. 8.14(a) and 8.14(b) are taken in $\mathbf{g}^* = \pm\mathbf{g}^*$: On the reversal of sign of $\mathbf{g}^*$, oscillation of contrast is reversed.

Contrasts which look like dislocations and misleading contrasts of dislocations

Contrasts which beginners likely take as dislocations are bend contours (see Section 6.3). Bend contours of low indices (bend contours associated with low indexed (hkl) planes) are broad as shown by A and B in Fig. 8.15, so that they are unlikely to be mistaken as dislocations; however, bend contours of high indices (bend contours associated with high indexed (hkl) planes) are narrow as shown by C and D in Fig. 8.15 and may be mistaken as dislocations.

In order to distinguish between bend contours and real defects such as dislocations, it is recommended to tilt a specimen slightly: In Figs. 8.15(a) and 8.15(b), bend contours A, B, C and D are displaced

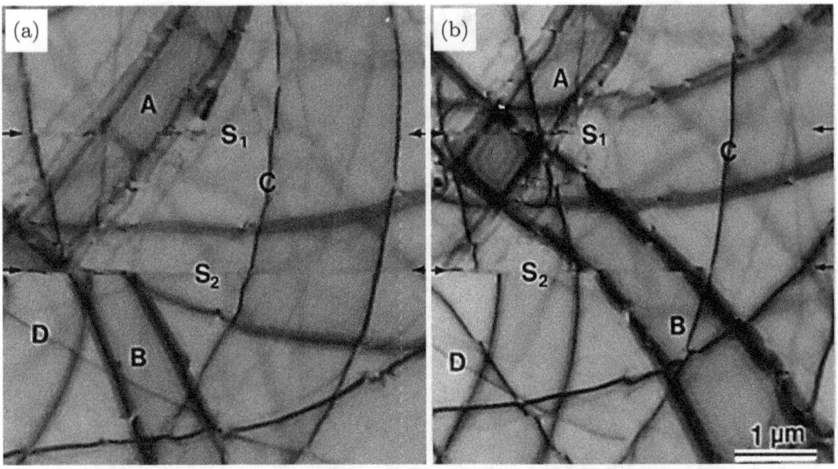

Fig. 8.15: Bend contours which are difficult to distinguish from dislocations. Low-order bend contours such as A and B are unlikely taken as dislocations by mistake, but high-order bend contours such as C and D may be taken dislocation lines by mistake. When the foil is tilted slightly, contours A, B, C and D are displaced largely, while groups of dislocations indicated by S_1 and S_2 (slip bands) remain unchanged.

largely, while groups of dislocations indicated by S_1 and S_2 (slip bands) remain unchanged.

When double-image of a dislocation appears, this does not necessarily mean that the dislocation consists of more than one dislocation. When two-beam condition (see Section 6.1) is not satisfied perfectly and two strong reflections (for instance $\mathbf{g}_1^*$ and $\mathbf{g}_2^*$ in Fig. 8.16(a)) are excited simultaneously, the positions of images stemming from the respective reflections, i.e., $\mathbf{g}_1^*$, $\mathbf{g}_2^*$ are different, albeit small, so that they appear as if two dislocations are present. To distinguish this, a two-beam condition must be satisfied perfectly as shown in Fig. 8.16(b).

However, in some cases, even when the two-beam condition is satisfied perfectly, a single dislocation appears as if two dislocations. One case is when $\mathbf{g}^* \cdot \mathbf{b} = 2$ is satisfied, especially when $\mathbf{s}_g^* = 0$. Figure 8.17 shows an example. For dislocation ②, $\mathbf{g}^* \cdot \mathbf{b} = 2$, while for dislocation ①, $\mathbf{g}^* \cdot \mathbf{b} = 1$. Dislocation ① shows a single image, while dislocation ② shows a double image. Inversely, a double image can be used to determine the absolute value of $\mathbf{b}$. In addition, when

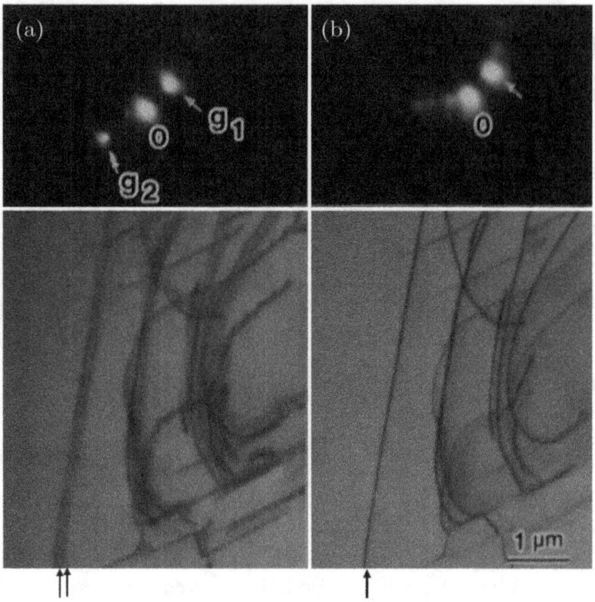

Fig. 8.16: When two strong diffraction spots ($\mathbf{g}_1^*$ and $\mathbf{g}_2^*$) are excited simultaneously, a dislocation may show double contrast (a). When the specimen is tilted in such a way that only $\mathbf{g}_1^*$ is excited, the double contrast becomes single contrast (b).

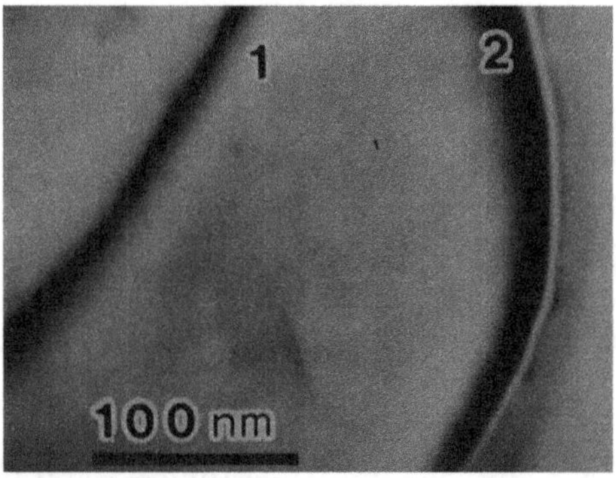

Fig. 8.17: When imaged in $\mathbf{g}^* \cdot \mathbf{b} = 2$, a dislocation shows double contrast like ②. Dislocation ① shows single contrast, for which is $\mathbf{g}^* \cdot \mathbf{b} = 1$.

an elastic anisotropy is very high, a dislocation may show a double image even when $\mathbf{g}^* \cdot \mathbf{b} = 0$ (see for detail Section 9.1.4).

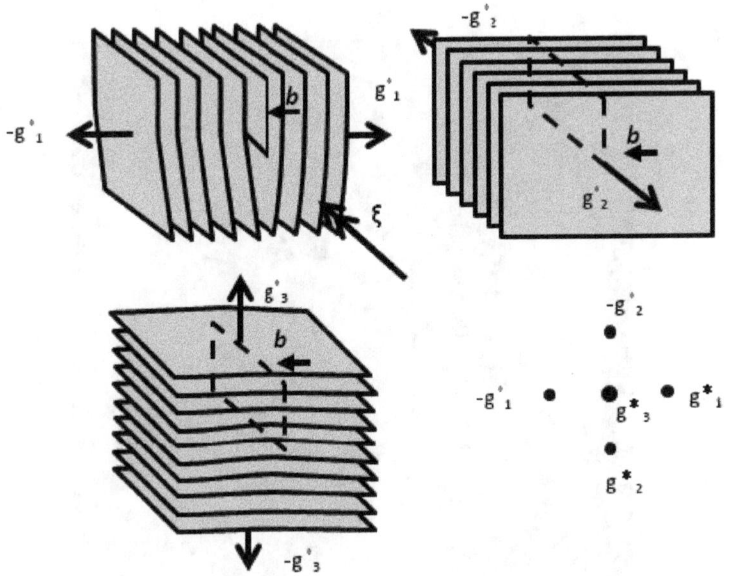

Fig. 8.18: In $\mathbf{g}_1^*$ a dislocation appears strongly, but in $\mathbf{g}_2^*$ it is invisible. In $\mathbf{g}_3^*$ the lattice planes are open downwards slightly, producing a residual contrast.

> **Point 8.3.**
>
> (1) Take images in as many as different reflections ($\pm\mathbf{g}^*$), and confirm that all the images are consistent.
> (2) Dislocations often make up a network. In such a case, confirm that Kirchhoff's rule (see Theorem 3.3) holds.

8.5 Supplement

Edge dislocation

In two-beam approximation, planes corresponding to $\mathbf{g}_3^*$ are bent slightly, opening downward (Fig. 8.18) (see also Fig. 8.2).

Distinguishing between a high-order bend contour and a dislocation

In a hard crystal such as SiC bend contours are straight, so that sometimes it is difficult to distinguish between a dislocation and bend

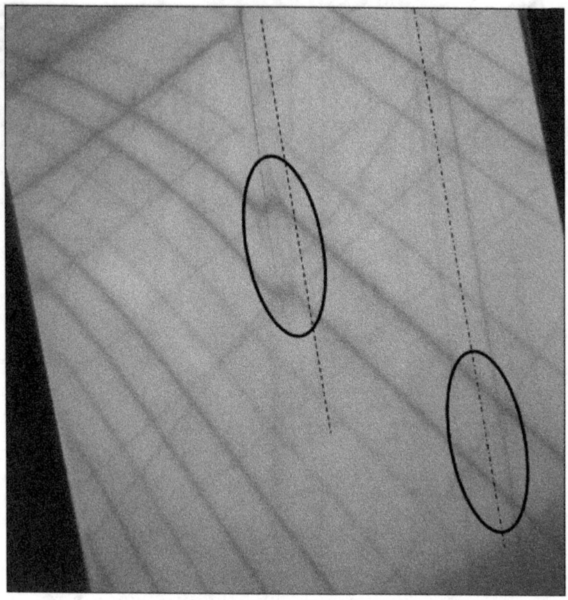

Fig. 8.19: A dislocation line and bend contours. Bend contrasts, for instance ① and ②, are straight where they cross linear contrast (b), while they stagger where they cross (a).

contours as shown in Fig. 8.19. Many straight lines are observed: Among them one dislocation exists. Can readers tell which straight lines are bend contours and which is the dislocation? The answer is a line parallel to a broken line is the dislocation and the others are all bend contours. How this is identified a dislocation? Points to distinguish between dislocations and bend contours are as follows. Straight lines intersecting the dislocation (indicated by a broken line) are all bend contours. Where these bend contours intersect the dislocation under question, the bend contours are bent locally. The bending results from interaction of the strain around bend contour and the strain field of dislocation. Figure 8.20 shows a little bit simpler example, in a Cu alloy, of interaction between a dislocation and a bend contour.

Let us suppose that a dislocation is present in a thin crystal which is bent gradually (Fig. 8.21(a)). For the sake of simplicity let us suppose that the dislocation is an edge dislocation (Fig. 8.21(b)). Now,

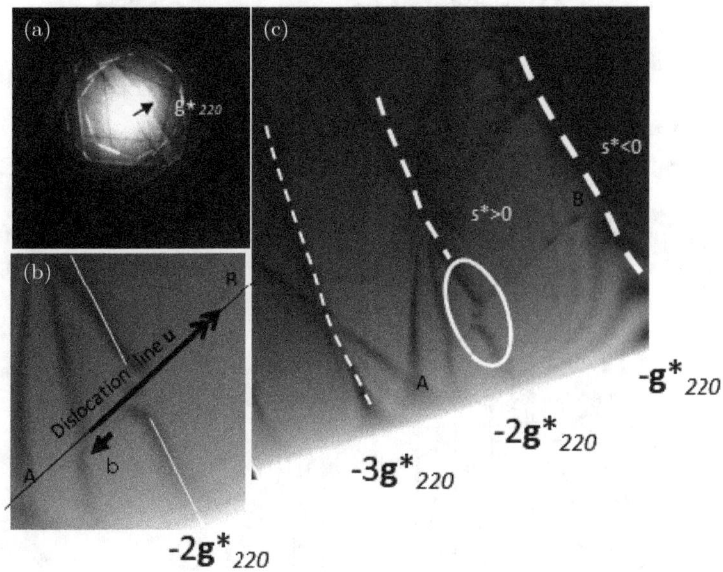

Fig. 8.20: (a) shows a diffraction pattern where $\mathbf{g}^* = \mathit{220}$ satisfies Bragg's condition. In (b) where dislocation AB (the direction is $\mathbf{u}$) intersects a bend contour of $-2\mathbf{g}^*_{\mathit{220}}$ (shown by white line), the bend contour is split. (c) shows the general view at a lower magnification.

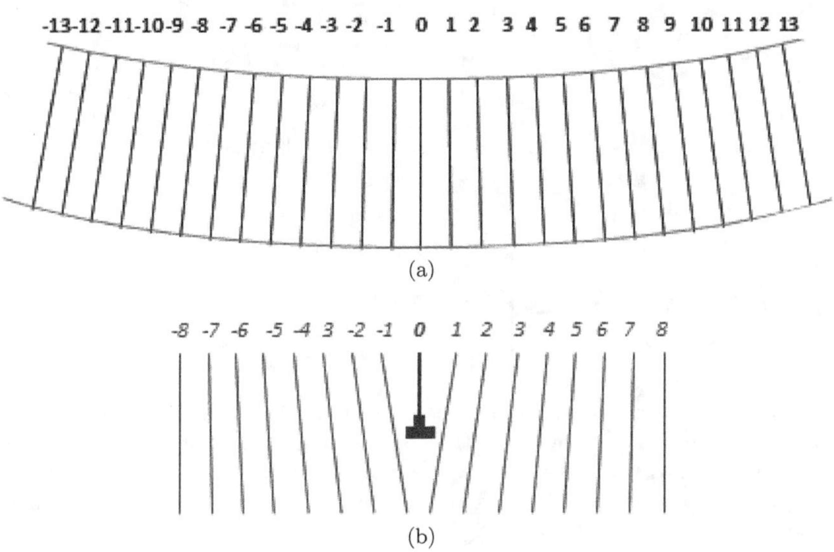

Fig. 8.21: (a) Lattice planes in a gradually curved matrix. (b) Rotation of lattice planes near an edge dislocation.

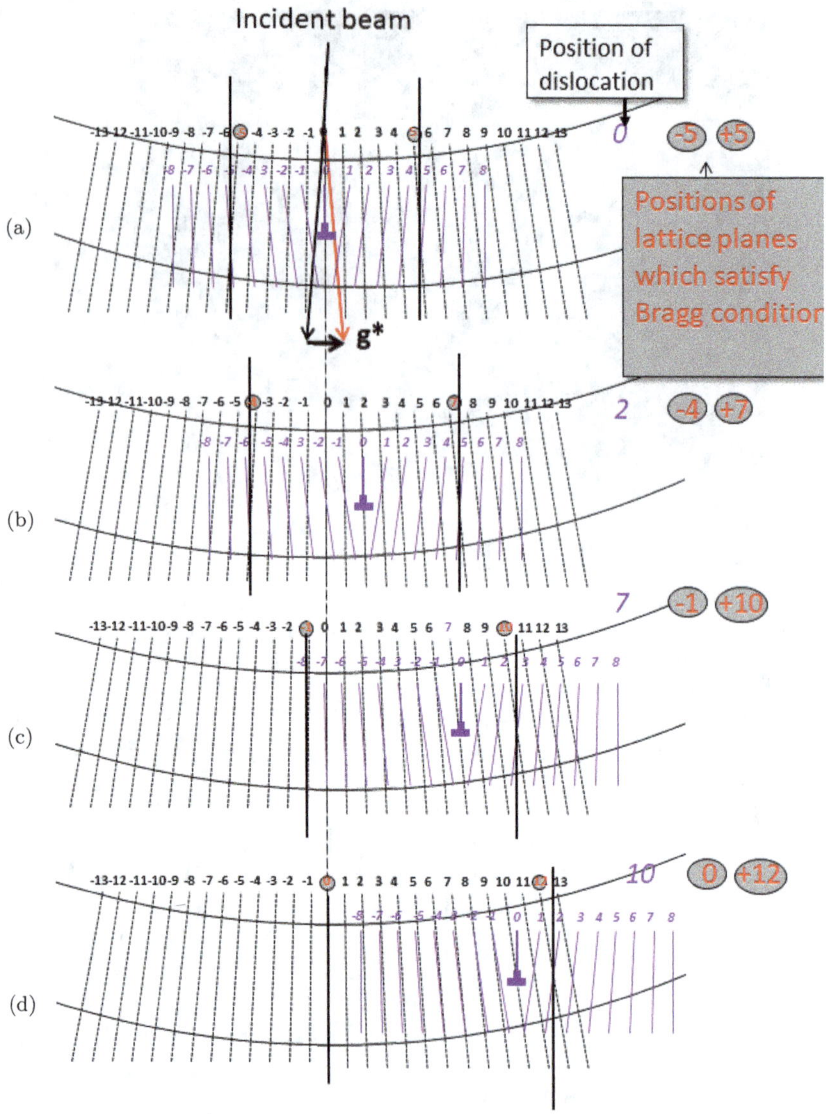

Fig. 8.22: In (a) the dislocation is located at lattice plane 0 of the matrix. Bending of the lattice planes near the dislocation cancels that of lattice planes of matrix, so that two lattice planes −5 and +5 of the matrix satisfy Bragg's condition. As a result, the bend contours split into two. In (b) a dislocation is located at 2.

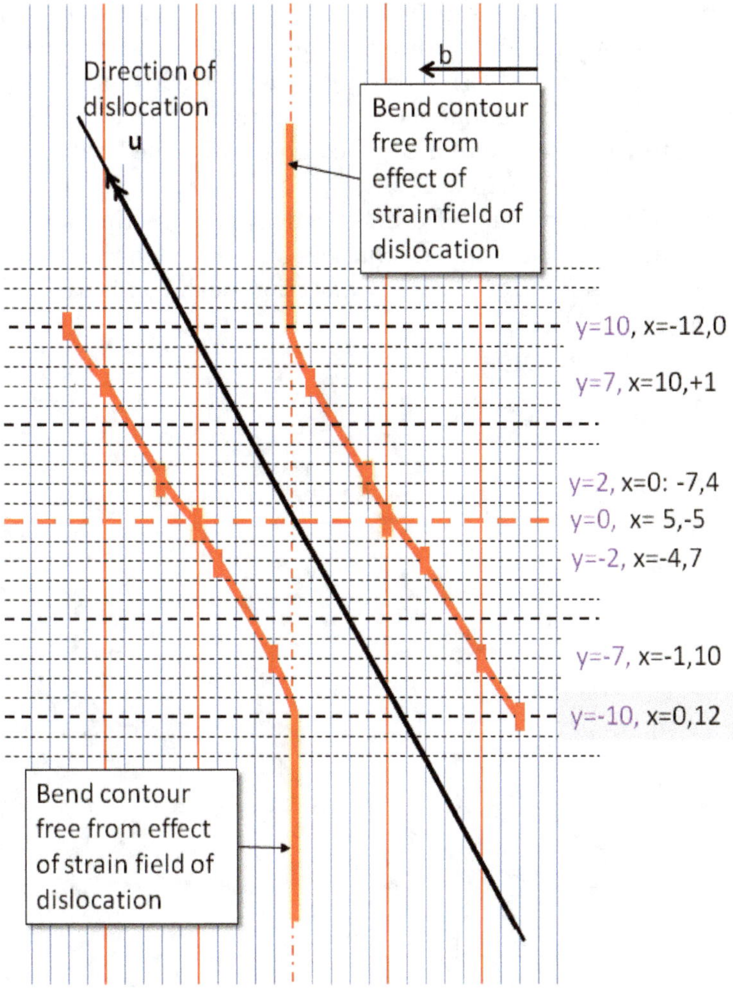

Fig. 8.23: Illustrating locations of the bend contours shown in Fig. 8.22.

suppose that the dislocation is located at a lattice site denoted by 0. As Fig. 8.22(a) shows, an elastic strain of the matrix, which bends the lattices of matrix downward, while the strain field of dislocation bends the lattice upward: They cancel each other. As a result, lattices 5 and −5 become vertical. Let us assume that when a lattice is vertical, the incident beam satisfies the Bragg's condition for $+\mathbf{g}^*$ which reflects the incident beam to the right. The position at which the Bragg's condition is satisfied (locally), i.e., $\mathbf{s}_g^* = 0$ corresponds to the bend contour. When the location of the dislocation changes to lattices 2, 7 and 10 (corresponding to Figs. 8.22(b)–8.22(d)), the location of the bend contour moves to the right. This situation corresponds to the case where a bend contour is inclined with respect to a dislocation (Fig. 8.23).

If behavior of splitting of a bend contour intersecting a dislocation is examined in detail, its Burgers vector can be determined. This will be described in detail in a section dealing with LACBED (Sections 11.2 and 11.3).

Chapter 9

Dark Field Weak-Beam Method, Stereographic Analysis

9.1 Dark Field Weak-Beam Method

Principle of dark field weak-beam method

So far, we have confined ourselves to images formed under condition where $\mathbf{s}_g^*$ is zero or slightly positive. In contrast to this, in the dark field weak-beam (DFWB) method, contrast formed under $|\mathbf{s}_g^*| \gg 0$ is important.

Let us consider $\theta \gg \theta_B$, that is, $|\mathbf{s}_g^*| \gg 0$ in Fig. 8.5. In this case, column ① does not satisfy the Bragg's condition, while column ② does. Therefore, that region which produces contrast of a dislocation is smaller and closer to the dislocation core. In other words, a sharper image is formed at a position closer to the actual dislocation core.

Figure 9.1 shows how BF and DF images of a dislocation change on increasing $\mathbf{s}_g^*$. Evidently the image of dislocation becomes sharper on increasing $\mathbf{s}_g^*$; however, BF image of dislocation diminishes significantly. For $|\mathbf{s}_g^*| \gg 0$, the diffracted intensity is weak (see Section 6.3). In other words, the transmitted intensity does not diminish very much. Thus, in the BF mode a weak (albeit sharp) image is superimposed on the background of high intensity.

By contrast, in the DF mode under $|\mathbf{s}_g^*| \gg 0$, the diffracted intensity is weak. In other words, a sharp image is superimposed on the

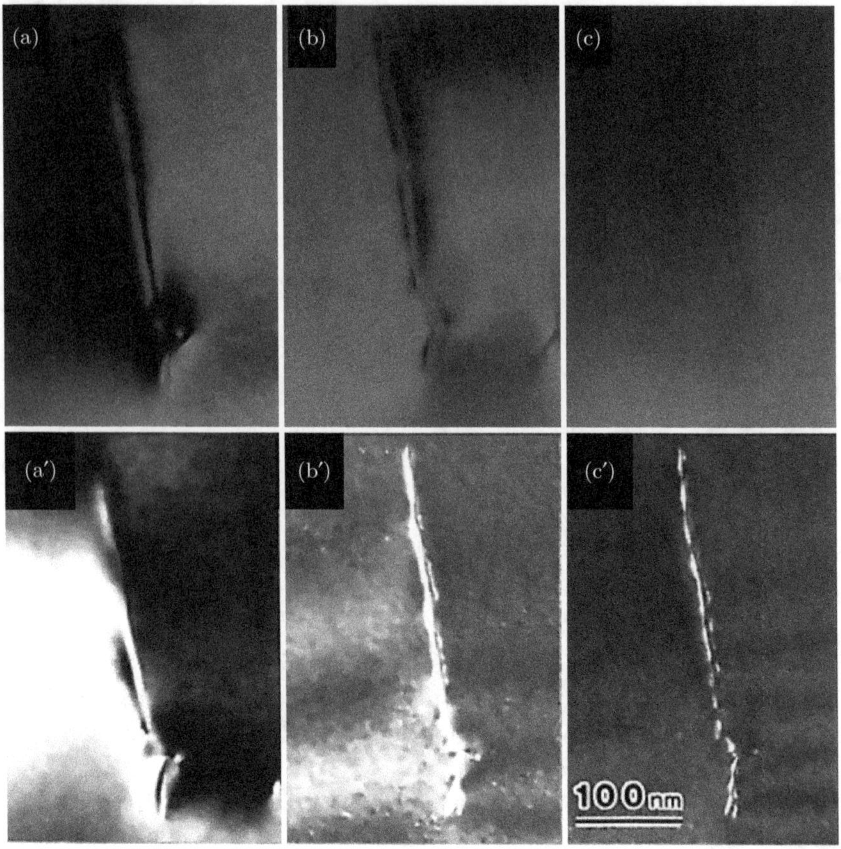

Fig. 9.1: (a)–(c) show BF images with $\mathbf{g}^*$, $2\mathbf{g}^*$ and $3\mathbf{g}^*$ excited, respectively. (a′)–(c′) show DF images corresponding to (a)–(c). $\mathbf{s}_g^*$ for $\mathbf{g}^*$ increases in the order of (a′)→(b′)→(c′). With increasing $\mathbf{s}_g^*$ the image of dislocation becomes weaker but sharper. (c′) is the condition recommended for weak-beam imaging.

weak background, producing a sharp image of dislocation. Therefore, it is recommended to observe in the DF mode in the WB method. In contrast, BF imaging with $|\mathbf{s}_g^*| \gg 0$ as shown in Fig. 9.1(c)) is referred to as a high-order reflection imaging method. As will be described later, when trying to make $\mathbf{s}^*$ bigger for the first order $\mathbf{g}^*$, the second and the third order reflections are excited necessarily.

Procedures of DFWB method

The concrete procedures of DFWB method are as follows.

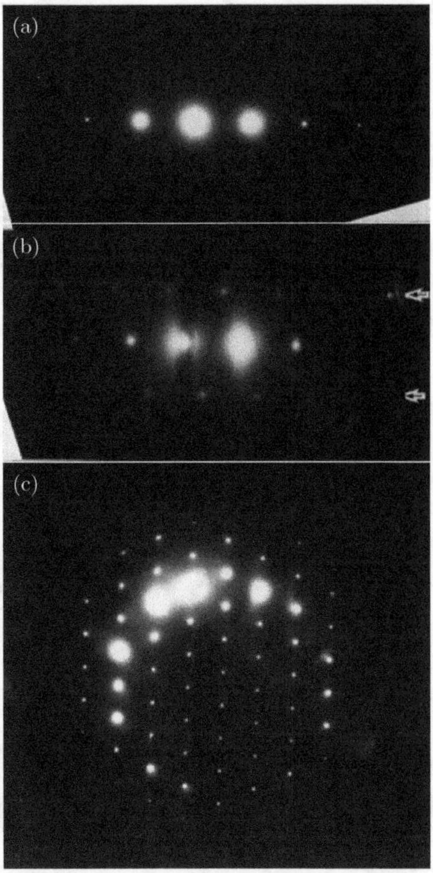

Fig. 9.2: (a) Only single systematic reflections are excited. (b) Effectively only systematic reflections are excited. (c) Non-systematic reflections are excited.

Point 9.1. Practice of DFWB method

(1) Tilt the specimen so that only $\pm\mathbf{g}^*_{hkl}$, $\pm 2\mathbf{g}^*_{hkl}$, $\pm 3\mathbf{g}^*_{hkl}$ are excited, as shown in Fig. 9.2(a)).
(2) In BF mode, tilt the specimen so that $\mathbf{g}^*$ satisfies Bragg's condition (Fig. 9.3(a)).
(3) Switch to DF mode (without tilting the specimen), and deflect the incident beam (using a deflection coil) in such a way that the direct spot is now located at the position where the original $-\mathbf{g}^*$ was located. Then, $3\mathbf{g}^*$ satisfies the Bragg's condition,

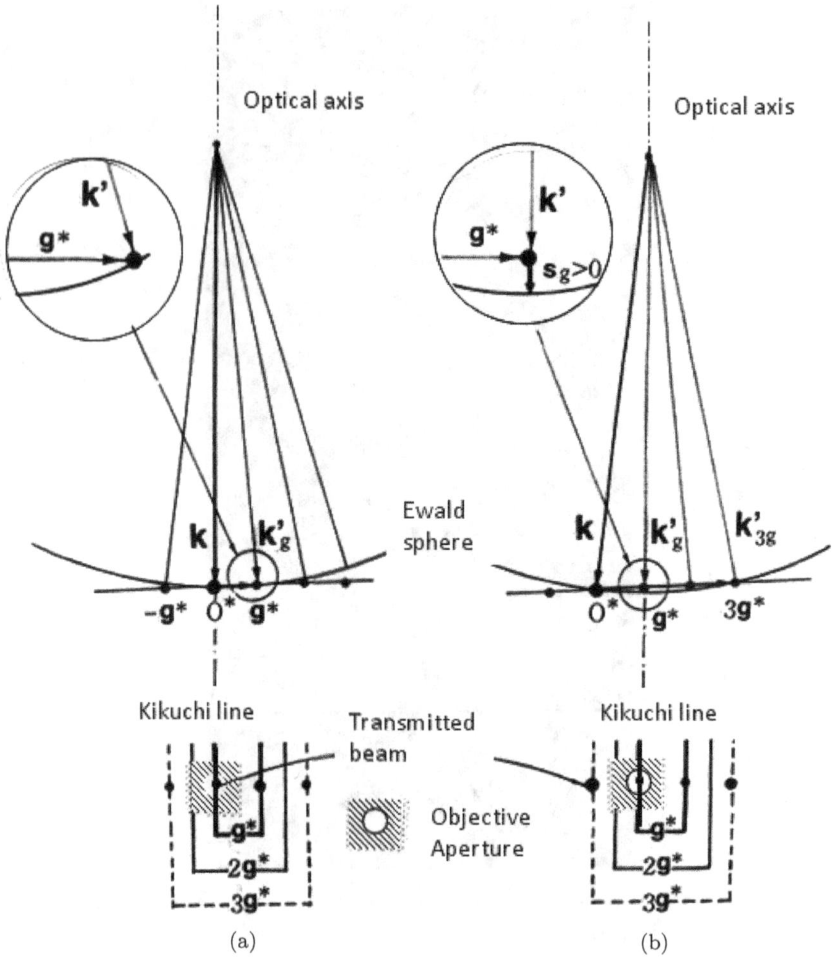

Fig. 9.3: How to set $\mathbf{g}^*/3\mathbf{g}^*$ for DFWB. In (a) the Ewald sphere intersects $\mathbf{g}^*$, so that $\mathbf{s}_g^*$ for $\mathbf{g}^*$ is zero. When the incident beam ($\mathbf{0}^*$) is tilted to the original position of $-\mathbf{g}^*$ (b), the Ewald sphere intersects $3\mathbf{g}^*$, so that $\mathbf{s}_g^*$ for $\mathbf{g}^*$ increases. In this case, the diffracted beam ($\mathbf{g}^*$) is along the optical axis of microscope.

resulting in $\mathbf{s}_g^* \gg 0$ for $\mathbf{g}^*$ (Fig. 9.3(b)). Set an objective aperture to the optical axis (where the direct spot was located before) using a deflection coil so that the diffracted beam $\mathbf{g}^*$ passes through the objective aperture. This imaging mode, in which an objective aperture is set to $\mathbf{g}^*$, while $3\mathbf{g}^*$ is excited, is denoted by $\mathbf{g}^*/3\mathbf{g}^*$.

When only one reflection $\mathbf{g}^*_{hkl}$ and its multiples by integers ($n \times \mathbf{g}^*_{hkl}$ where n is integers) are excited, this situation is referred to as systematic reflection. (a) illustrates an ideal systematic reflection and $\mathbf{g}^*_{hkl}$, $\pm 2\mathbf{g}^*_{hkl}$ and $\pm 3\mathbf{g}^*_{hkl}$ are referred to as the first order, the second order and the third order reflections, respectively. In (b), although weak reflections other than the systematic ones (indicated by arrows) appear, they are weak, so that this situation is taken as an effective systematic reflection. By contrast, (c) is an example of non-systematic reflection. It is noted that reflecting condition of (a)(b) corresponds to Fig. 5.11(c) and that of (c) corresponds to Fig. 5.11(b).

Benefit of $\mathbf{g}^*/3\mathbf{g}^*$ method is that the conventional BF imaging is switched into the DFWB imaging quickly by switching BF mode to DF mode.

Figure 9.4 shows a concrete example. Figure 9.4(a) shows a conventional two beam reflecting condition and corresponds to Fig. 9.3(a): First order reflection $\mathbf{g}^*$ is excited. Figure 9.4(b) shows $\mathbf{g}^*/3\mathbf{g}^*$ condition and corresponds to Fig. 9.4(b): Third order Kikuchi line lies just on the spot of $3\mathbf{g}^*$. Setting $\mathbf{g}^*$ to the optical axis of the lens indicated by and setting an objective aperture to $\mathbf{g}^*$, the DFWB condition of $\mathbf{g}^*/3\mathbf{g}^*$ is realized.

Point 9.2.
DFWB method is restricted to a very thin foil specimen, so that Kikuchi lines do not appear. However, when thickness contours run parallel to the edge of specimen, adjust the diffraction condition at a rather thick region and then move to a thinner region. A region where complex bend contours are observed is not appropriate for DFWB observation.

Exercise 9.1. Suppose that nth order reflection is excited, as shown in Fig. 9.3(b) ($n = 3$ in this particular case). Show that $|\mathbf{s}^*_{\mathrm{g}}|$ is given by the following equation.

$$|\mathbf{s}^*_{\mathrm{g}}| = \frac{(n-1)\,|\mathbf{g}^*|^2}{2\,|\mathbf{k}|} \tag{9.1}$$

Point 9.3.
$|\mathbf{s}^*_{\mathrm{g}}| \gg 0.2\,\mathrm{nm}^{-1}$ is recommended. As can be seen from Eq. (9.1), even if $\mathbf{g}^*/3\mathbf{g}^*$ is satisfied, $|\mathbf{s}^*_{\mathrm{g}}|$ is not necessarily large enough when $\mathbf{g}^*$ is small.

(a)

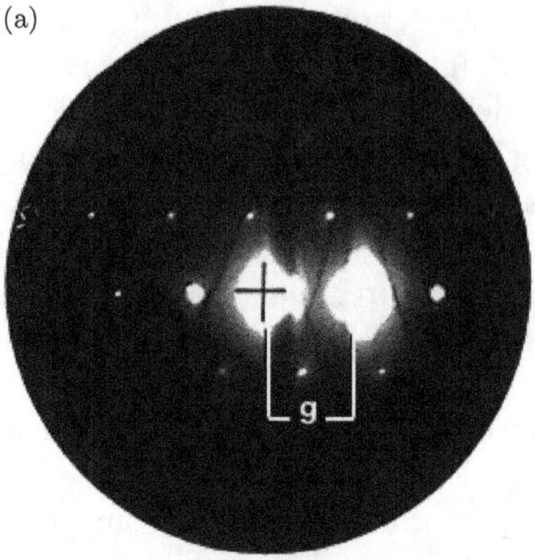

(b)

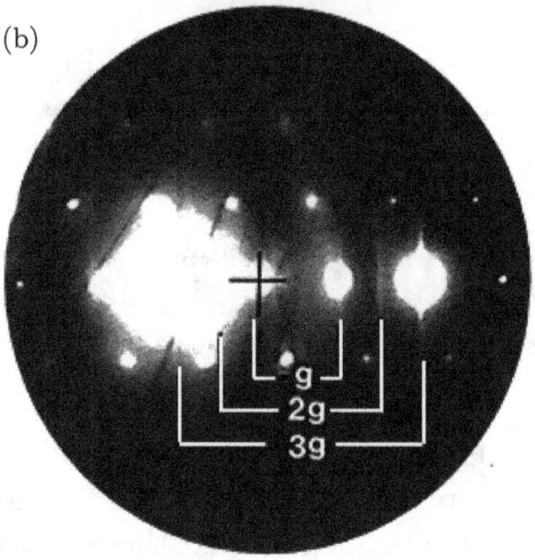

Fig. 9.4: (a) and (b) show diffraction patterns which correspond to Figs. 9.3(a) and 9.3(b), respectively. In (a) the first-order Kikuchi pair is located at spots of $\mathbf{0}^*$ and $\mathbf{g}^*$. In (b) the third-order Kikuchi pair is located at spots of $\mathbf{0}^*$ and $3\mathbf{g}^*$. + indicates the optical axis.

In practice, when a DFWB micrograph is taken, besides the length of $\mathbf{g}^*$, brightness (and hence an exposure time) and drift[1] of a specimen are important factors to determine the picture quality. It is recommended to take pictures with different length of $\mathbf{g}^*$ without clinging to $\mathbf{g}^*/3\mathbf{g}^*$ condition.

[1]Due to mechanical and/or thermal instability a specimen keeps moving. Needless to say, when a drift is present during exposure, the image is blurred.

Observing a dissociated dislocation

DFWB method is most effectively applied to observation of a dissociated dislocation. For instance, a dislocation on a glide plane in FCC is dissociated into two Shockley partials (see Section 3.4). Dislocations in ordered alloys and intermetallic compound are dissociated into more than one superlattice partial dislocations (super-partials) (see Section 3.5).

Figure 9.5 shows an example of a dissociated dislocation in FCC alloy. This dislocation lies on (*111*) plane. In (a) two partials are visible clearly, while in (b) and (c) only one of the two partials, which makes up the dissociated dislocation, is visible. In (a) $\mathbf{b}_T // \mathbf{g}^*$ ($\mathbf{b}_T$: Burgers vector of the perfect dislocation), so that the Burgers vectors of partials dislocations $\mathbf{p}_1$ and $\mathbf{p}_2$, that is, $\mathbf{b}_{p1}$ and $\mathbf{b}_{p2}$, make an angle of 30° with $\mathbf{g}^*$, while in (b) and (c), $\mathbf{b}_{p1}$ and $\mathbf{b}_{p2}$ are perpendicular

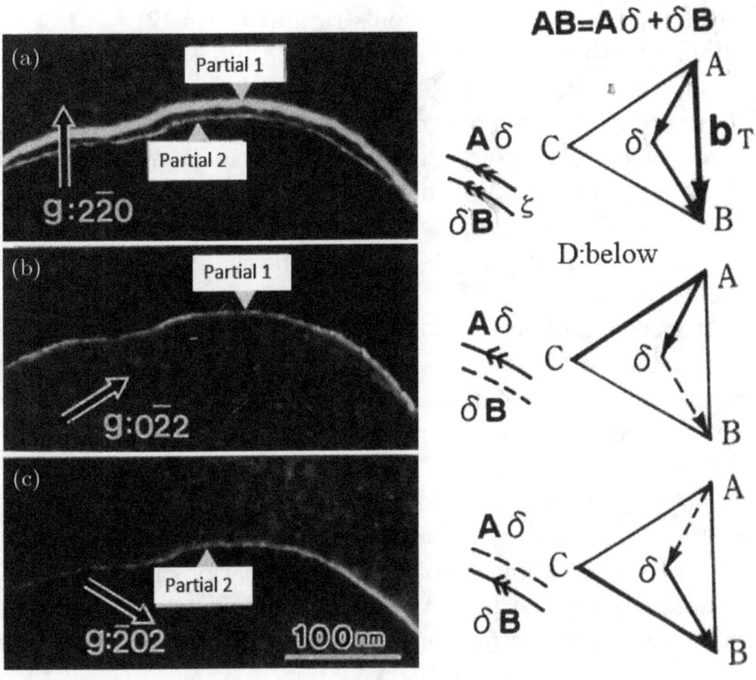

Fig. 9.5: DFWB images of a dissociated dislocation in FCC crystal. In (a) both partials $\mathbf{b}_{p1}$ and $\mathbf{b}_{p2}$ are visible clearly. In (b) $\mathbf{b}_{p2}$ is perpendicular to $\mathbf{g}^*$, so that $\mathbf{g}^* \cdot \mathbf{b}_{p2} = 0$ and $\mathbf{b}_{p2}$ is invisible. (c) Same is true for $\mathbf{b}_{p1}$ (by courtesy of Dr. Y. Kaneko).

to $\mathbf{g}^*$, so that $\mathbf{g}^* \cdot \mathbf{b}_\mathrm{p} = 0$, and contrast disappears. Also, in all of (a)(b)(c) contrast of the SF is invisible. Displacement vector of the SF is $\mathbf{R}_\mathrm{F} = \frac{1}{3}[111]$, resulting in $\mathbf{g}^* \cdot \mathbf{R}_\mathrm{F} = 0$ for $\mathbf{g}^*_{2\bar{2}0}$, $\mathbf{g}^*_{0\bar{2}2}$ and $\mathbf{g}^*_{\bar{2}02}$.

From separation between two images in Fig. 9.5(a) (Δ_obs) a true separation Δ of the dissociated dislocations is given by

$$\Delta_\mathrm{obs} = \left\{\Delta^2 + \frac{(a_1+a_2)^2}{a_1^2 a_2^2} + \frac{2\,(a_1+a_2)}{a_1 a_2}\Delta - \frac{4\Delta}{a_1}\right\}^{1/2} \tag{9.2}$$

Here,

$$a_i = -s_g \left[\frac{\mathbf{g}^*}{2\pi} \cdot \left\{\mathbf{b}_{ip} + \frac{\mathbf{b}_{ipe}}{2(1-\nu)}\right\}\right] \tag{9.3}$$

where $\mathbf{b}_{i\mathrm{p}}$ is the Burgers vector of ith partial, $\mathbf{b}_{i\mathrm{pe}}$ is the edge component of $\mathbf{b}_{i\mathrm{p}}$ and ν is the Poisson's ratio ($i = 1$ or 2).

Obtaining values of Δ using Eqs. (9.2) and (9.3), the stacking-fault energy is given by Eq. (3.5).[2] Figure 9.6 shows an example.

In order to distinguish between a dissociated dislocation and a dislocation dipole, it is recommended to image in DFWB mode using $+\mathbf{g}^*$ and $-\mathbf{g}^*$. A dislocation dipole is made up by a pair of closely positioned parallel dislocations with $\pm\mathbf{b}$, as shown in Fig. 9.7. In an edge dislocation dipole, stable positions of the constituent dislocations are situated at $\theta = 45°$.

In the case of a dissociated dislocation, on the reversal of the sign of $\mathbf{g}^*$, the contrast of the partial dislocations is reversed, however, the separation remains unchanged in $+\mathbf{g}^*$ and $-\mathbf{g}^*$, as shown in Fig. 9.8. By contrast, in the case of a dipole, images in $+\mathbf{g}^*$ and $-\mathbf{g}^*$ are quite different as shown in Fig. 9.9. This can be accounted for by considering that the image of a dislocation appears either at right or left side of the dislocation core (see Section 8.2). For a dissociated dislocation, positions of images of partial dislocations in $+\mathbf{g}^*$ and $-\mathbf{g}^*$ is displaced slightly, however, the directions and amounts of the displacement are the same for $+\mathbf{g}^*$ and $-\mathbf{g}^*$, as shown in Figs. 9.10(a)

[2]For a stricter treatment, anisotropic elasticity theory must be used where orientation dependence of elastic constants is taken into consideration.

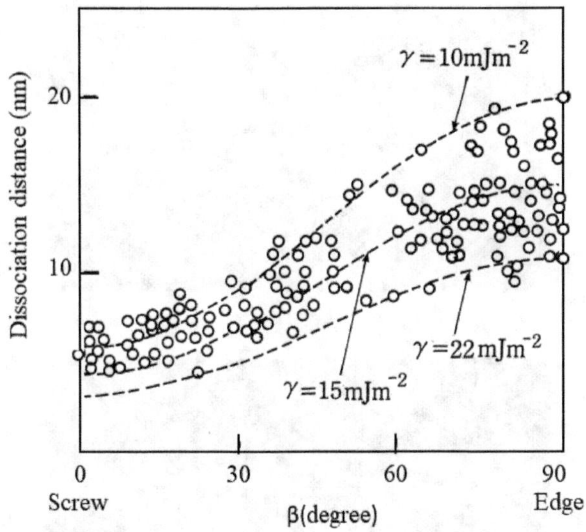

Fig. 9.6: The dissociation distance (Δ) as plotted against β, an angle the Burgers vector of the total dislocation ($\mathbf{b}_T$) makes with the dislocation line, for dissociated dislocations in an Fe–15at.%Cr–10at.%Ni alloy. Broken lines are theoretical curves calculated based on Eq. (3.5) (by courtesy of Dr. Y. Kaneko).

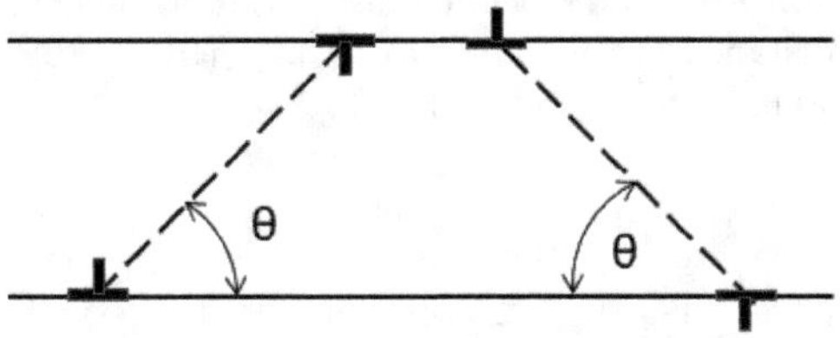

Fig. 9.7: Dislocation dipoles. A pair of dislocations with $+\mathbf{b}$ and $-\mathbf{b}$ lie parallel with each other.

and 9.10(b). As a result, the separation remains unchanged. By contrast, for a dislocation dipole, images of dislocations are displaced to either inside or outside, depending on the sign of $\mathbf{g}^*$, resulting in a large difference in the separation when imaged in $+\mathbf{g}^*$ and $-\mathbf{g}^*$, as shown Figs. 9.10(c) and 9.10(d). This situation is true for an ordinary BF image (dislocation ④ in Fig. 8.6 is a dislocation dipole and behavior of the contrast on the reversal of $\mathbf{g}^*$ is the same as Fig. 9.9).

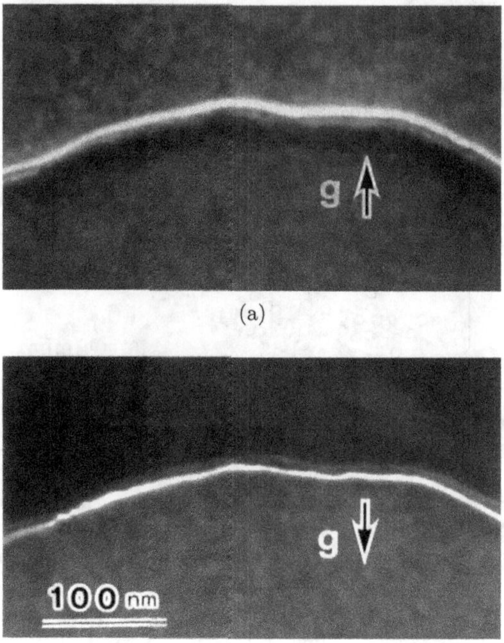

(a)

(b)

Fig. 9.8: DFWB images of a dissociated dislocation as imaged in $+\mathbf{g}^*$ and $-\mathbf{g}^*$. The dissociation distance is the same in (a) and (b). Note that either of the partials appears strongly in (a) and (b).

In a dissociated dislocation, one of the two partial dislocations appears strongly (see Figs. 9.5(a) and 9.8). Besides, as shown in Figs. 9.8(a) and 9.8(b), the partials appearing strongly change on the reversal of the sign of $\mathbf{g}^*$. This is accounted for as follows. As illustrated schematically in Figs. 9.10(a) and 9.10(b), displacement of lattices in a region between the two partials cancels each other, since the displacement of the respective (two) partials are opposite. By contrast, outside the dissociated dislocation, such a cancellation of lattice displacement does not occur. Therefore, contrast of a (partial) dislocation appears stronger inside than outside.

Similar discussion is applicable to a dislocation dipole. In this case, in contrast to a dissociated dislocation, inside the dipole the displacement resulting from the two dislocations strengthens each other, so that stronger contrast appears inside than outside. B and

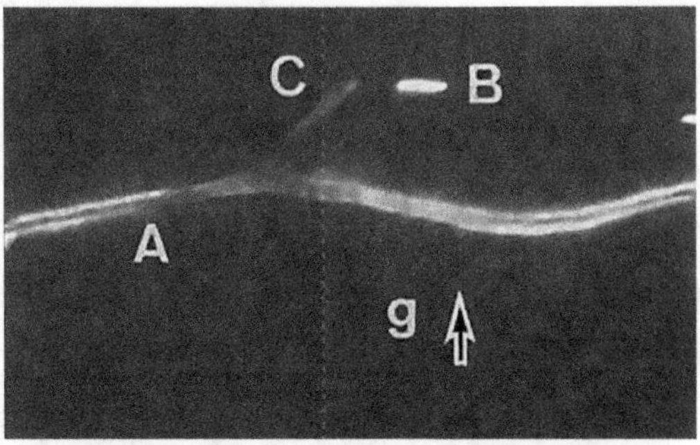

(a)

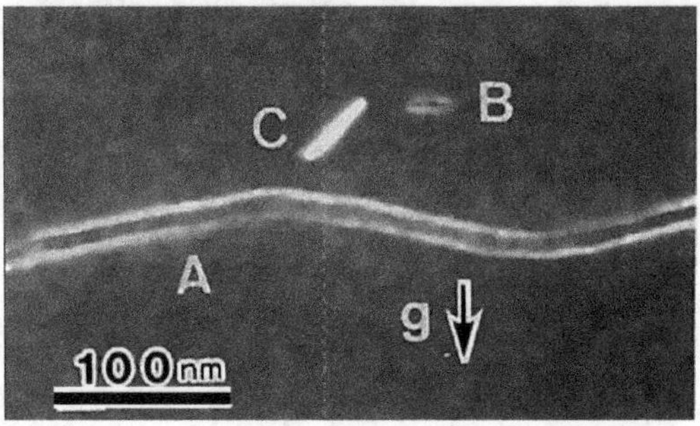

(b)

Fig. 9.9: Dislocation dipoles imaged in $+\mathbf{g}^*$ and $-\mathbf{g}^*$. Separation of two dislocations labelled by A is not the same in (a) and (b). Same is true for dipoles B and C.

C shown in Fig. 9.9 are dipoles and this feature is especially strongly manifested.

Application to elastically anisotropic crystals

DFWB method is applicable to an elastically anisotropic crystals, to which the conventional BF imaging to determine Burgers vector is not applicable. Figure 9.11 shows BF images of dislocations in $\mathbf{g}^*_{121}$ and $\mathbf{g}^*_{1\bar{2}1}$ in β-CuZn which has an extremely high elastic anisotropy A$\{= 2C_{44}/(C_{11} - C_{12})\} = 10$.[3] It has been concluded by a separate experiment that the Burgers vector **b** of these dislocations are [111].

[3]For elastically isotropic crystals $A = 1$.

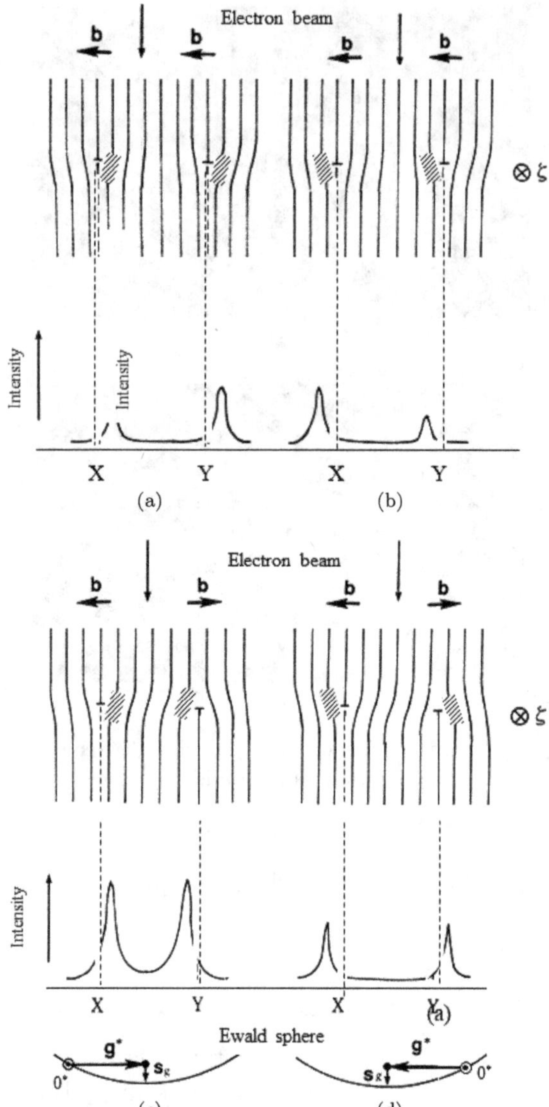

Fig. 9.10: Explanation of Figs. 9.8 and 9.9. For a dissociated dislocation images of the two partials shift to the same direction on the reversal of the sign of $\mathbf{g}^*$ ($\mathbf{g}^*$ versus $-\mathbf{g}^*$), so that the separation between the images of the two partials remains unchanged between (a) and (b). For a dipole, images of the two dislocations making up the dipole shift to the reverse directions with each other (c) and (d). That is, in (c) they shift to inside, while in (d) outside. As a result, the separation in $+\mathbf{g}^*$ and $-\mathbf{g}^*$ is not the same. (c) and (d) redrawn from Cherns *et al.* (1980).

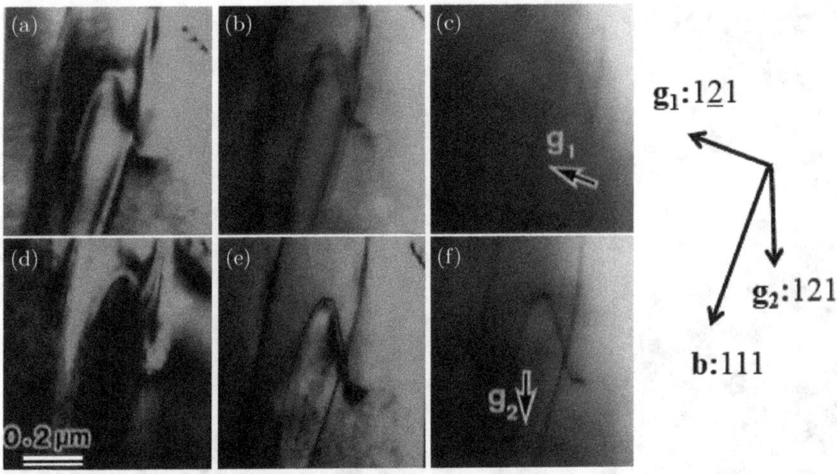

Fig. 9.11: BF images of dislocations ($\mathbf{b}$=[111]) in β-CuZn imaged in high-ordered reflections. The elasticity anisotropy of β-CuZn is very high. (a)–(c) are imaged in $\mathbf{g}^*_{1\bar{2}1}$, while (d)–(f) are imaged in $\mathbf{g}^*_{121}$. $\mathbf{B}$=[101], so that $\mathbf{g}^*_{1\bar{2}1}$ and $\mathbf{g}^*_{121}$ are crystallographically equivalent. However, for (a)–(c) $\mathbf{g}^* \cdot \mathbf{b} = 0$, while for (d)–(f) $\mathbf{g}^* \cdot \mathbf{b} \neq 0$ (reproduced with permission from Saka (1984)).

Therefore, in (a)(b)(c) $\mathbf{g}^*_{1\bar{2}1} \cdot \mathbf{b} = 0$, while in (d)(e)(f) $\mathbf{g}^*_{121} \cdot \mathbf{b} \neq 0$. In (a)(d) $\mathbf{g}^*_{121}$ and $\mathbf{g}^*_{1\bar{2}1}$ are excited, and in (b)(e) $2\mathbf{g}^*_{121}$ and $2\mathbf{g}^*_{1\bar{2}1}$ are excited and in (c)(f) $3\mathbf{g}^*_{121}$ and $3\mathbf{g}^*_{1\bar{2}1}$ are excited. In other words, $\mathbf{s}^*_\mathrm{g}$ for $\mathbf{g}^*$ increases in the order of (a)(d)→(b)(e)→(c)(f).

When the dislocations are imaged in BF mode with the first and the second order $\mathbf{g}^*$ excited, dislocations for which $\mathbf{g}^*_{1\bar{2}1} \cdot \mathbf{b} = 0$ are visible ((a)(b)), so that an invisibility criterion $\mathbf{g}^* \cdot \mathbf{b} = 0$ is not applicable.[4] However, when imaged with $3\mathbf{g}^*$ excited (in a high order reflection imaging mode), the dislocations are invisible in (c), while in (f) they are visible apparently. In other words, the conventional $\mathbf{g}^* \cdot \mathbf{b} = 0$ is applicable.

Similarly to a high order reflection imaging mode, in DFWB method an invisibility criterion $\mathbf{g}^* \cdot \mathbf{b} = 0$ is applicable. Figures 9.12(a)–9.12(d) and 9.12(a')–9.12(d') show an example of contrast experiment carried out on dislocations both in the high order

[4] In practice, when imaged in $2\mathbf{g}^*$, for $\mathbf{g}^* \cdot \mathbf{b} = 0$ (b) the double image of dislocations is symmetric, while in $\mathbf{g}^* \cdot \mathbf{b} \neq 0$ (e), the double image of dislocations is asymmetric, so that it is possible to distinguish between $\mathbf{g}^* \cdot \mathbf{b} = 0$ and $\mathbf{g}^* \cdot \mathbf{b} \neq 0$.

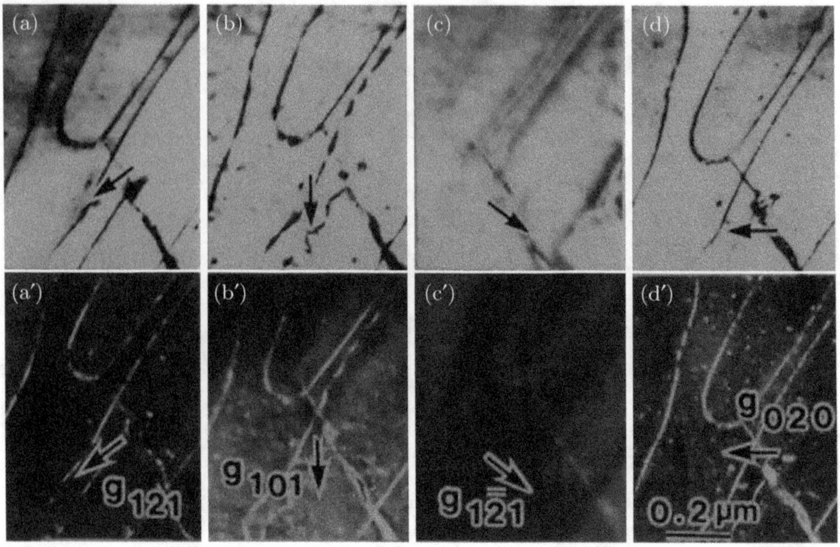

Fig. 9.12: Contrast experiments carried out on a dislocation in β-CuZn using the high-order reflection method and DFWB method. In $\mathbf{g}^* = 1\bar{2}1$, (c) and (c′), the dislocation is invisible, that is $\mathbf{g}^* \cdot \mathbf{b} = 0$ (reproduced with permission from Saka (1984)).

reflection imaging mode and the DFWB mode. In both cases dislocations are invisible in $\mathbf{g}^*_{1\bar{2}1}$(c), indicating that the invisibility criterion $\mathbf{g}^* \cdot \mathbf{b} = 0$ is applicable. In practice it is recommended to use both the high order reflection imaging mode and the DFWB method.

Determining the absolute value of Burgers vector

In the methods to determine Burgers vector described in Sections 8.1 and 8.2, the absolute values of the Burgers vectors cannot be determined. However, DFWB method allows one to determine the absolute value of Burgers vector. Figures 9.13(a) and 9.13(b) show DFWB images ($\mathbf{s}^*_g \gg 0$) of a dissociated dislocation piercing a foil of an FCC alloy. When a specimen is wedge-shaped, thickness contours appear. In a DFWB method, spacing of the thickness contours is quite small, since $\mathbf{s}^*_g$ is large (see Section 6.12).

In (a) at the emerging point of the dislocation, two extra contours are observed at the left side (indicated by arrows). Similarly, in (b)

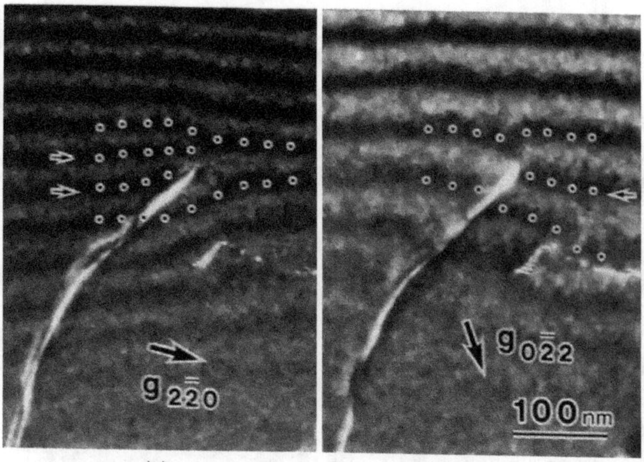

(a) (b)

Fig. 9.13: Determining the absolute value of Burgers vector $|\mathbf{b}|$ using disturbance of thickness contours. (a) and (b) show DFWB images in $\mathbf{g}^*_{2\bar{2}0}$ and $\mathbf{g}^*_{0\bar{2}2}$, respectively. In (a) at the emergent point of the dislocation, two extra thickness contours are present at the left side, while in (b) one extra thickness contour is present at the right side.

one extra contour is observed at the right side. Number of extra contours N is related with $\mathbf{b}$ and $\mathbf{g}^*_{hkl}$ by the following relationships.

Theorem 9.1:
$\mathbf{b} \cdot \mathbf{g}^*_{hkl} = \mathrm{N}$ or $\mathrm{b_x}h + \mathrm{b_y}k + \mathrm{b_z}l = \mathrm{N}$, where $\mathbf{b} = [\mathrm{b_x}, \mathrm{b_y}, \mathrm{b_z}]$.

Theorem 9.2:
Place a wedge-shaped specimen in such a way that the thinner part of the specimen points upward. When extra contours appear in the left side, $\mathrm{N} < 0$. When extra contours appear in the right side, $\mathrm{N} < 0$.

Exercise 9.2. The dislocation shown in Fig. 9.13 is a dissociated dislocation in an FCC alloy. The foil surface is (111) and $\mathbf{b}$ is nearly in the plane of foil. Assuming that $\mathrm{b}_z = 0$ determine $\mathbf{b}$.[5]

Figures 9.14(a) and 9.14(b) show a BF image ($\mathbf{s}^*_\mathrm{g} \gg 0$) and a DFWB image ($\mathbf{s}^*_\mathrm{g} \gg 0$) of a dissociated dislocation in an FCC alloy. In the BF image in (a) the dislocation image is accompanied

[5]By applying Eq. (9.4) for three different $\mathbf{g}^*$, $\mathbf{b}$ is determined uniquely.

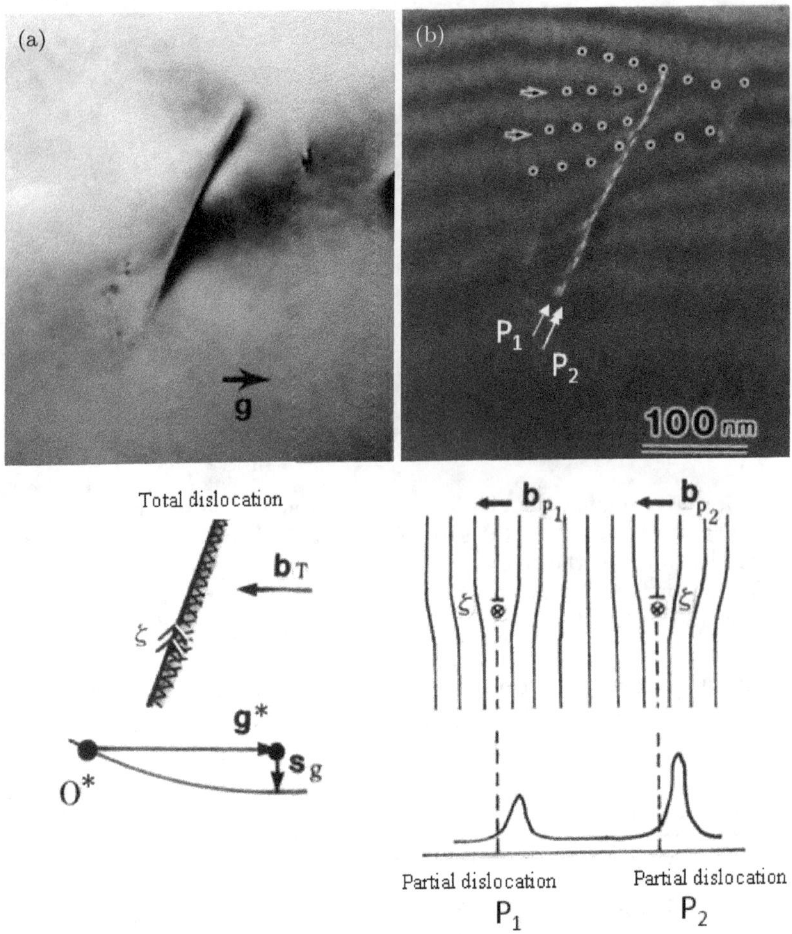

Fig. 9.14: (a) BF image and (b) DFWB image of a dissociated dislocation in FCC crystal imaged in $\mathbf{s}_g^* > 0$. In (a) a tail is observed at the right side of dislocation. In (b) the right-sided partial appears (P_2) strongly (see Fig. 9.10(a)).

with a tail at the right side. This is in keeping with the right-sided partial dislocation (P_2) appearing stronger. Therefore, from Theorem 8.5, $\mathbf{g}_{hkl}^* \cdot \mathbf{b} < 0$. On the other hand, Theorem 9.2 says that when $\mathbf{g}^* \cdot \mathbf{b} = \mathrm{N} < 0$, extra fringes should appear at the left side of the dislocation. Indeed, in (b) two extra fringes appear at the left side of the dislocation. The results on **b** obtained from (a) and (b) are in keeping with each other.

9.2 Stereographic Observation

In order to obtain a stereograph, it is necessary to observe an identical object along two different directions. Figure 9.15(a) illustrates optical paths into right and left eyes when we observe an object.

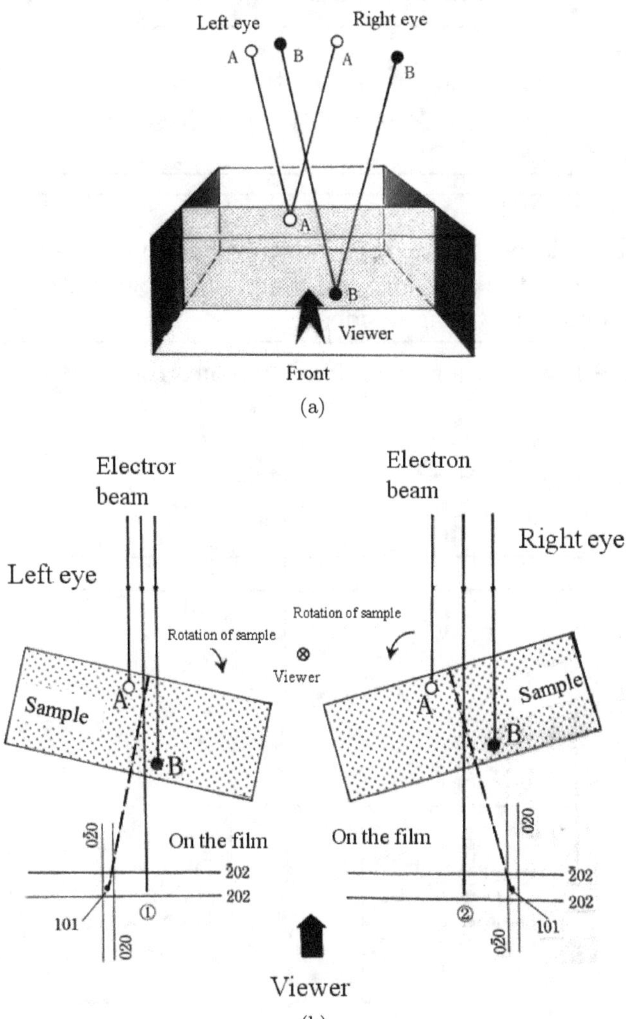

Fig. 9.15: How to take a stereo-picture. (a) Ray paths for eyes. (b) The corresponding situation for TEM is achieved by tilting the sample as shown in (b).

Difference in the angle of optical ray paths which enter the right and left eyes makes the object appear three-dimensionally as shown in Fig. 9.15(a).

In order to take a stereomicrograph, a pair of micrographs must be taken with the electron beam tilted by ~10°, as shown in Fig. 9.15(b). For a crystal the pair of micrographs must be taken under exactly the same diffracting condition: Kikuchi map is useful for this purpose.

Point 9.4.
Setting $\mathbf{g}^*$ and $\mathbf{s}_g^*$ to the same diffracting conditions along a Kikuchi pair at (①) and (②), take a pair of micrographs, as shown in Fig. 9.16.

Theorem 9.3:
In practice, place a micrograph taken at the right position ① of a Kikuchi pair in the left side, and place a micrograph taken at the left position ② in the right side.

Figure 9.17 shows an example of stereomicrograph.

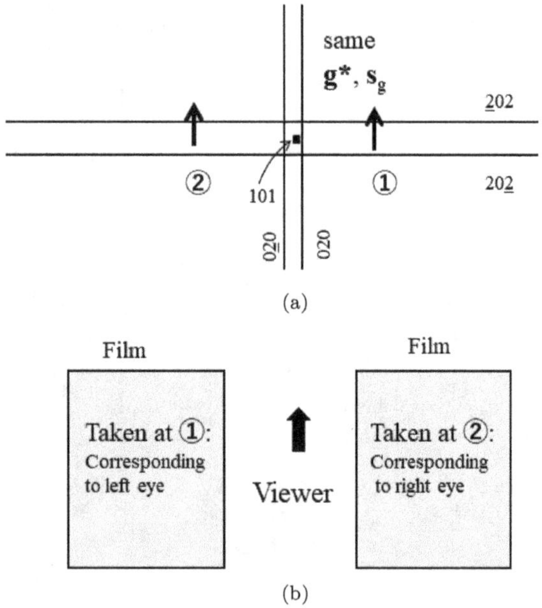

Fig. 9.16: Explanation of Theorem 9.3. Place a micrograph ① taken at the right side of a Kikuchi map at the left side, and a micrograph ② taken at the left side of a Kikuchi map at the right side.

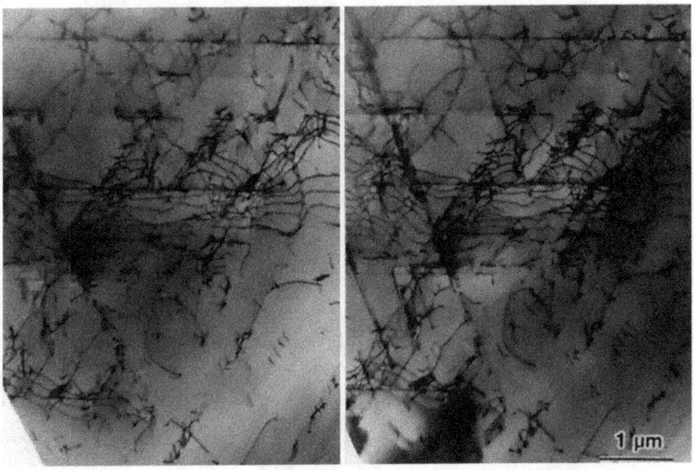

(a) (b)

Fig. 9.17: Example of a stereomicrograph. Place a paper vertically between (a) and (b).

9.3 Trace Analysis

Determining directions

In order to determine the direction of a straight dislocation line ST, it is necessary to observe the dislocation along at least two directions. For instance, when a dislocation line ST is observed along B_1, only its projection S_1T_1 is observed. When observed along B_2, S_2T_2 is obtained. As can be seen from Fig. 9.18, the dislocation line ST is an intersection of those planes whose normal lines are $\mathbf{n}_1$(shaded with light grey) and $\mathbf{n}_2$ (shaded with dark grey), that is a zone axis of $\mathbf{n}_1$ and $\mathbf{n}_2$. The direction of ST can be determined with methods described in Fig. 2.13.

Determining planes

Let us suppose that a dislocation loop lies on a plane (for instance, a slip plane) and we try to determine the plane. Determine orientations of two segments A_1A_2 and B_1B_2 which make as large an angle as possible with each other (ideally 90°) as shown in Fig. 9.19(a), then determine the plane containing both A_1A_2 and B_1B_2 using the

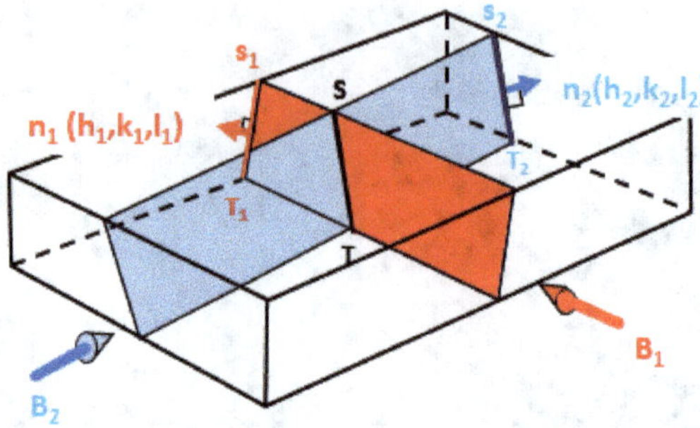

Fig. 9.18: ST indicates the position of the actual dislocation. ST is projected to S_1T_1 when viewed along B_1, while projected to S_2T_2 when viewed along B_2. An intersection of a plane on which S_1T_1 lies ($\mathbf{n}_1$) with that on which S_2T_2 lies ($\mathbf{n}_2$) is ST.

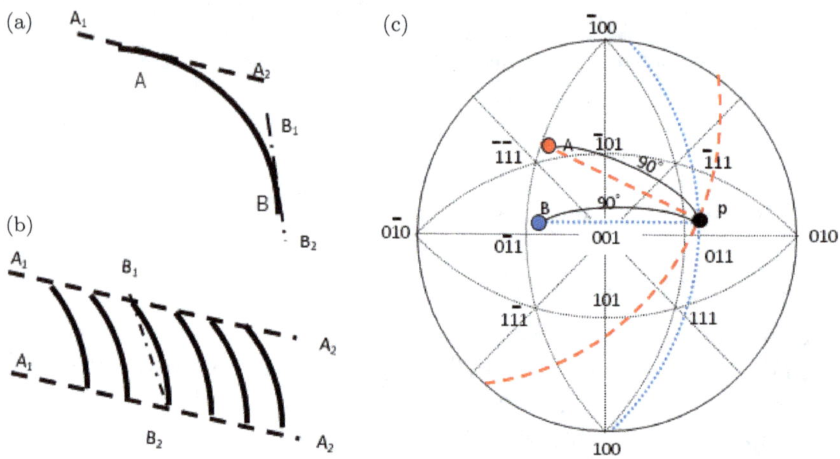

Fig. 9.19: A plane on which both of A and B lie is the plane P on which the dislocation(s) lies (see also Fig. 2.13).

method described in Section 9.3.1. Alternatively, when the dislocations are piled up on a single slip plane (Fig. 9.19(b)), an intersection with specimen A_1A_2 and lines connecting emergence points of dislocations at the upper and lower surfaces can be used.

9.4 Effects of Surface

TEM samples are thin foils, so that dislocations are inevitably affected by surfaces. In order to avoid this, it is hoped to use as

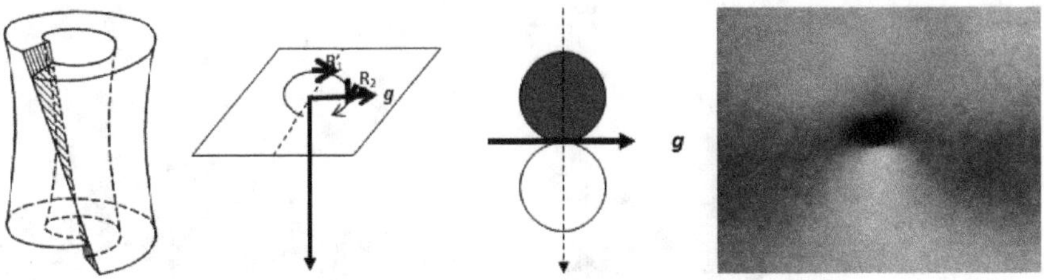

(a) (b) (c) (d)

Fig. 9.20: (a) Eshelby twist. (b) Eshelby twist induces displacements parallel to the surface. (c) at $\mathbf{R}_2$, $\mathbf{R}_2 \cdot \mathbf{g}^* = 0$, so that no contrast appears, while at $\mathbf{R}_1$, $\mathbf{R}_1 // \mathbf{g}^*$, so that contrast appears. (c) shows an example.

thick a specimen as possible. Readers refer to Saka (2017) for behavior of dislocations in a thin foil. Here, let us constrict ourselves to effects of surface on images of a dislocation.

Eshelby twist

Let us consider a pure screw dislocation perpendicular to the foil surface. In this case, diffracting vectors available for imaging (for instance $\mathbf{g}_1^*$, $\mathbf{g}_2^*$ and $\mathbf{g}_3^*$) are all perpendicular to the Burgers vector of the screw dislocation $\mathbf{b}$. Therefore, according to Theorem 8.1, the (screw) dislocation should be invisible.

However, a screw dislocation twists material near the surfaces to relax stress as shown in Fig. 9.20(a). This phenomenon is referred to as Eshelby twist. This produces displacement near the surfaces parallel to the plane of foil as indicated by $\mathbf{R}$ in Fig. 9.20(b). At places where $\mathbf{R}$ is perpendicular to $\mathbf{g}^*$, i.e., $\mathbf{R}_2$, no contrast appears, however, where $\mathbf{R}$ is parallel to $\mathbf{g}^*$, i.e., $\mathbf{R}_1$, contrast appears. As a result, a pair of black-and-white contrast appears as shown in Fig. 9.20(c).

Slip traces

When dislocations move in a foil specimen (after it is thinned), traces of its movement appear at surfaces. Figure 9.21 shows historic pictures when Hirsch *et al.* first observed dislocation motion. At first sight, there may appear no difference between (a) and (b). However, a careful inspection shows that in (b) a straight contrast bent by 90° appears. Hirsch *et al.* succeeded in confirming the existence of dislocation.

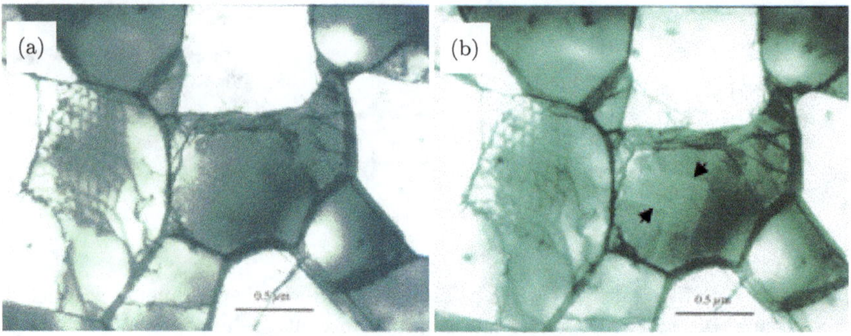

Fig. 9.21: Reproduced from Humphreys (2002).

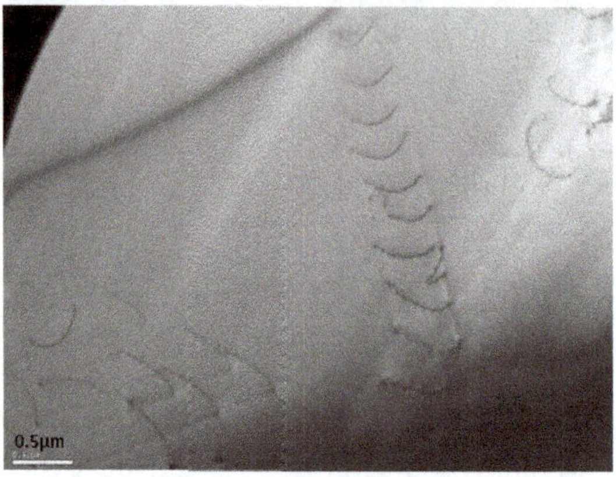

Fig. 9.22: Example of surface traces of dislocation motion.

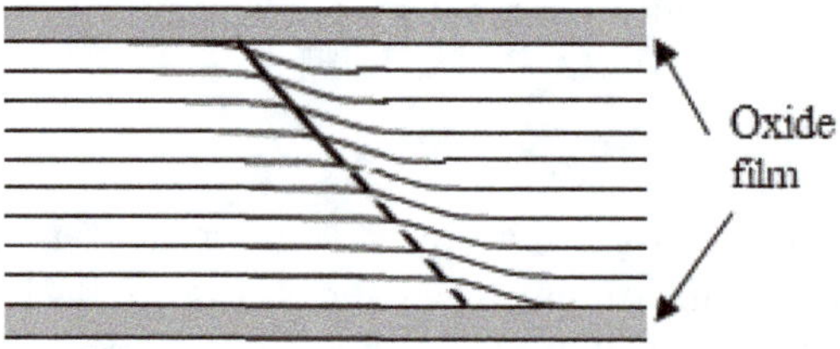

Fig. 9.23: Mechanism by which slip traces appear.

Figure 9.22 shows an example of motion of dislocations in a Cu alloy. The dislocations bow out to the right, indicating that the dislocations move to the right. The dislocations pierce the foil, leaving pairs of traces. Mechanisms by which contrast appears at the intersections of dislocations and foil surface are considered as follows. Surfaces of a foil specimen are covered with thin oxide layers and the surfaces are bent by dislocation (Fig. 9.23). Step formed as a result of motion of a dislocation is too small ($\sim$0.1 nm) to produce visible contrast.

Chapter 10

Diffraction Contrast in STEM

10.1 General Theory

In scanning transmission electron microscope (STEM), an electron beam with a small diameter d (electron probe) illuminates a small area. By scanning this probe, a wide area can be observed. In other words, the resolving power of STEM is determined by d, i.e., by the illuminating lens system. The probe has a semi-angle of $2\theta_C$. This means that direction of the incident electron beam is diverted (Fig. 10.1).

Advantages and drawbacks of STEM over CTEM can be summarized as follows:

(1) In STEM a thicker specimen can be observed than in CTEM.
(2) In STEM bend contours are suppressed. This is of particular benefit in observing a complexly bent specimen.
(3) Under appropriate condition, $\mathbf{g}^* \cdot \mathbf{b}$ and $\mathbf{g}^* \cdot \mathbf{R}$ criteria, which are established in CTEM, are applicable.
(4) Combining energy-dispersive X-ray spectroscopy (EDX) and electron energy-loss spectroscopy (EELS) with STEM facilitates detection of segregation in a local small area.
(5) However, it is to be noted that in the aforementioned benefit (2) information on strain is lost, which is a drawback.

In STEM if the probe is fixed, no image is obtained. In order to obtain an image, the probe must scan the specimen, acquiring

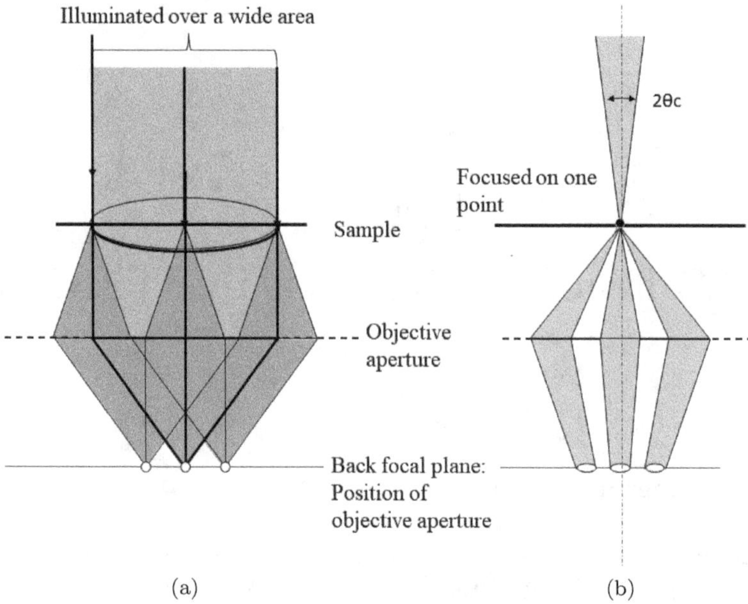

Fig. 10.1: Ray paths of CTEM (a) and STEM (b).

Table 10.1: Features of illuminating beam CTEM and STEM.

	CTEM	STEM
Illuminated area	Wide	Narrow
Semi-angle	Small	Large
Incident electron beam	Parallel	Diverted

a signal with a synchronized detector (Fig. 10.2). Since the size of display is not different significantly from one display to another, the smaller is the area (L × L) over which the probe scans, the larger the final magnification is. The resolution is determined by the size of the probe *d*. When the probe is fixed at one point, an electron diffraction pattern from that (small) area is obtained. However, the incident electron beam is not parallel, so that sharp spots like those in CTEM are not obtained; the diffraction spots become disks. This is referred to as a diffraction disk. The area illuminated by a probe is small, so that it can be regarded as an unstrained crystal. In other

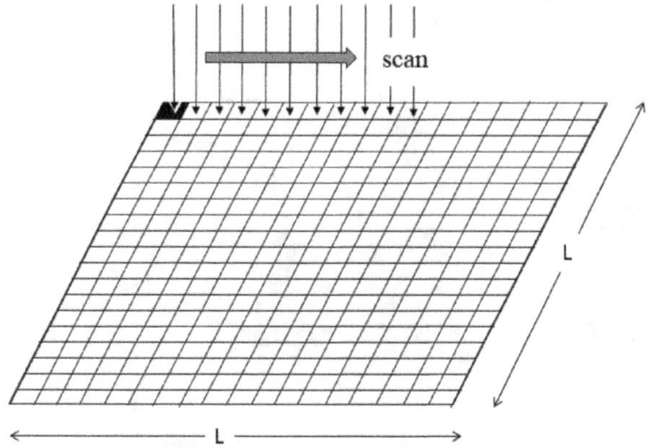

Fig. 10.2: STEM image is formed by scanning an electron probe.

words, crystallographic information from the small area is obtained. This method is referred to as convergent-beam electron diffraction (CBED). CBED will be treated in Section 11.1.

10.2 Diffraction Contrast in STEM

Two methods are used to detect electrons which transmit a specimen. One is to place a detector at the optical axis of the transmitted beam. This corresponds to BF imaging in CTEM. For this reason, it is customarily referred to as BF imaging (in STEM). The other method is to use an annular detector. In this case the transmitted electron beam is not detected: This is a kind of DF imaging and referred to as annular dark field imaging, the detail of which will be described in Section 10.4. If the position of the transmitted disk is displaced from the optical axis in such a way that the diffraction disk now falls on the so-called BF detector, the image obtained corresponds to DF image in CTEM (Fig. 10.3(b)). In order to distinguish this type of (DF) imaging from a genuine BF imaging, hereinafter let us call a BF detector an optical-axis detector (OAD).

The distance between a specimen and a detector is referred to as a camera length. When the camera length is small, both of the transmitted and diffracted beams are detected by the OAD. This situation corresponds to Fig. 6.1(a) in CTEM and no diffraction

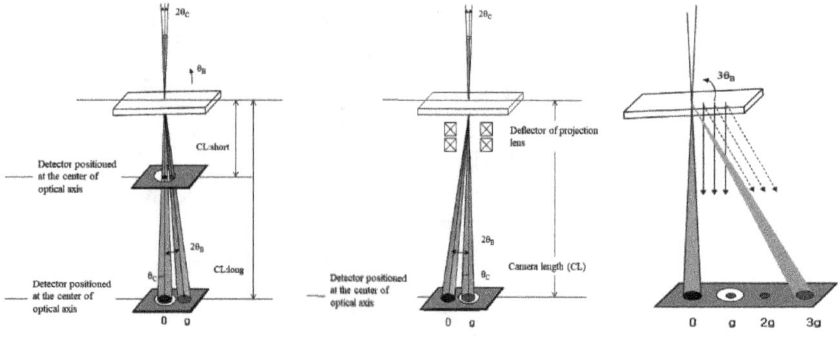

(a) (b) (c)

Fig. 10.3: (a) Illustrates a ray path for BF imaging in STEM. When the camera length is short, both of the incident disk and diffraction disk reach a detector, so that diffraction contrast is not produced. In order to obtain diffraction contrast, the camera length must be long enough to separate the incident disk and diffraction disk. (b) Illustrates a ray path for DF imaging in STEM. By the deflecting coil of a projector lens, the diffracted electron cone is set to the center of the optic-axis centered (OAD) detector. (c) Illustrates a ray path for DFWB imaging in STEM (reproduced from Iwata and Saka (2017)).

contrast is produced. When the camera length is long enough, it is possible to separate the diffracted beams from the transmitted one: When an OAD is set to the transmitted beam, BF image is obtained (Fig. 10.3(a)) and when set to the diffracted beam, DF image is obtained (Fig. 10.3(b)). Since the position of an OAD is fixed, in order to bring the diffracted beam to an OAD, the deflector of projection lens is used. This situation corresponds to Figs. 6.3(b), 6.3(b′) and 6.3(b″) in CTEM. In Figs. 6.3(b), 6.3(b′) and 6.3(b″), the diffracted beam is tilted by $2\theta_B$, so that the picture quality (resolution) is poor. In STEM, the resolution is determined by the diameter of an electron probe d which illuminates locally a sample and is not affected by an optical path of electrons after it transmits the sample. In other words, a handy DF imaging method in STEM (Fig. 10.3(b)) is comparable to a strict DF imaging method in CTEM. Also, $\mathbf{g}^*$ is not reversed. Furthermore, deflecting the incident beam by $2\theta_B$ in CTEM is fairly tedious, which is skipped in STEM.

In order to observe a weak beam image, a specimen is tilted so as to satisfy $3\mathbf{g}^*$ and then the diffracted beam of $\mathbf{g}^*$ is deflected to OAD by projector lens (PL). This is $\mathbf{g}^*/3\mathbf{g}^*$ imaging mode.

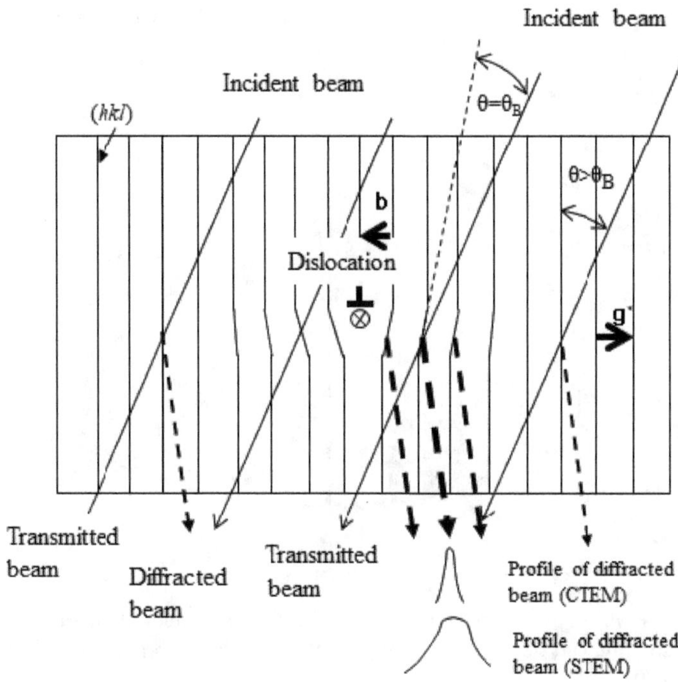

Fig. 10.4: Diffraction of electron beams around a dislocation in CTEM (a) and STEM (b). In STEM the diffracted beam is diverted (reproduced from Iwata and Saka (2017)).

10.3 Resolution of Dislocations Observed by Diffraction Contrast in STEM

Let us discuss on the resolution of diffraction contrast in STEM. Figure 10.4 corresponds to Fig. 8.5 redrawn for STEM.

Figure 8.5 illustrates schematically mechanism by which the diffraction contrast of dislocation is formed. In STEM, however, an image of dislocation widens, as illustrated in Fig. 10.4. The width of an image of dislocation is determined by the diameter of the incident electron probe d and a semi-angle $2\theta_B$. Ideally both the diameter d and the semi-angle $2\theta_B$ are small. Practically in order to make d smaller keeping brightness, it is necessary to have a large value of $2\theta_B$. Inversely, in order to have a smaller $2\theta_B$, d must be larger. In order to cope with this *antinomy*, it is cordial to use a brighter

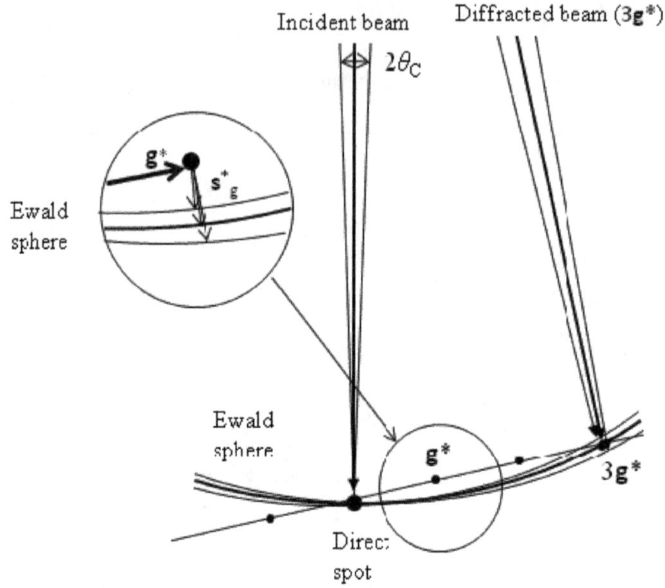

Fig. 10.5: Ewald sphere construction in STEM. The incident beam has a semi-angle θ_c, so that the deviation parameter $\mathbf{s}_g^*$ is diverted (reproduced from Iwata and Saka (2017)).

electron gun. For this purpose, STEM equipped with a field emission gun (FE-STEM) has been developed.

In CTEM a profile of dislocation image is sharp as schematically shown by broken lines, while in STEM it widens because of effects caused by a spot size d and a semi-angle $2\theta_\mathrm{B}$. Excepting this, mechanisms of image formation in CTEM and STEM are essentially the same (Fig. 10.5). Figure 10.6(a) shows BF images of dislocations in a Cu–Al alloy imaged in the so-called two-beam condition. The dislocations are imaged in $\mathbf{g}^* \cdot \mathbf{b} = 2$, so that the images of the dislocations show a strong-weak pair as in the case of dislocation 2 in Fig. 8.17. When the dislocation is imaged in DFWB method, it is evident that it is dissociated into two Shockley partials. Furthermore, on the reversal of sign of $\mathbf{g}^*$, the strong-weak pair is also reversed. This is perfectly in keeping with CTEM (see Fig. 9.8). Resolution of dislocation images in DFWB is determined by a probe size d. Figure 10.7 shows an example of a dissociated dislocation in Si observed using FE-STEM.

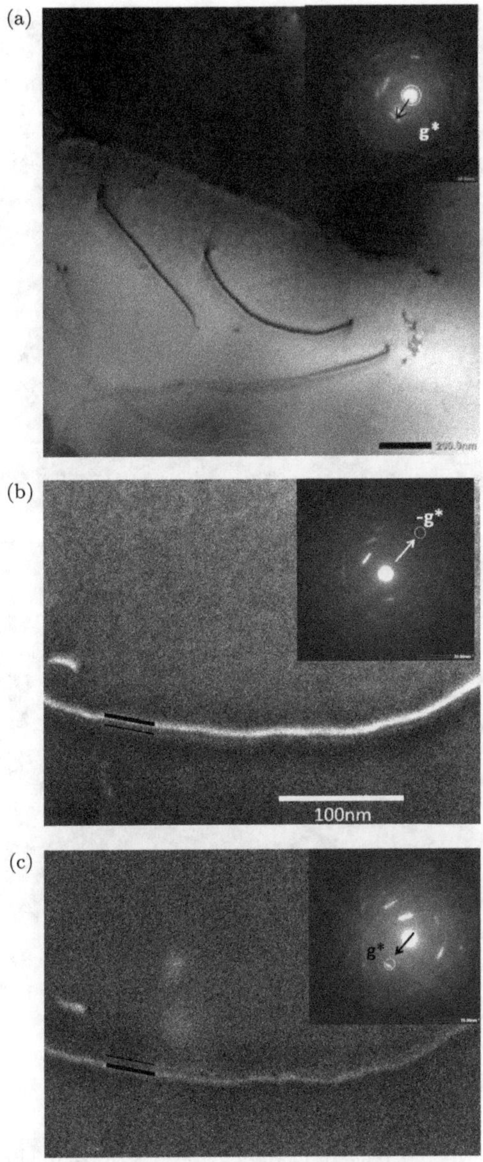

Fig. 10.6: (a) shows a BF image of a dissociated dislocation in a Cu–Al alloy imaged in $\mathbf{g}^* \cdot \mathbf{b} = 2$. The dislocation shows a double image, but this results from a diffraction effect and does not show that individual partials are resolved (see dislocation 2 in Fig. 8.17). (b) and (c) show DFWB imaged in $+\mathbf{g}^*$ and $-\mathbf{g}^*$, respectively. It is noted that intensities of the two partials are reversed on the reversal of the sign of $\mathbf{g}^*$ (reproduced from Iwata and Saka (2017)).

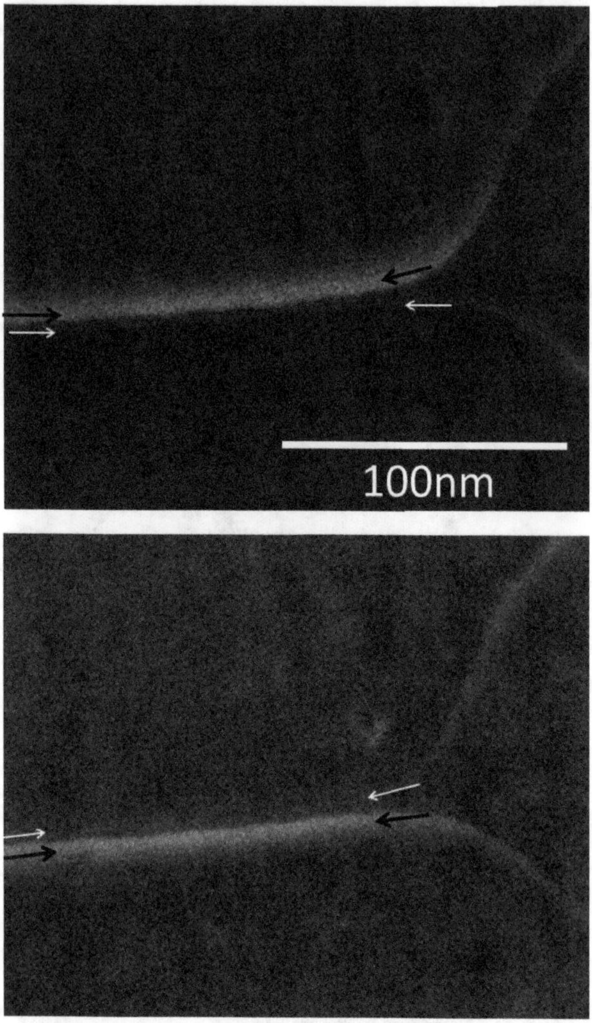

Fig. 10.7: DFWB images of a dislocation in Si as observed in an FE-STEM (unpublished work by Iwata and Saka).

In summary, in CTEM it is necessary to deflect the incident electron beam by a deflecting coil of condenser lens (CL), while in STEM what is necessary is just to deflect the diffracted beam by a projection lens (PL) to OAD. Deflecting the incident beam by CL in CTEM is rather tedious, while deflecting the diffracted beam by PL is much easier (Table 10.2).

Table 10.2: How to take DFWB image in $\mathbf{g}^*/3\mathbf{g}^*$ condition.

	CTEM	STEM
Step 1	Tilt the sample so as to satisfy Bragg's condition for $\mathbf{g}^*$.	Tilt the sample so as to satisfy Bragg's condition for $3\mathbf{g}^*$.
Step 2	Deflect the incident beam to the position of $-\mathbf{g}^*$ (Now Bragg's condition for $3\mathbf{g}^*$ is satisfied).	Set the camera length long enough.
Step 3	Insert an objective aperture with small size to $\mathbf{g}^*$.	Bring the $\mathbf{g}^*$ disk to the center of the detector (OAD) by deflector of the projector lens.
Step 4	Record the image on a film or TV.	Start scanning of the incident probe.

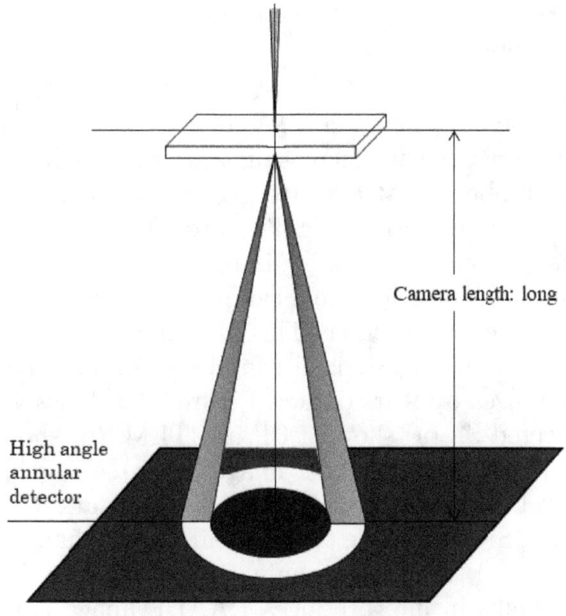

Fig. 10.8: Principle of HAADF.

10.4 HAADF

An imaging method using an annular aperture with a large camera length is referred to as high angle annular dark field (HAADF) method (Fig. 10.8). The advantage of this method is that it enables

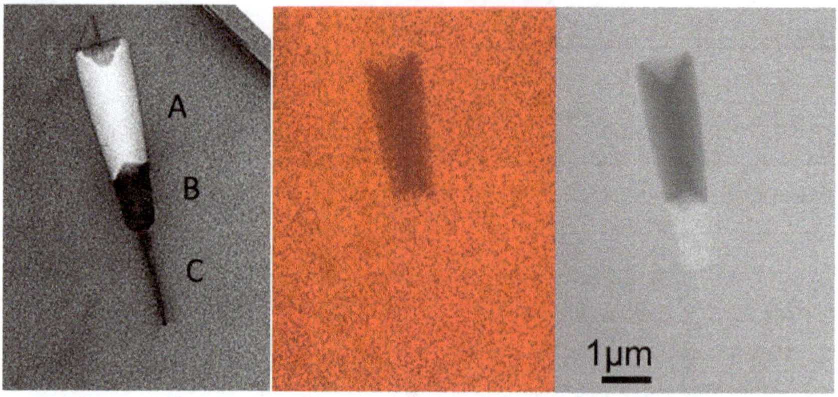

(a) (b) (c)

Fig. 10.9: (a) BF image of a void formed in Si formed by a laser injection. (b) EDX map around the void. Red area shows Si. (c) HAADF image. The void shows a dark contrast.

one to distinguish mass of material. As Fig. 4.4 shows, an atomic scattering factor decreases with increasing scattering angle and increases with atomic number Z. At a scattering angle larger than a certain value, the atomic scattering factor is proportional to Z^2. In other words, intensity of electrons falling in the annular aperture increases with increasing Z. The region in a sample which consists of material with higher Z appears brighter. This imaging method is referred to as Z-contrast. Alternatively, in a sample consisting of single value of Z, a thinner area appears darker. Figure 10.9 shows voids formed in a laser-injected Si. (a) shows a BF in CTEM, (b) shows a Si-map of EDX and (c) shows a HAADF image. In (b) **A** is deficient in Si, that is, a void. In (a) **A** appears bright and in (c) dark. **B** and **C** show no features in a Si-map, while they appear darker than matrix in (a) and brighter in (c). Detailed examination shows that **B** and **C** are dislocations. Thus, in interpreting a HAADF image it is important to be careful not to be confused with a diffraction contrast which may be involved accidentally.

Advantages and drawbacks of STEM

In STEM complex bend contours in a complicatedly bent sample disappear. However, inversely, this means that information concerning distribution of a strain in the sample also disappears.

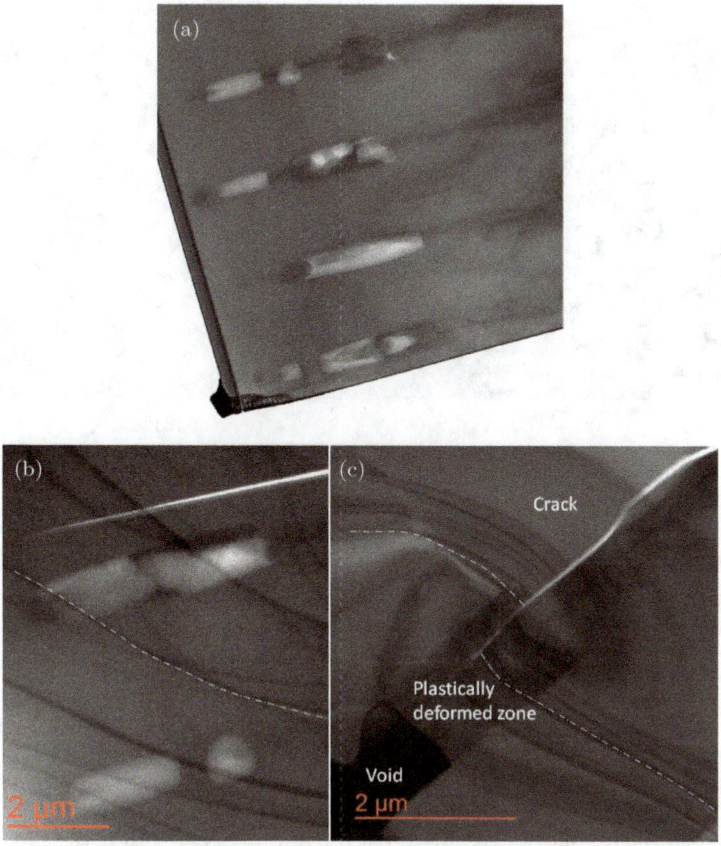

Fig. 10.10: (a) STEM BF image of a void formed in Si by a laser injection. (b) and (c) CTEM BF images. In (b) bend contours are not disturbed even near voids, while in (c) bend contours near a plastic zone in the vicinity of void are disturbed severely; the Si matrix is distorted severely. Black contrast in the left bottom of (c) is a void which is covered by protective tungsten. In (b) and (c) typical bend contours are shown by white chain lines (reproduced from Kawaguchi *et al.* (2019)).

Figures 10.10(a) and 10.10(b) show STEM and CTEM images of voids which are formed in Si by a laser. In CTEM image (Fig. 10.10(b)), bend contours are evident around voids, while in STEM image (Fig. 10.10(a)) no bend contours are visible. When only images of voids are to be observed, the bend contours are awkward. However, bend contours provide information concerning strains inside a sample. Thus, in (a) it may appear that strain is

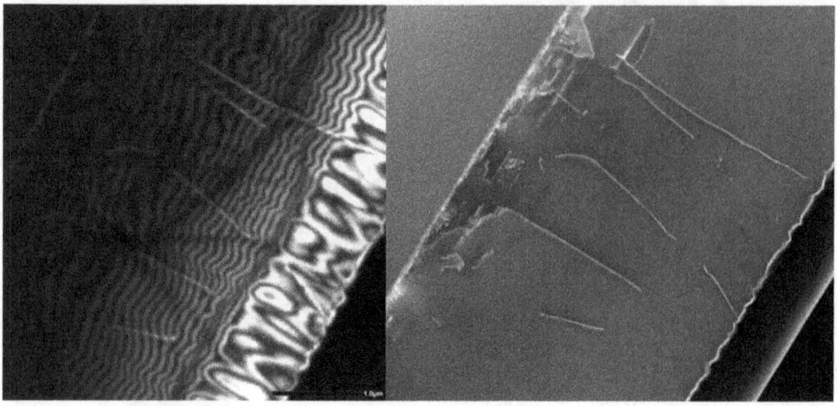

(a) (b)

Fig. 10.11: DFWB images of dislocations in GaN imaged in CTEM (a) and STEM (b) (by courtesy of Dr. H. Iwata).

absent in the sample, since no bend contour are observed, while in (b) it is evident that a strain is present since bend contours are observed. Figure 10.10(c) shows a plastically deformed zone near a void: Evidently around the void severe elastic strains are present, that is, bend contours are observed. Thus, information of strain is lost in STEM.

Similar thing can be said about thickness contours. Figures 10.11(a) and 10.11(b) show dislocations in GaN, imaged in CTEM and STEM, respectively. Fringes parallel to the edge of the sample in Fig. 10.11(a) are thickness contours. In Fig. 10.11(b), such fringes are not observed, giving an impression that dislocation images appear more distinguished. However, these thickness fringes provide very important information on Burgers vectors of dislocations (see Section 9.1.5): In STEM such information is lost.

Chapter 11

CBED and LACBED

11.1 Introduction

Figure 11.1 illustrates a path of electron beam cone, with a semi-angle θ_c, which is reflected by (hkl) plane. Let us assume a two-beam condition for the sake of simplicity. The electron beam has a semi-angle of θ_c in STEM mode, and correspondingly diffraction disks are formed instead of diffraction spots in CTEM. Electron beams passing through the centers of these disks (red lines) satisfy the Bragg's condition perfectly ($\mathbf{s}_g^* = 0$) (Fig. 11.2). Green lines correspond to $\mathbf{s}_g^* > 0$ and blue lines to $\mathbf{s}_g^* < 0$. It is noted that when a reciprocal point is inside the Ewald sphere, $\mathbf{s}_g^* > 0$ and when outside the Ewald sphere, $\mathbf{s}_g^* < 0$ (Fig. 5.14). In other words, information from $\mathbf{s}_g^* > 0$ to $\mathbf{s}_g^* < 0$ is contained in a single diffraction disk (Fig. 11.3(a)). The intensity profile in this disk along x-axis is shown in Fig. 11.3(b). Such a profile is referred to as a rocking curve. Rocking curves are classified into two: (1) a crystal sample is slightly rocked with the direction of the incident electron beam fixed (crystal-rocking curve), or (2) the direction of the incident beam is rocked slightly with the orientation of crystal sample fixed (beam-rocking curve).

When the incident beam is along a zone axis, the deviation parameters $\mathbf{s}_g^*$ differ both along x and y-axes, the rocking curve is two-dimensional (Fig. 11.3(c)).

Detailed analysis of the two-dimensional rocking curves provides precise determination of lattice constants, point group and space

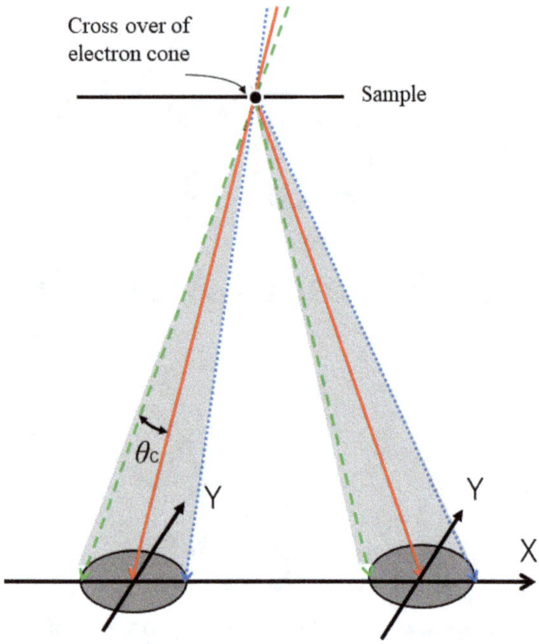

Fig. 11.1: Ray path of electron beam cone with θc.

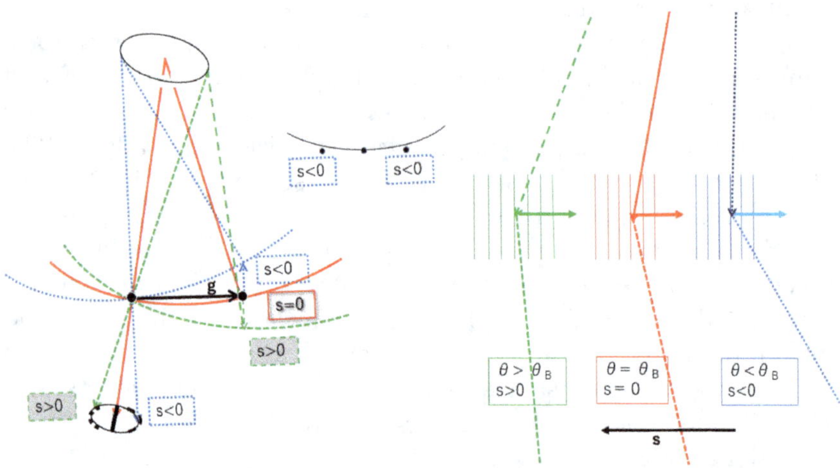

Fig. 11.2: Ewald construction for electron beam cone.

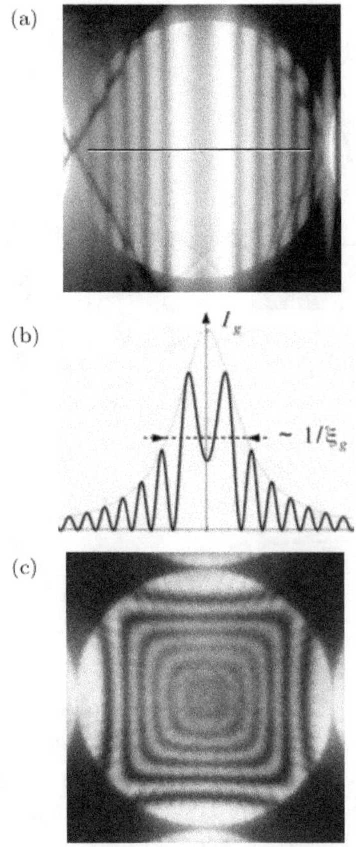

Fig. 11.3: Rocking curves (Reproduced with permission from Tanaka and Tsuda, 2014).

group etc. However, this is out of scope of this book and readers refer to Tanaka *et al.*

11.2 Determining Nature of a Dislocation by LACBED

In CBED, information of diffraction (i.e., in reciprocal space), that is, a beam-rocking curve is obtained over a small area (in real space) on which an electron cone converses. On the other hand, the electron probe stays on a certain point on a sample, so that information of

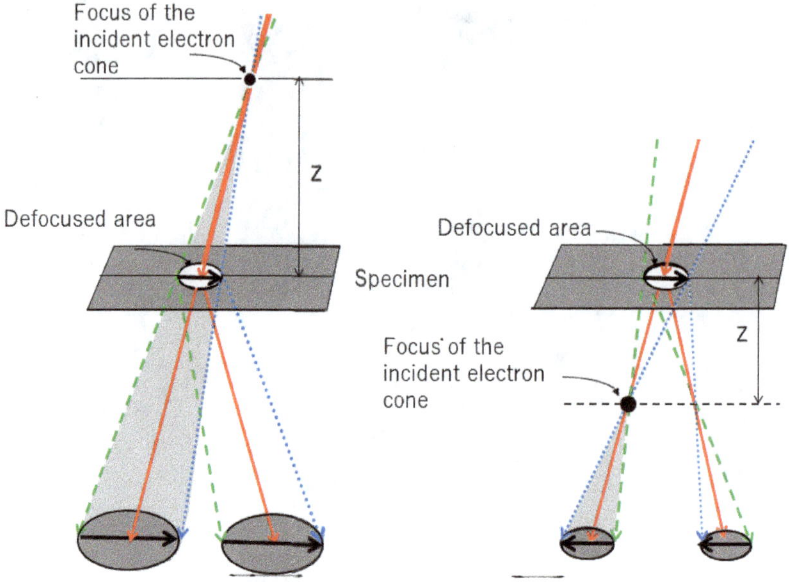

Fig. 11.4: Two modes of illumination in LACBED.

sample in real space, that is, image of a sample, is not obtained. (In order to obtain information in real space, the probe must scan the sample (see Fig. 10.1). By contrast, in large angle convergent beam electron diffraction (LACBED) the focus of an electron probe is shifted up or down by a distance z. Then, a little bit wider area on a sample (indicated by a circle) is illuminated, so that information both in real and reciprocal spaces over this defocused area is obtained. The focal point of the electron probe may be above (Fig. 11.4(a)) or below (Fig. 11.4(b)). In the case of above, the sample and its image point to the same direction, while in the case of below they point to the opposite directions. In this book, "above" position is employed.

Thus, if a lattice defect like a dislocation is present in the defocused area of LACBED, that area where Bragg's condition is satisfied locally is displaced for the reason similar to that described in Figs. 8.21(a) and 8.21(b). Such a local area is referred to as a bend contour in CTEM since this is a crystal-rocking curve. By contrast, in LACBED it is referred to as a Bragg's line since this is a beam-rocking curve. The direction of the displacement is determined by Burgers vector of the dislocation. The reason for this displacement

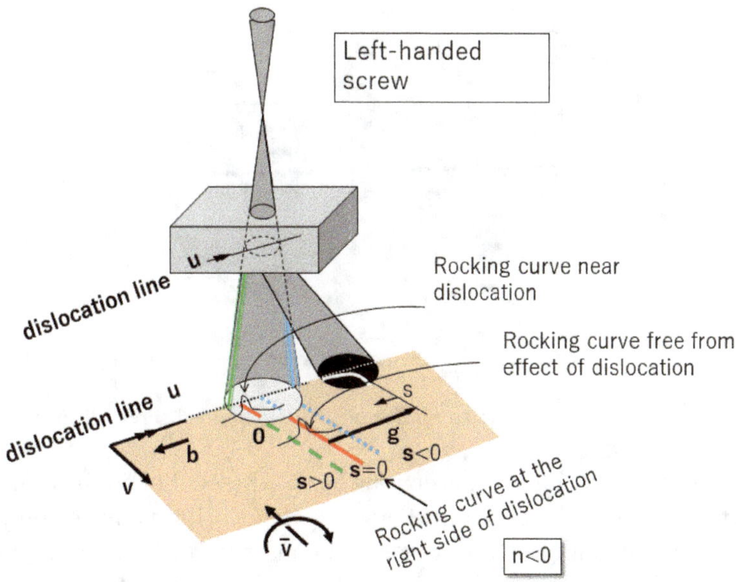

Fig. 11.5: Principle of LACBED.

is in principle the same as that for Figs. 8.21–8.23. In Figs. 8.22(a)–8.22(d), deviation from the Bragg's condition is caused by elastic bending of a specimen (i.e., bend contours) while in LACBED a specimen (in real space) is not bent but the incident beam has a cone-like shape which results in rocking of the incident electron beam. In other words, Figs. 8.22(a)–8.22(d) represent crystal-rocking, while LACBED stems from beam-rocking.

Figure 11.5 illustrates the principles of LACBED.

Now let us consider that a left-handed screw dislocation lies in the direction of **u**.[1] By definition **u** and **b** are antiparallel (see Fig. 2.11 in "Classical theory of crystal dislocations — From iron to gallium nitride" by Saka). Now let us pay attention to the right-sided area of the dislocation. Here, an upper part of the crystal is

[1]So far, in this book, the sense of direction of a dislocation has been denoted by ζ. In Chapters 10 and 11, it will be denoted by **u**, since ζ is easy to confuse with ξ, which is customarily used to denote an extinction distance in diffraction theory of crystals (see Section 6.4, Eqs. (6.8) and (6.9)).

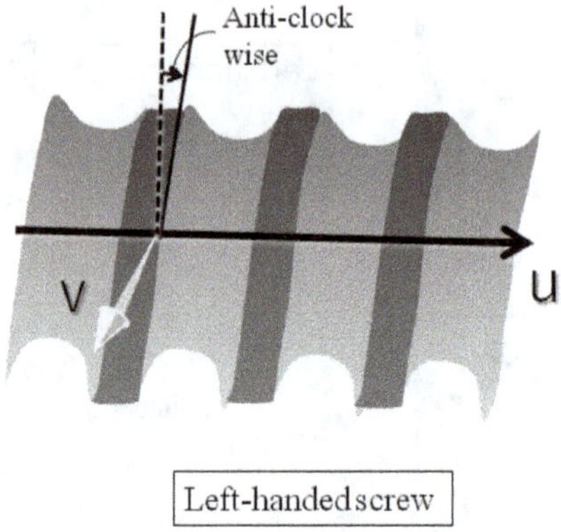

Fig. 11.6: Bending of lattice planes near a left-handed screw dislocation.

displaced by **b** while the lower part is fixed (see Theorem 3.2). Distortion of lattice planes around a screw dislocation is quite complex: Readers are strongly recommended to construct models in Appendix G to confirm distortions of right-handed and left-handed screw dislocations.

Bending of lattices in the right side of **u** is clockwise as shown in Fig. 11.6. It may appear strange that the bending of lattices along a left-handed screw is clockwise. However, this can be rationalized as follows. In Fig. 11.6, the rotation axis of bending of lattice planes is $\bar{\mathbf{v}}$, which is opposite to **v**. Since the co-ordinate system employed is right-hand (RH), the *proper* rotation axis should be **v**. Then, the rotation is anti-clockwise: This is in agreement with the rotation of lattice planes around a left-handed screw.

Now let us consider a case where both **u** and $|\mathbf{b}|$ point to the same direction, that is, either $\mathbf{u} = \mathbf{b}$ or $\mathbf{u} = -\mathbf{b}$ (Fig. 11.5). A ray which passes the center of an electron cone (drawn in red) satisfies the Bragg's condition, i.e., $\mathbf{s}_g^* = 0$ as illustrated in Fig. 11.7(a). For a ray drawn in green, $\mathbf{s}_g^* > 0$ while for a ray drawn in navy blue, $\mathbf{s}_g^* < 0$ (when a reciprocal point is inside Ewald sphere $\mathbf{s}_g^* < 0$, while outside Ewald sphere $\mathbf{s}_g^* > 0$ (see Fig. 5.14)).

Around a left-handed screw dislocation, lattice planes rotate clockwise with respect to $\bar{\mathbf{v}}$, which points toward the paper. That is, the lattice planes rotate as shown by solid lines in Fig. 11.8. Then, the ray at the center (red), which satisfies the Bragg's condition remote from the dislocation, does not satisfy the Bragg's condition

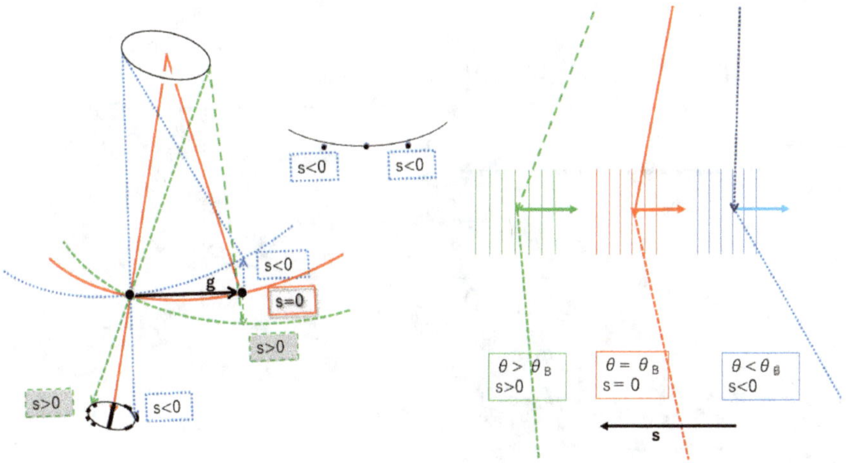

Fig. 11.7: Appearance of Bragg's line in perfect crystal.

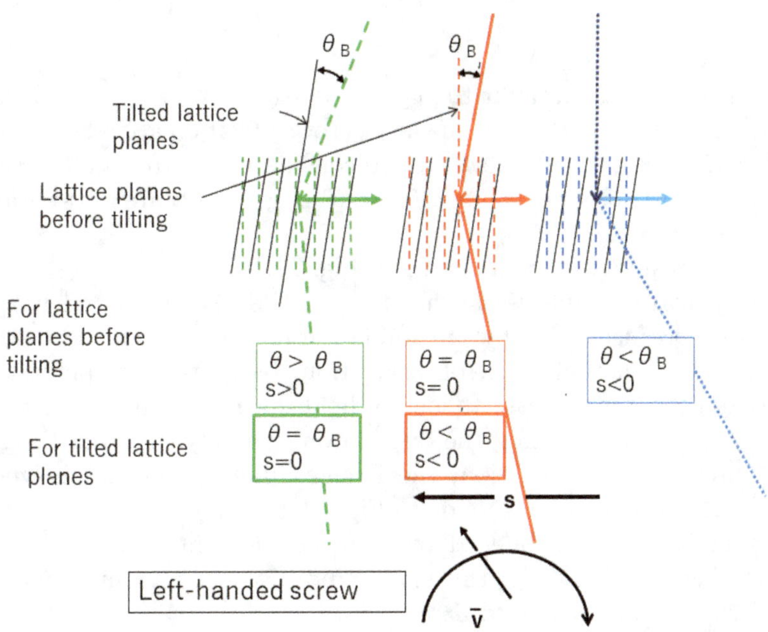

Fig. 11.8: Displacement of Bragg's line due to tilting of lattice planes.

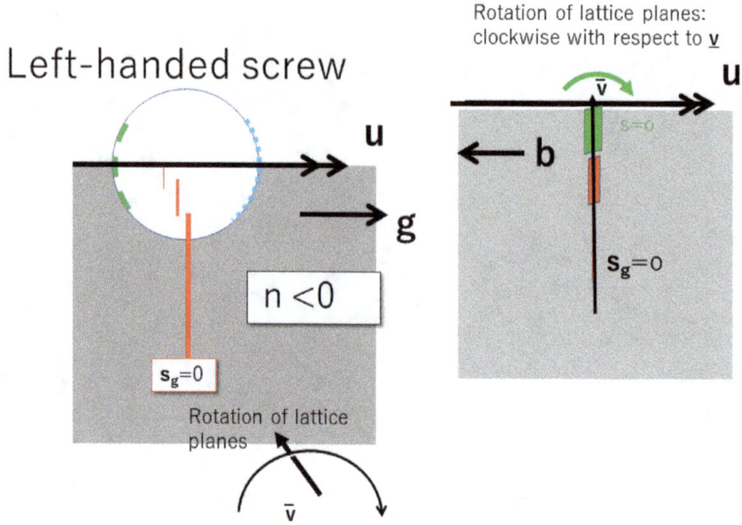

Fig. 11.9: Bending of Bragg's line near a dislocation line.

any longer. Instead a ray path in green satisfies the Bragg's condition. In other words, the Bragg's line moves toward left on approaching the dislocation (Fig. 11.9). Now, we are considering a left-handed screw dislocation. Therefore, $\mathbf{g}^* \cdot \mathbf{b} < 0$. That is, $\mathbf{g}^*$ and $\mathbf{b}$ are anti-parallel with each other.

These results are summarized in Fig. 11.10.

A typical example is shown in Fig. 11.11. White round images are voids formed in Si by ion implantation. As may be seen from Fig. 11.11, an actual image of dislocation (by diffraction contrast) is so weak that it is not easy to locate the dislocation. Thus, such voids or etch pits help locating the dislocation. Anyway, where a Bragg's line intersects a dislocation, the Bragg's line is split as shown in Fig. 11.11(b); this corresponds to Fig. 11.10(d).

Bragg's lines are indexed by focusing disks in Fig. 11.12(a) to spots. Then, by indexing this spotty diffraction by a usual manner, the Bragg's line is indexed $5\bar{3}3$.

Number of splitting of a Bragg's line, n, is given by $\mathbf{g}^* \cdot \mathbf{b} = n$ (Fig. 11.13). This is referred to as Cherns–Preston's rule.

Figure 11.14 shows results obtained on a dislocation in 4H-SiC.

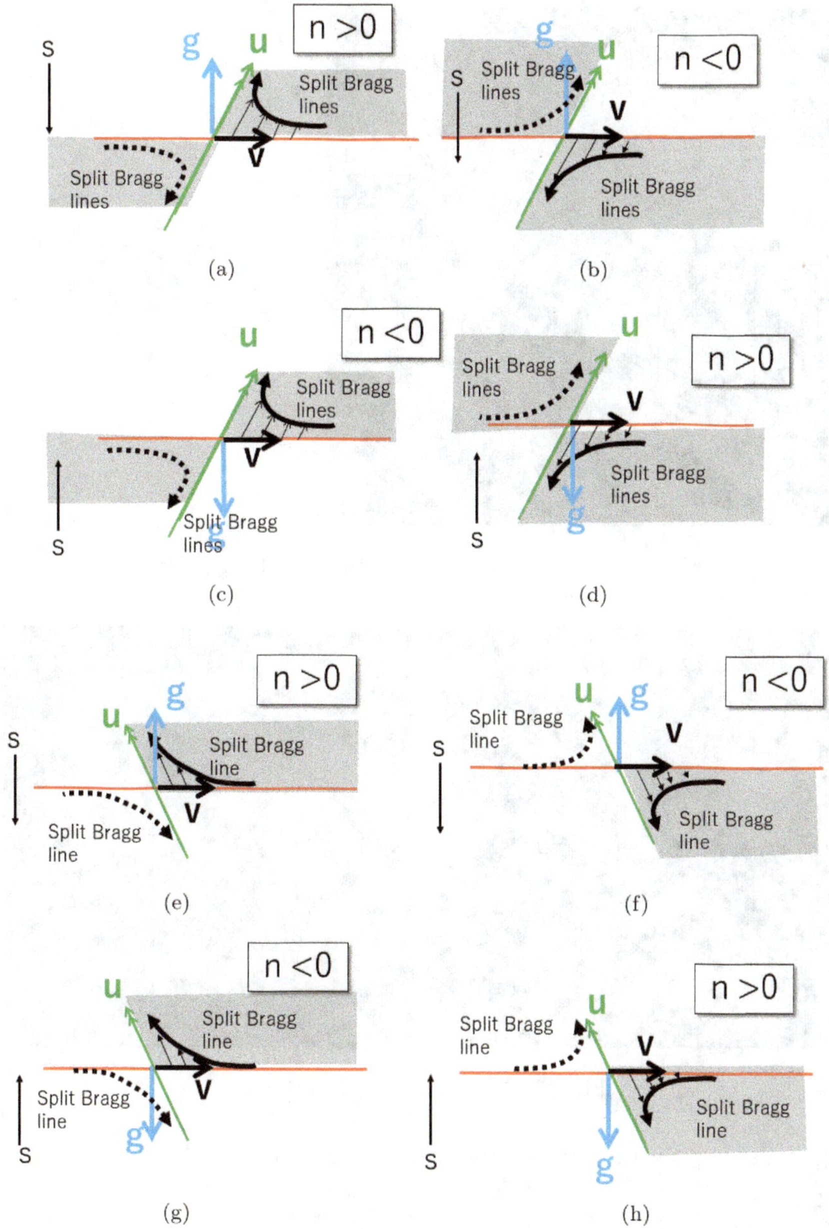

Fig. 11.10: Splitting of a Bragg's line when it intersects a dislocation line.

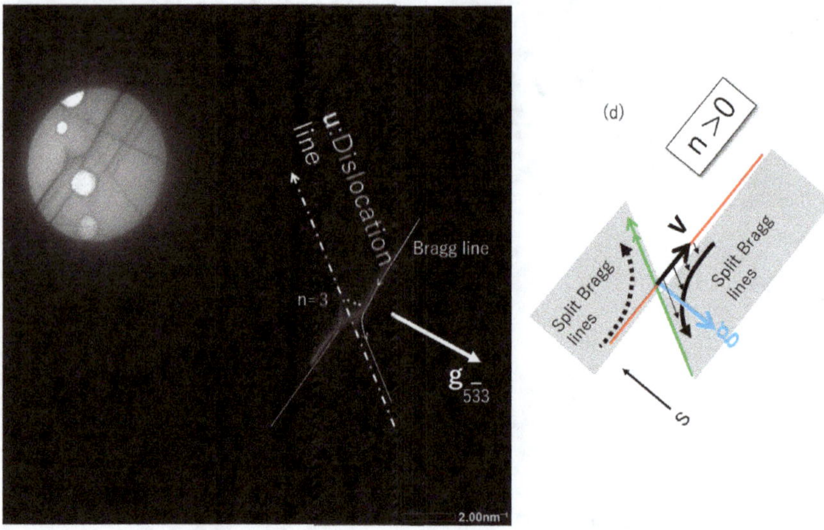

Fig. 11.11: An example of splitting of a Bragg's line.

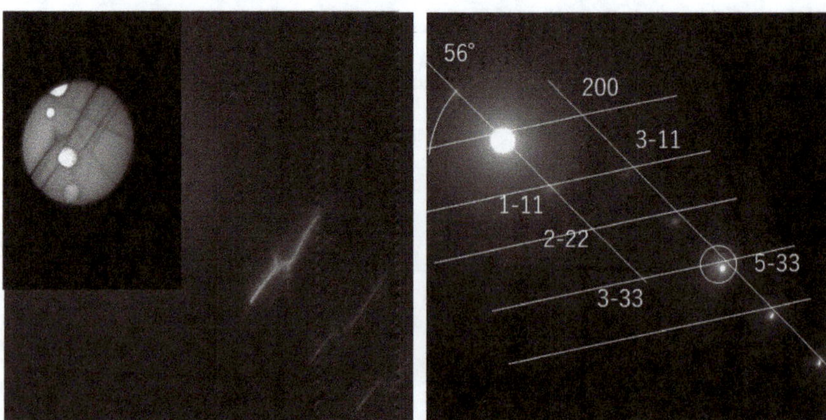

Fig. 11.12: Indexing of Bragg line.

$\mathbf{g}_1^* = \mathit{10} \cdot \mathit{11} \quad -\mathit{1}\mathrm{u} + \mathit{0}\mathrm{v} + \mathit{11}\mathrm{w} = 13$ (red)
$\mathbf{g}_2^* = \mathit{01} \cdot \mathit{14} \quad \mathit{0}\mathrm{u} + \mathit{1}\mathrm{v} + \mathit{14}\mathrm{w} = 14$ (green)
$\mathbf{g}_3^* = \mathit{0}\bar{\mathit{1}} \cdot \bar{\mathit{1}}\bar{\mathit{5}} \quad \mathit{0}\mathrm{u} - \mathit{1}\mathrm{v} - \mathit{15}\mathrm{w} = -15$ (blue)
$\mathbf{g}_4^* = \mathit{10} \cdot \bar{\mathit{1}}\bar{\mathit{2}} \quad \mathit{1}\mathrm{u} + \mathit{0}\mathrm{v} - \mathit{12}\mathrm{w} = -14$ (yellow)
$\mathbf{g}_5^* = \mathit{21} \cdot \bar{\mathit{9}} \quad \mathit{2}\mathrm{u} + \mathit{1}\mathrm{v} - \mathit{9}\mathrm{w} = 13$ (purple)

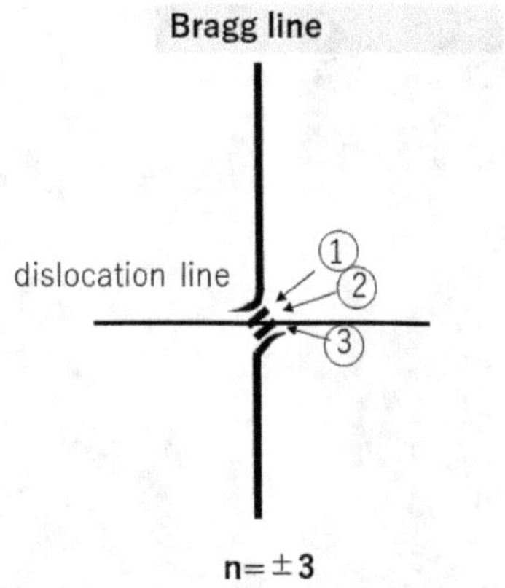

Fig. 11.13: Splitting of a Bragg's line.

From these simultaneous equations, $\mathbf{b} = [\bar{2}01]$. This corresponds to $\mathbf{b} = \mathbf{c} + 2\mathbf{a}$.

11.3 Advantages and Drawbacks of Different Methods to Determine Absolute Values of Burgers Vectors

Three methods are available to determine absolute values of Burgers vectors:

(1) Making use of shift of thickness contours (Section 9.1.5)

This method employs DFWB method, so that high resolution observation of dislocations can be carried out at the same time. However, it is desirable that the thickness contours are parallel to the edge of a wedge-shaped sample; a complexly bent sample is not appropriate.

(2) LACBED (Fig. 11.15(a), Section 11.2)

Bragg's lines are twisted where they intersect a dislocation. The illuminated area is narrow enough to be considered flat. The Bragg's lines correspond to maxima in the beam-rocking curve, where the incident electron beam is rocked. In this sense, this is a compromise of the real and reciprocal spaces. Diffraction contrast of a dislocation

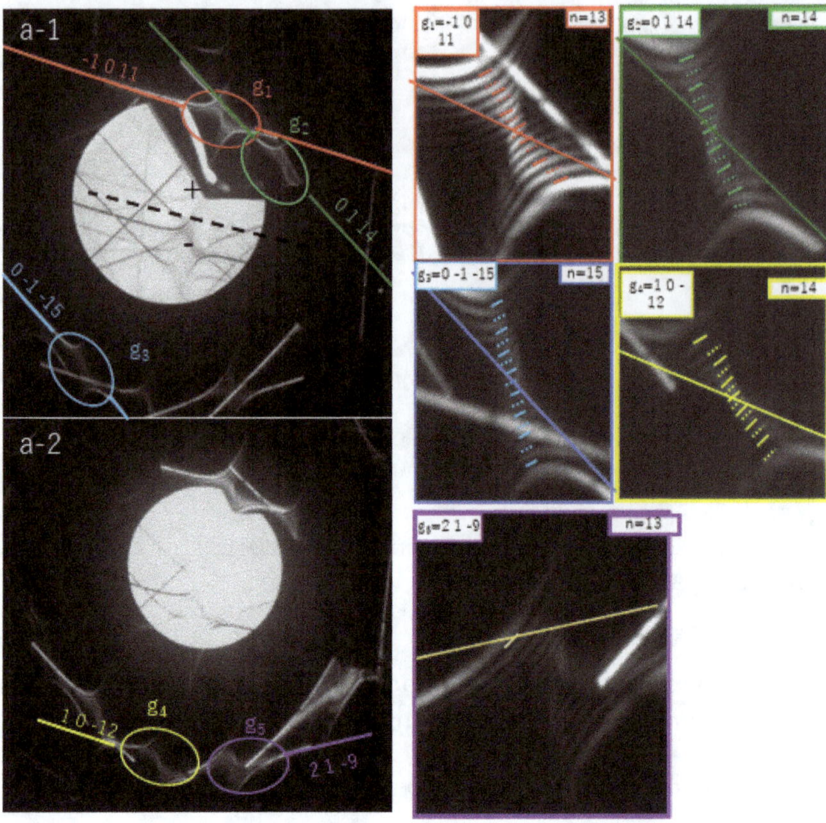

Fig. 11.14: Results obtained on a dislocation in 4H-SiC (Reproduced from Onda *et al.*, 2005, pp. 489–495).

is not strong enough, it is convenient to make use of an etch pit or a void to locate the dislocation.

(3) Spiecker–Jäger's method (Fig. 11.5(b))

This method makes use of a crystal-rocking curve, where the crystal is rocked. Wide area can be observed. In addition, DF imaging can be used, if necessary. Here, Cherns–Preston's rule is applicable. Figure 11.16 shows an enlarged micrograph of an encircle area in Fig. 8.19. The number of splitting where a dislocation crosses bend

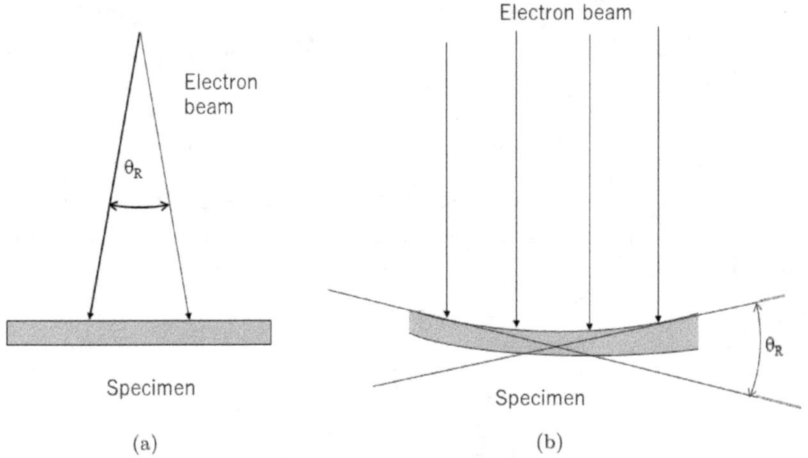

Fig. 11.15: LACBED (a) and bend contour method (b). In LACBED a specimen is flat and the incident electron beam rocks in a range from 0 to θ_R, while in a bend contour method the incident beam is parallel and the specimen rocks from 0 to θ_R.

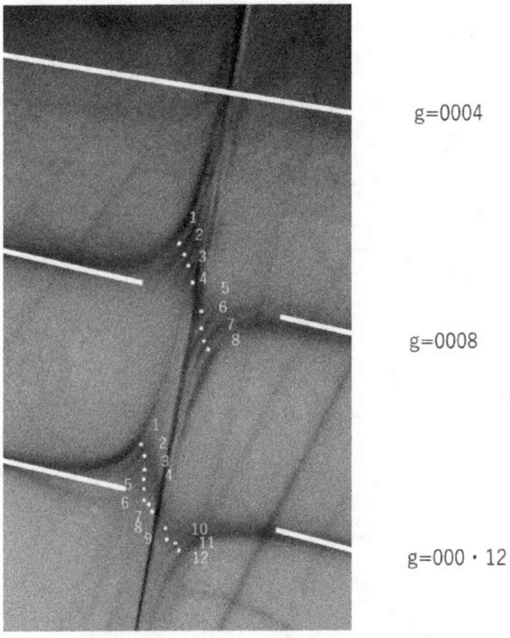

Fig. 11.16: Example of splitting of bend contour. When a dislocation intersects the bend contour for $\mathbf{g}^* = \mathit{000 \cdot 8}$, the bend contour splits into 8, while when intersect the bend contour of $\mathbf{g}^* = \mathit{000 \cdot 12}$, it splits into 12. This indicates that the Burgers vector of the dislocation under question is uv1.

contours, $n = 8$ for $\mathbf{g}^* = \mathit{0008}$ and $n = 12$ for $\mathbf{g}^* = \mathit{000{\cdot}12}$. This indicates that $\mathbf{g}^* \cdot \mathbf{b} = 8$ (for $\mathbf{g}^* = \mathit{0008}$) and $\mathbf{g}^* \cdot \mathbf{b} = 12$ (for $\mathbf{g}^* = \mathit{000{\cdot}12}$). Making $\mathbf{b} = [\mathrm{uvw}]$,

$$\mathit{0}\mathrm{u} + \mathit{0}\mathrm{v} + \mathit{8}\mathrm{w} = 8, \;\; \mathit{0}\mathrm{u} + \mathit{0}\mathrm{v} + \mathit{12}\mathrm{w} = 12.$$

We obtain $w = 1$. Needless to say, it is necessary to observe splitting of bend contours with different indices to determined u and v.

Appendices

Appendix A

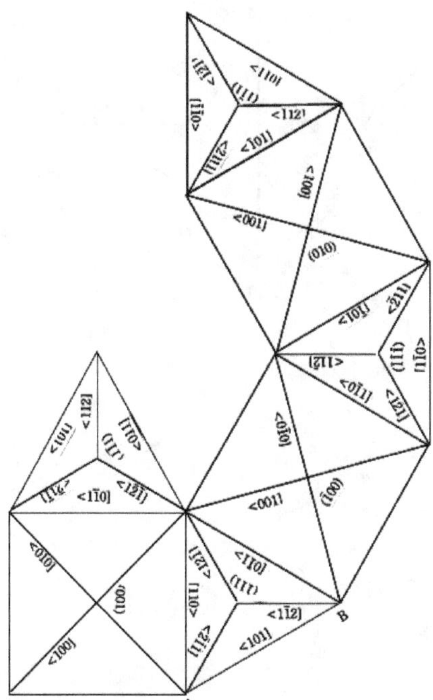

Fig. A.1: Three-dimensional model of a cubic crystal bounded with {*100*} and {*111*} planes. AB in A.1 and A.2 is common. Vector such as <100] indicates that this vector points to left and the one such as [1$\bar{1}$0> indicates that this vector points to right.

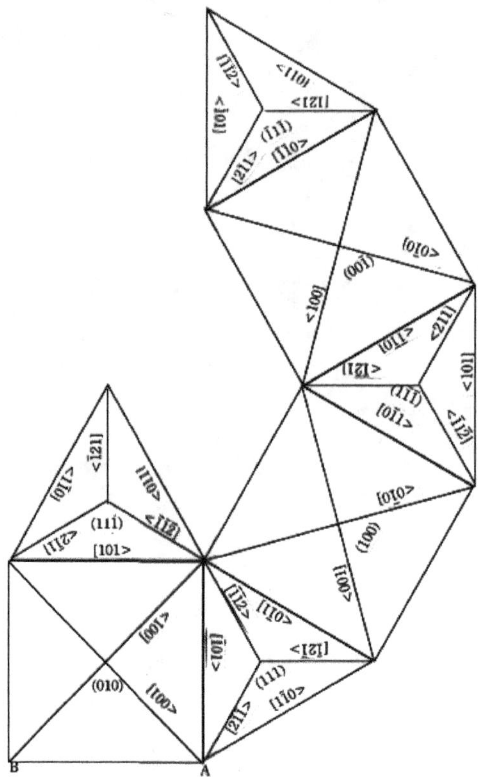

Fig. A.2.

Appendix B

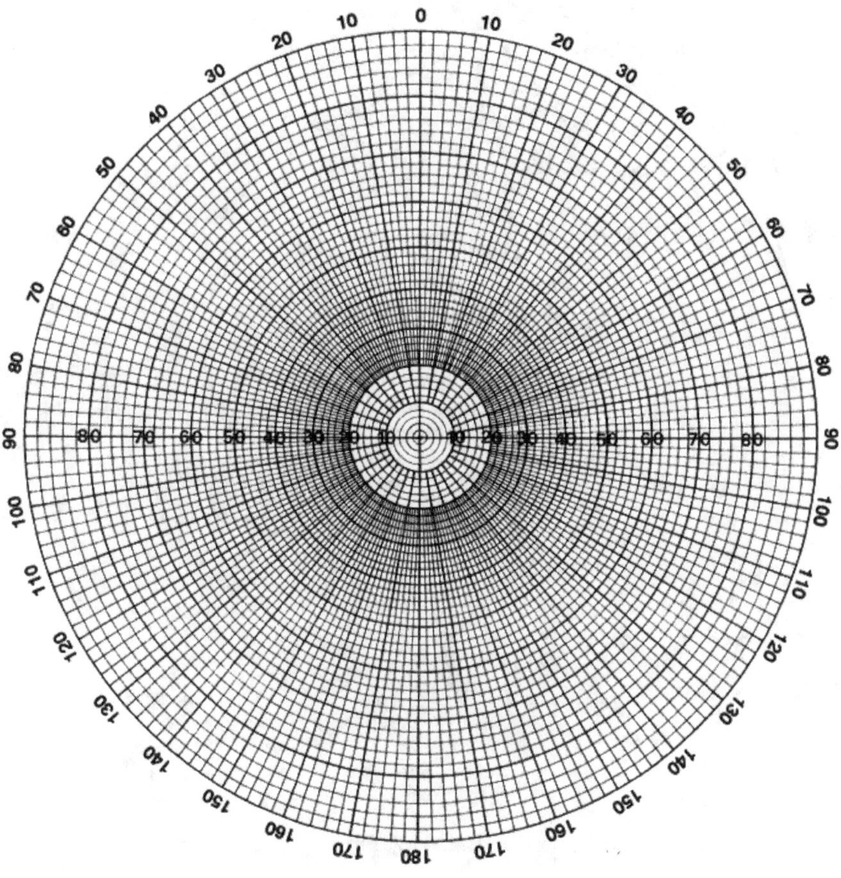

Fig. B.1: Polar net.

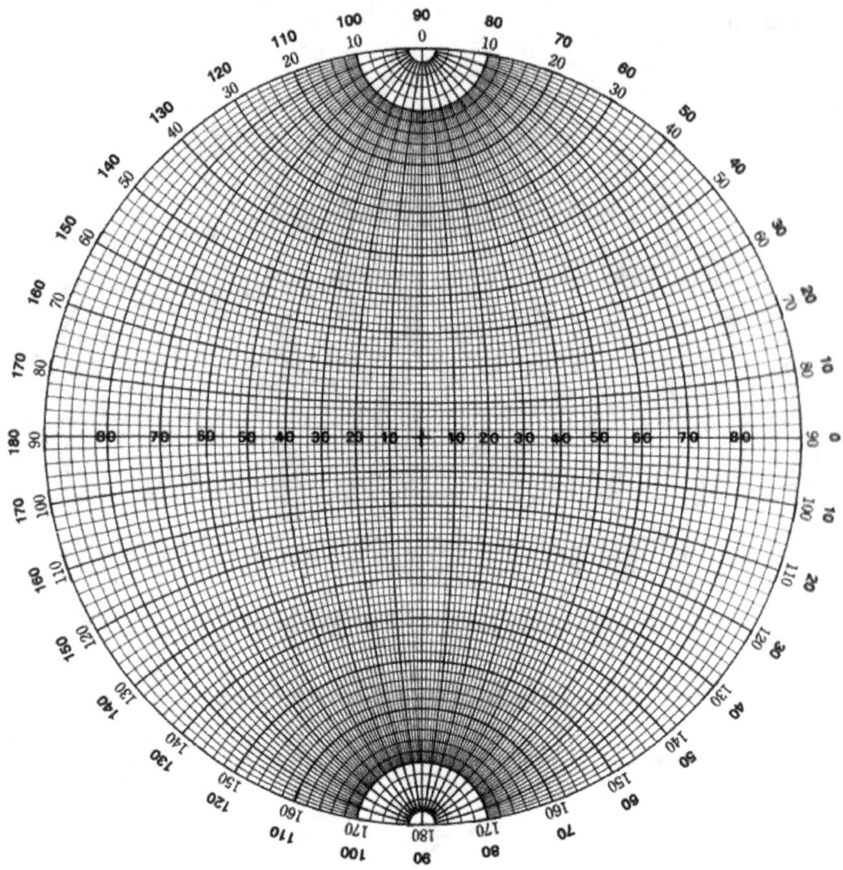

Fig. B.2: Wulff net.

Appendix C

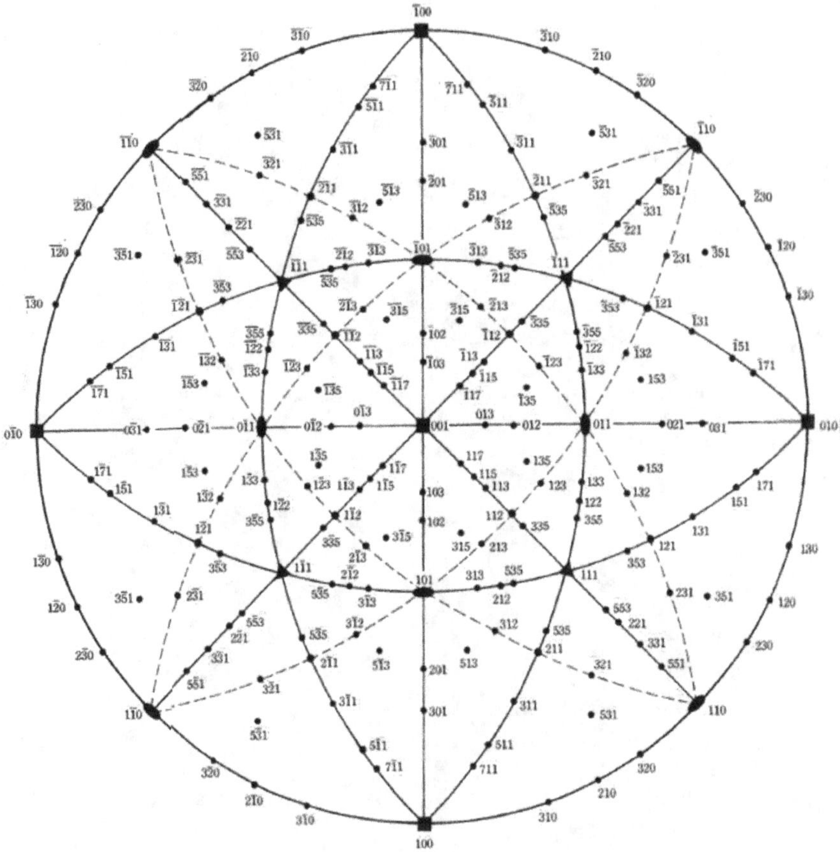

Fig. C.1: The 001 stereographic projection of a cubic crystal.

Appendix D

HKL	hkl	Values of angles between planes (and between directions) in cubic crystal (degrees*)						
100	100	0	90					
	110	45	90					
	111	54.73						
	210	26.57	63.43	90				
	211	35.27	65.96					
	221	48.18	70.53					
	310	18.43	71.56	90				
	311	25.23	72.45					
	320	33.68	56.32	90				
	321	36.76	57.68	74.5				
110	110	0	60	90				
	111	35.27	90					
	210	18.43	50.77	71.57				
	211	30	54.73	73.13	90			
	221	19.47	45	76.37	90			
	310	26.57	47.87	63.43	77.08			
	311	31.48	64.77	90				
	320	11.32	53.97	66.9	78.68			
	321	19.1	40.9	55.47	67.8	79.1		
111	111	0	70.53					
	210	39.28	75.03					
	211	19.47	61.87	90				
	221	15.8	54.73	78.9				
	310	43.1	68.58					
	311	29.5	58.52	79.98				
	320	36.82	80.78					
	321	22.2	51.88	72.02	90			
210	210	0	36.87	53.13	66.42	78.47	90	
	211	24.1	43.08	56.78	79.48	90		
	221	26.57	41.82	53.4	63.43	72.39	90	
	310	8.13	31.85	45	64.9	73.57	81.87	
	311	19.28	47.6	66.13	82.25			
	320	7.12	29.75	41.92	60.25	68.15	75.63	82.88
	321	17.02	33.22	53.3	61.43	68.98	83.13	90
211	211	0	33.55	48.18	60	70.53	80.4	
	221	17.42	35.27	47.12	65.4	74.2	82.2	
	310	25.35	49.8	58.92	75.03	82.58		
	311	10.02	42.4	60.5	75.75			
	320	25.07	37.62	55.55	63.08	83.5		
	321	10.9	29.2	40.2	49.1	56.93		
		70.9	77.4	83.73	90			
221	221	0	27.27	38.95	63.62	83.62	90	
	310	32.52	42.45	58.2	65.07	83.95		
	311	25.23	47.28	59.83	72.45	84.23		
	320	22.4	42.3	49.67	68.3	79.35	84.7	
	321	11.48	27.02	36.7	57.68	63.55	74.5	
		79.44	84.88					
310	310	0	25.85	36.87	53.13	72.55	84.27	90
	311	17.55	40.28	55.1	67.68	79.02	90	
	320	15.25	47.87	52.13	58.25	74.75	79.9	
	321	21.62	32.32	40.48	47.47	53.73	59.53	
		65	85.15	90				
311	311	0	35.1	50.48	62.97	84.78		
	320	23.1		41.18	54.12	65.28	75.47	85.2
	321	14.77	36.32	49.87	61.08	71.2	80.73	
320	320	0	22.62	46.18	62.52	67.38	72.08	90
	321	15.5	27.18	35.38	48.15	53.62	58.75	68.25
		72.75	77.15	85.75	90			
321	321	0	21.78	31	38.22	44.42	50	60
		64.62	69.07	73.4	81.78	85.9		

*Values of angles are indicated in the decimal system.

Fig. D.1: Angles between crystallographic planes $(HKL\}$ and (hkl) and directions ⟨HKL⟩ and $\langle hk\ell \rangle$ in crystals of cubic system. The values are represented in the decimal system, that is, 74.5 means $74^\circ 30'$ in the sexagesimal system.

Appendix E

Crystal system:

Angle between $(h_1k_1l_1)$ and $(h_2k_2l_2)$, ϕ
Angle between $[u_1v_1w_1]$ and $[u_2v_2w_2]$, ρ

Cubic:

$$\cos\phi = \frac{h_1h_2 + k_1k_2 + l_1l_2}{\{(h_1^2 + k_1^2 + l_1^2)(h_2^2 + k_2^2 + l_2^2)\}^{1/2}}$$

$$\cos\rho = \frac{u_1u_2 + v_1v_2 + w_1w_2}{\{(u_1^2 + v_1^2 + w_1^2)(u_2^2 + v_2^2 + w_2^2)\}^{1/2}}$$

Tetragonal:

$$\cos\phi = \frac{\frac{1}{a^2}(h_1h_2 + k_1k_2) + \frac{1}{c^2}l_1l_2}{\left\{\left(\frac{1}{a^2}(h_1^2 + k_1^2) + \frac{1}{c^2}l_1^2\right)\left(\frac{1}{a^2}(h_2^2 + k_2^2) + \frac{1}{c^2}l_2^2\right)\right\}^{1/2}}$$

$$\cos\rho = \frac{a^2(u_1u_2 + v_1v_2) + c^2w_1w_2}{\{(a^2(u_1^2 + u_1^2) + c^2w_1^2)(a^2(u_2^2 + u_2^2) + c^2w_2^2)\}^{1/2}}$$

Orthorhombic:

$$\cos\phi = \frac{\frac{1}{a^2}h_1h_2 + \frac{1}{b^2}k_1k_2 + \frac{1}{c^2}l_1l_2}{\left\{\left(\frac{1}{a^2}h_1^2 + \frac{1}{b^2}k_1^2 + \frac{1}{c^2}l_1^2\right)\left(\frac{1}{a^2}h_2^2 + \frac{1}{b^2}k_2^2 + \frac{1}{c^2}l_2^2\right)\right\}^{1/2}}$$

$$\cos\rho = \frac{a^2u_1u_2 + b^2v_1v_2 + c^2w_1w_2}{\left\{\left(a^2u_1^2 + b^2v_1^2 + c^2w\right)\left(a^2u_2^2 + b^2v_2^2 + c^2w_2^2\right)\right\}^{1/2}}$$

Hexagonal:

$$\cos\phi = \frac{h_1h_2 + k_1k_2 + \frac{1}{2}(h_1k_2 + k_1h_2) + \frac{3}{4}\frac{a^2}{c^2}l_1l_2}{\left\{\left(h_1^2 + k_1^2 + h_1k_1 + \frac{3}{4}\frac{a^2}{c^2}l_1^2\right)\left(h_2^2 + k_2^2 + h_2k_2 + \frac{3}{4}\frac{a^2}{c^2}l_2^2\right)\right\}^{1/2}}$$

$$\cos\rho = \frac{u_1u_2 + v_1v_2 - \frac{1}{2}(u_1v_2 + v_1u_2) + \frac{c^2}{a^2}w_1w_2}{\left\{\left(u_1^2 + v_1^2 - u_1v_1 + \frac{c^2}{a^2}w_1^2\right)\left(u_2^2 + v_2^2 - u_2v_2 + \frac{c^2}{a^2}w_2^2\right)\right\}^{1/2}}$$

Rhombohedral: Convert to corresponding hexagonal indices *) and use the above two formulae.

Monoclinic:

$$\cos\phi = \frac{\frac{1}{a^2}h_1h_2 + \frac{1}{b^2}k_1k_2\sin^2\beta + \frac{1}{c^2}l_1l_2 - ac\,(w_1u_2 + w_2u_1)\cos\beta}{\left\{\left(\frac{1}{a^2}h_1^2 + \frac{1}{b^2}k_1^2\sin^2\beta + \frac{1}{c^2}l_1^2 - \frac{2h_1l_1}{ac}\cos\beta\right) \times \left(\frac{1}{a^2}h_2^2 + \frac{1}{b^2}k_2^2\sin^2\beta + \frac{1}{c^2}l_2^2 - \frac{2h_2l_2}{ac}\cos\beta\right)\right\}^{1/2}}$$

$$\cos\rho = \frac{a^2u_1u_2 + b^2v_1v_2 + c^2w_1w_2 - \frac{1}{ac}\,(l_1h_2 + l_2h_1)\cos\beta}{\left\{\left(a^2u_1^2 + b^2v_1^2 + c^2w_1^2 + 2acu_1w_1\cos\beta\right) \times \left(a^2u_2^2 + b^2v_2^2 + c^2w_2^2 + 2acu_2w_2\cos\beta\right)\right\}^{1/2}}$$

Triclinic:

$$\cos\phi = \frac{F}{A_{h_1k_1l_1}A_{h_2k_2l_2}}$$

where

$$\begin{aligned} F = {} & h_1h_2b^2c^2\sin^2\alpha + k_1k_2a^2c^2\sin^2\beta + l_1l_2a^2b^2\sin^2\gamma \\ & + abc^2\,(\cos\alpha\cos\beta - \cos\gamma)\,(k_1h_2 + h_1k_2) \\ & + ab^2c\,(\cos\gamma\cos\alpha - \cos\beta)\,(h_1l_2 + l_1h_2) \\ & + a^2bc\,(\cos\beta\cos\gamma - \cos\alpha)\,(k_1l_2 + l_1k_2) \end{aligned}$$

and

$$\begin{aligned} A_{hkl} = {} & \{h^2b^2c^2\sin^2\alpha + k^2a^2c^2\sin^2\beta + l^2a^2b^2\sin^2\gamma \\ & + 2hkabc^2\,(\cos\alpha\cos\beta - \cos\gamma) + 2hlab^2c\,(\cos\gamma\cos\alpha - \cos\beta) \\ & + 2kla^2bc\,(\cos\beta\cos\gamma - \cos\alpha)\}^{1/2} \end{aligned}$$

$$\cos\rho = \frac{L}{I_{u_1v_1w_1}I_{u_2v_2w_2}}$$

where

$$\begin{aligned} L = {} & a^2u_1u_2 + b^2v_1v_2 + c^2w_1w_2 + bc\,(v_1w_2 + w_1v_2)\cos\alpha \\ & + ac\,(w_1u_2 + u_1w_2)\cos\beta + ab\,(u_1v_2 + v_1u_2)\cos\gamma \end{aligned}$$

and

$$I_{uvw} = (a^2u^2 + b^2v^2 + c^2w^2 + 2bcvw\cos\alpha \\ +2cawu\cos\beta + +2cawu\cos\gamma)^{1/2}$$

*Relationship between

Miller indices in rhombohedral system $(h_Rk_Rl_R)$ and Miller indices in hexagonal system (hkl) is as follows:

$$h_R = \frac{1}{3}(2h+k+l), \quad k_R = \frac{1}{3}(-h+k+l), \quad l_R = \frac{1}{3}(-h-2k+l)$$

Appendix F

Crystal system	Lattice spacing of (hkl) plane
Cubic	
$a = b = c$	
$\alpha = \beta = \gamma = 90°$	$\frac{1}{d^2} = \frac{1}{a^2}\left(h^2 + k^2 + l^2\right)$
Tetragonal	
$a = b \neq c$	
$\alpha = \beta = \gamma = 90°$	$\frac{1}{d^2} = \frac{1}{a^2}\left(h^2 + k^2\right) + \frac{1}{c^2}l^2$
Orthorhombic	
$a \neq b \neq c$	$\frac{1}{d^2} = \frac{1}{a^2}h^2 + \frac{1}{b^2}k^2 + \frac{1}{c^2}l^2$
Hexagonal	
$a = b \neq c$	
$\alpha = \beta = 90°;\ \gamma = 120°$	$\frac{1}{d^2} = \frac{4}{3a^2}\left(h^2 + hk + k^2\right) + \frac{1}{c^2}l^2$
Rhombohedral	
$a = b = c$	
$\alpha = \beta = \gamma < 120° \neq 90°$	$\frac{1}{d^2} = \frac{1}{a^2}\frac{(1+\cos\alpha)\{(h^2+k^2+l^2)-(1-\tan^2\frac{1}{2}\alpha)(hk+kl+lh)\}}{1+\cos\alpha-2\cos^2\alpha}$
Monoclinic	
$a \neq b \neq c$	$\frac{1}{d^2} = \frac{1}{a^2}\frac{h^2}{\sin^2\beta} + \frac{1}{b^2}k^2 + \frac{1}{c^2}\frac{l^2}{\sin^2\beta} - \frac{2hk\cos\beta}{a c\sin^2\beta}$
$\alpha = \gamma = 90° \neq \beta$	
Triclinic	
$a \neq b \neq c$	
$\alpha \neq \beta \neq \gamma$	$\frac{1}{d^2} = \frac{1}{V^2}(s_{11}h^2 + s_{22}k^2 + s_{33}l^2 + 2s_{12}hk + 2s_{23}kl + 2s_{31}lk)$,

where

$$V^2 = a^2b^2c^2\left(1 - \cos^2\alpha - \cos^2\beta - \cos^2\gamma + 2\cos\alpha\cos\beta\cos\gamma\right)$$

and

$$s_{11} = b^2c^2\sin^2\alpha \quad s_{22} = a^2c^2\sin^2\beta \quad s_{33} = a^2b^2\sin^2\gamma$$
$$s_{12} = abc^2(\cos\alpha\cos\beta - \cos\gamma) \quad s_{23} = a^2bc(\cos\beta\cos\gamma - \cos\alpha)$$
$$s_{31} = ab^2c(\cos\gamma\cos\alpha - \cos\beta)$$

Appendix G

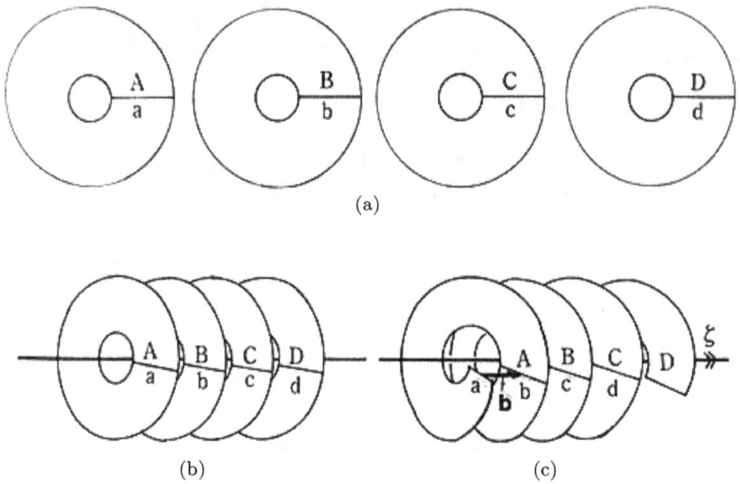

Fig. G.1: Structure of a screw dislocation.

Appendix H

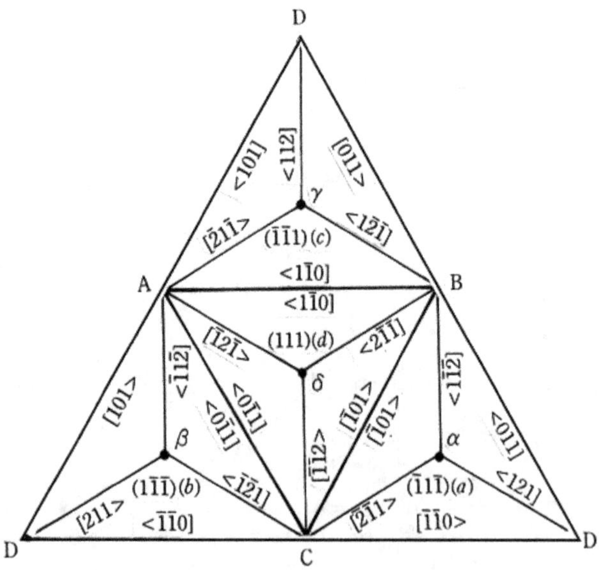

Fig. H.1: Thompson's tetrahedron.

Appendix I

Table I.1: Diffracting planes in crystals of cubic system.

				$\sqrt{N}$				$\sqrt{N}\sqrt{2}$	$\sqrt{N}/\sqrt{2}$	
h	k	l	$N = h^2 + k^2 + l^2$		BCC	FCC	Diamond	BCC	FCC	Diamond
1	0	0	1	1.000						
1	1	0	2	1.414	1.414			1.000		
1	1	1	3	1.732		1.732	1.732		1.000	1.000
2	0	0	4	2.000	2.000	2.000	2.000	1.414	1.155	1.155
2	1	0	5	2.236						
2	1	1	6	2.449	2.449			1.732		
2	2	0	8	2.828	2.828	2.828	2.828	2.000	1.633	1.633
3	0	0	9	3.000						
2	2	1	9	3.000						
3	1	0	10	3.162	3.162			2.236		
3	1	1	11	3.317		3.317	3.317		1.915	1.915
2	2	2	12	3.464	3.464	3.464		2.449	2.000	
3	2	0	13	3.606						
3	2	1	14	3.742	3.742			2.646		
4	0	0	16	4.000	4.000	4.000	4.000	2.828	2.309	2.309
4	1	0	17	4.123						
3	2	2	17	4.123						
4	1	1	18	4.243	4.243			3.000		
3	3	0	18	4.243	4.243			3.000		
3	3	1	19	4.359		4.359	4.359		2.517	2.517
4	2	0	20	4.472	4.472	4.472	4.472	3.162	2.582	2.582
4	2	1	21	4.583						
3	3	2	22	4.690	4.690			3.317		
4	2	2	24	4.899	4.899	4.899	4.899	3.464	2.828	2.828
5	0	0	25	5.000						
4	3	0	25	5.000						
5	1	0	26	5.099	5.099			3.605		
4	3	1	26	5.099	5.099			3.605		
5	1	1	27	5.196		5.196	5.196		3.000	3.000

(*Continued*)

Table I.1: (*Continued*)

h	k	l	$N = h^2 + k^2 + l^2$	$\sqrt{N}$				$\sqrt{N}\sqrt{2}$	$\sqrt{N}/\sqrt{2}$	
					BCC	FCC	Diamond	BCC	FCC	Diamond
3	3	3	27	5.196		5.196	5.196		3.000	3.000
4	3	2	29	5.386						
5	2	0	29	5.477						
5	2	1	30	5.477	5.477			3.873		
4	4	0	32	5.657	5.657	5.657	5.657	4.000	3.266	3.266
5	2	2	33	5.744						
4	4	1	33	5.745						
5	3	0	34	5.831	5.831			4.123		
4	3	3	34	5.831	5.831			4.123		
5	3	1	35	5.916		5.916	5.916		3.416	3.416
6	0	0	36	6.000	6.000	6.000	6.000	4.243	3.464	3.464
4	4	2	36	6.000	6.000	6.000	6.000	4.243	3.464	3.464

Table I.2: Occurrence of reflections for HCP crystal structure

Common allowed reflections			Common forbidden reflections (sometimes these occur by double reflection, depending on **B**.)
$10\bar{1}0$	$3\bar{2}11$	$20\bar{2}3$	0001
$1\bar{2}10$	$2\bar{2}01$	$\bar{2}4\bar{2}3$	0003
$20\bar{2}0$	$\bar{2}112$	$10\bar{1}4$	$1\bar{2}11$
$12\bar{3}0$	$01\bar{1}2$	$\bar{2}114$	$1\bar{2}13$
$30\bar{3}0$	0002	$10\bar{1}5$	$\bar{3}033$
$13\bar{4}0$	$11\bar{2}2$	$\bar{1}2\bar{1}6$	$1\bar{2}15$
$5\bar{1}\bar{4}0$	$21\bar{3}2$		
$01\bar{1}1$	$10\bar{1}3$		

Appendix J

FCC/L1_2

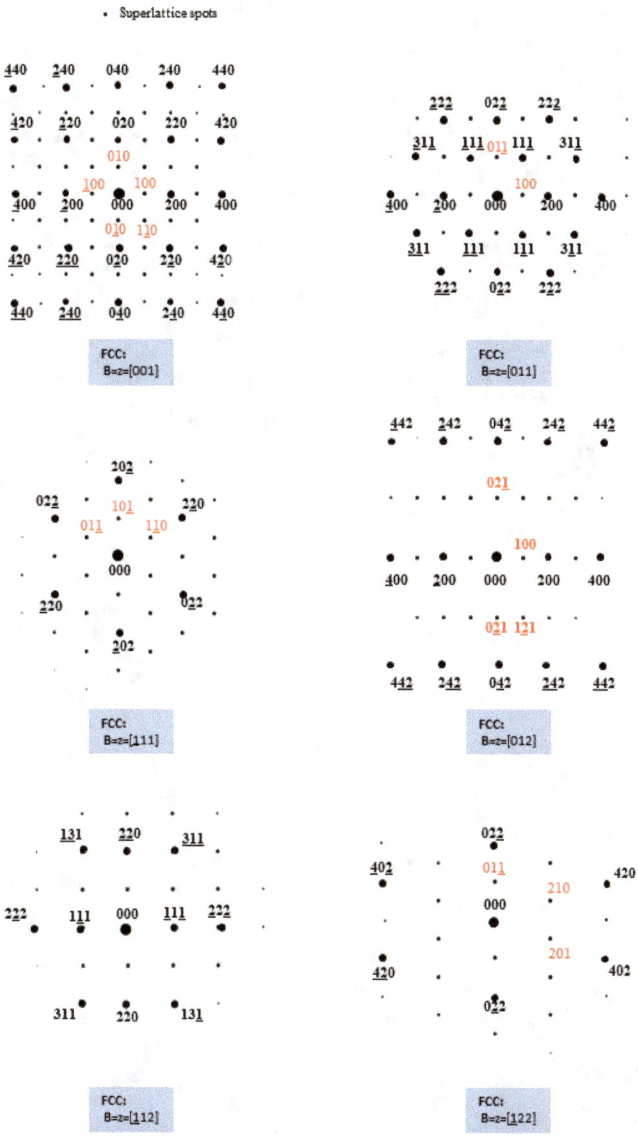

Fig. J.1: Some examples of electron diffraction patterns for the FCC crystal structure. Smaller circles indicate positions of the forbidden spots for the ordered FCC(L1_2).

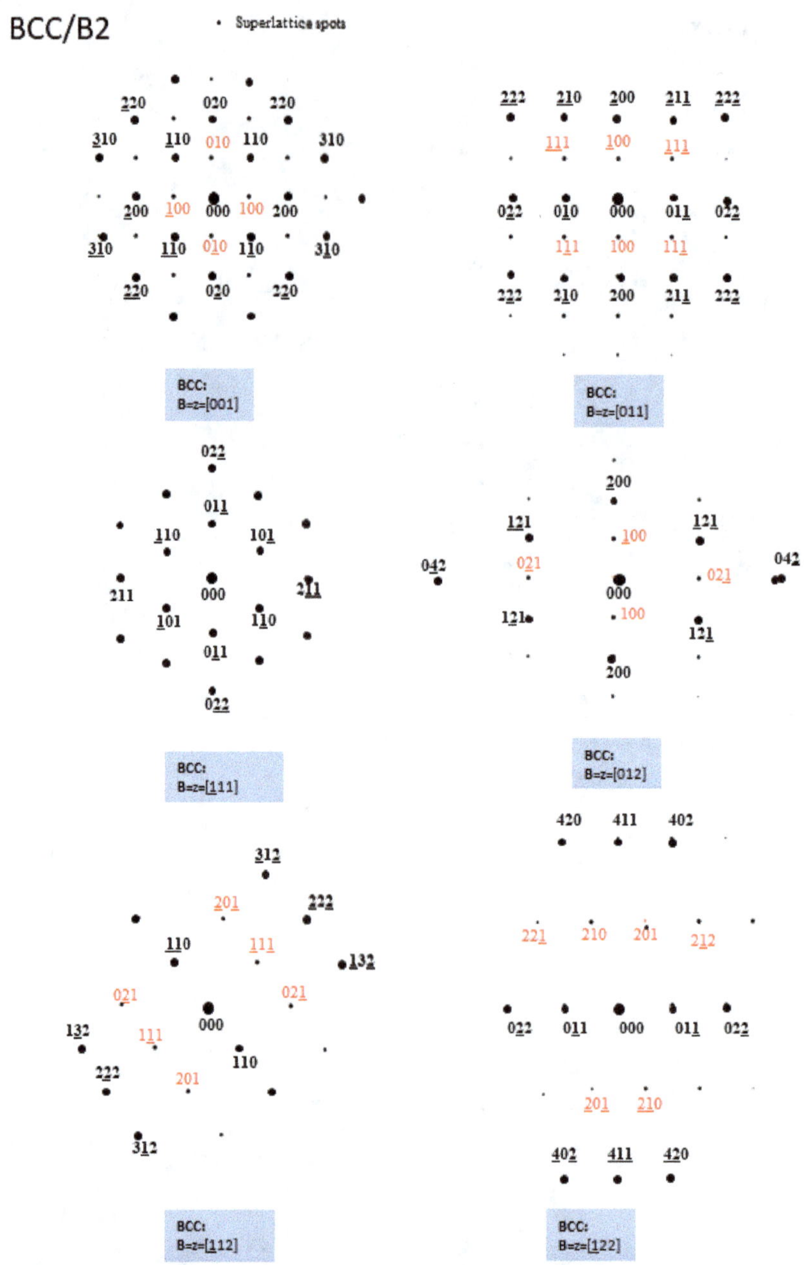

Fig. J.2: Some examples of electron diffraction patterns for the BCC crystal structure. Smaller circles indicate positions of the forbidden spots for the ordered BCC(B2).

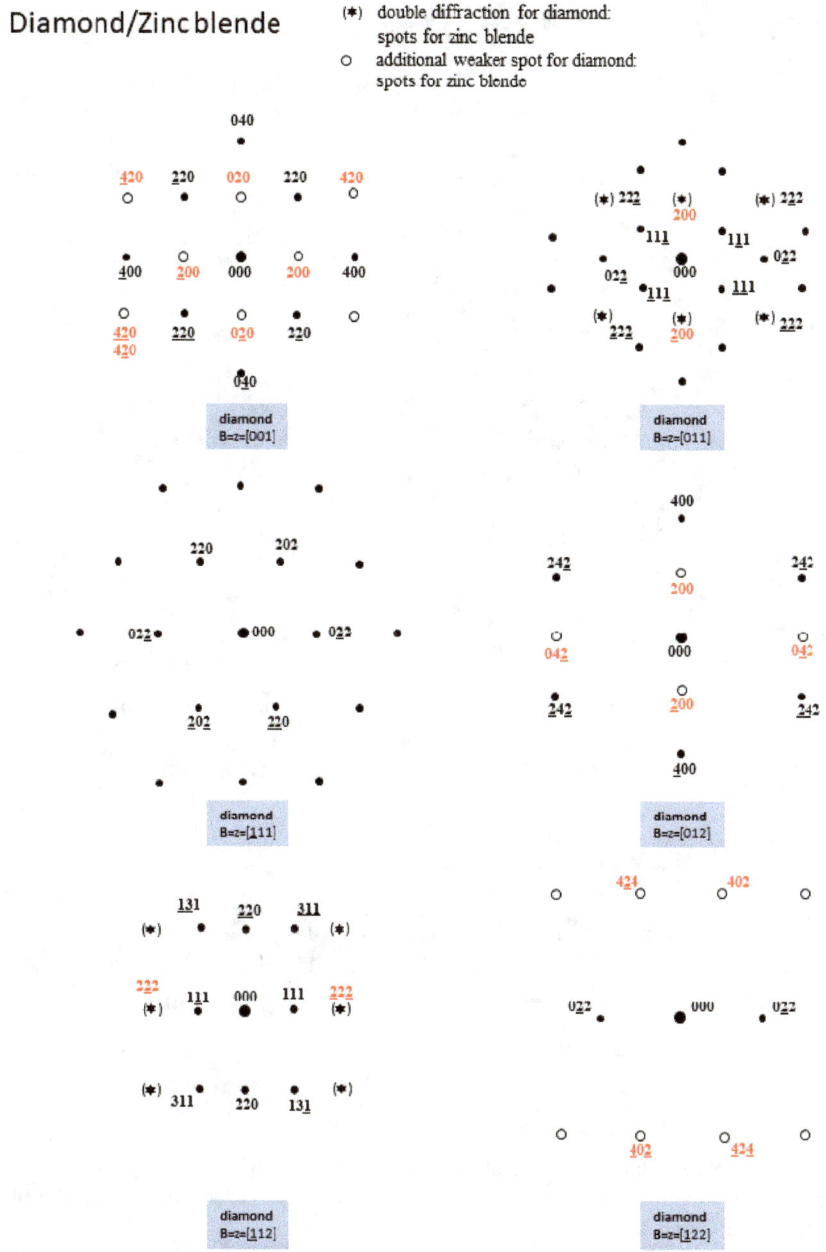

Fig. J.3: Some examples of electron diffraction patterns for the diamond crystal structure. ○ additional weaker spots for diamond, spots for zinc blende. ∗ spots that can occur due to double diffraction for diamond, spots for zinc blende.

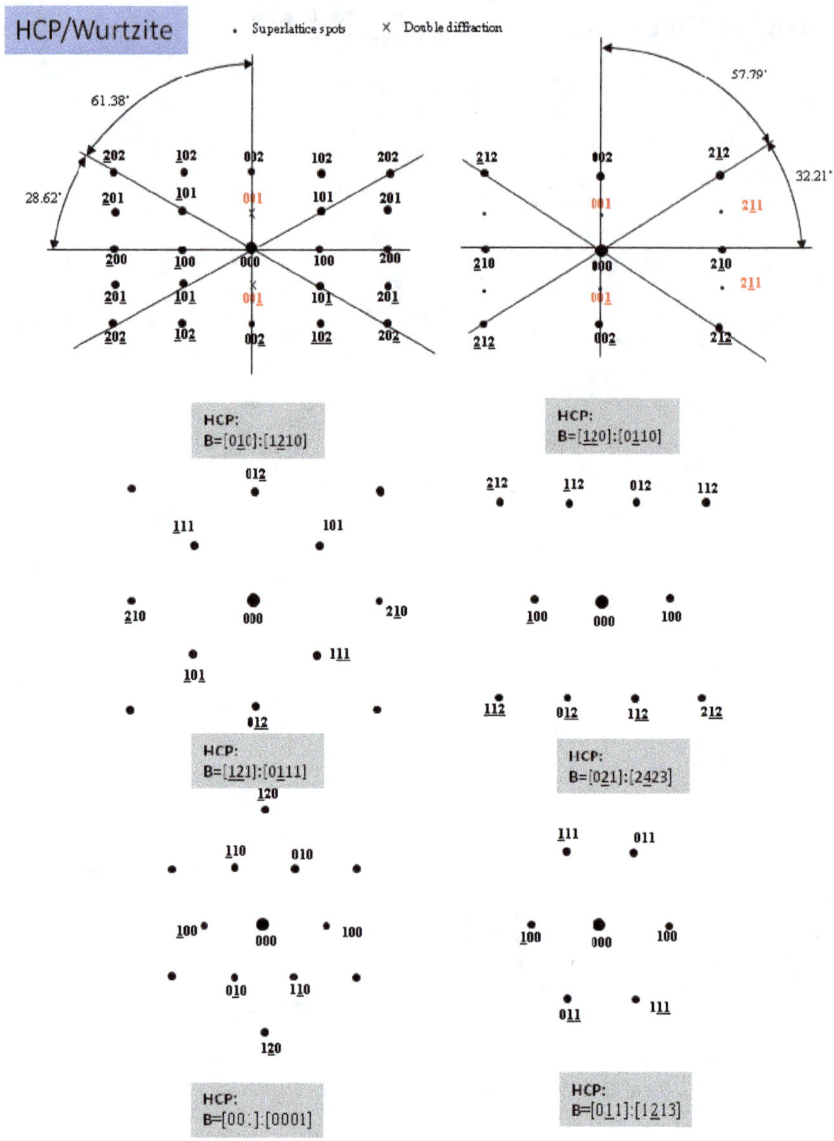

Fig. J.4: Some examples of electron diffraction patterns for HCP crystal structure. Smaller circles: spots for wurtzite, x: spots that can occur due to double diffraction for HCP and spots for wurtzite.

Appendix K

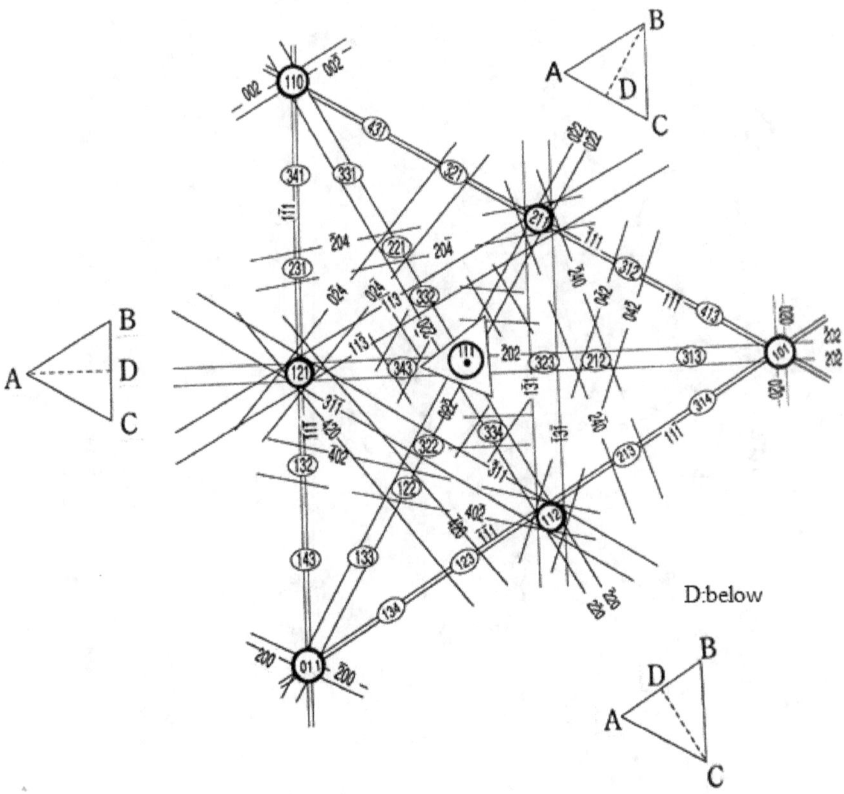

Fig. K.1: Kikuchi map around [111] of FCC crystal.

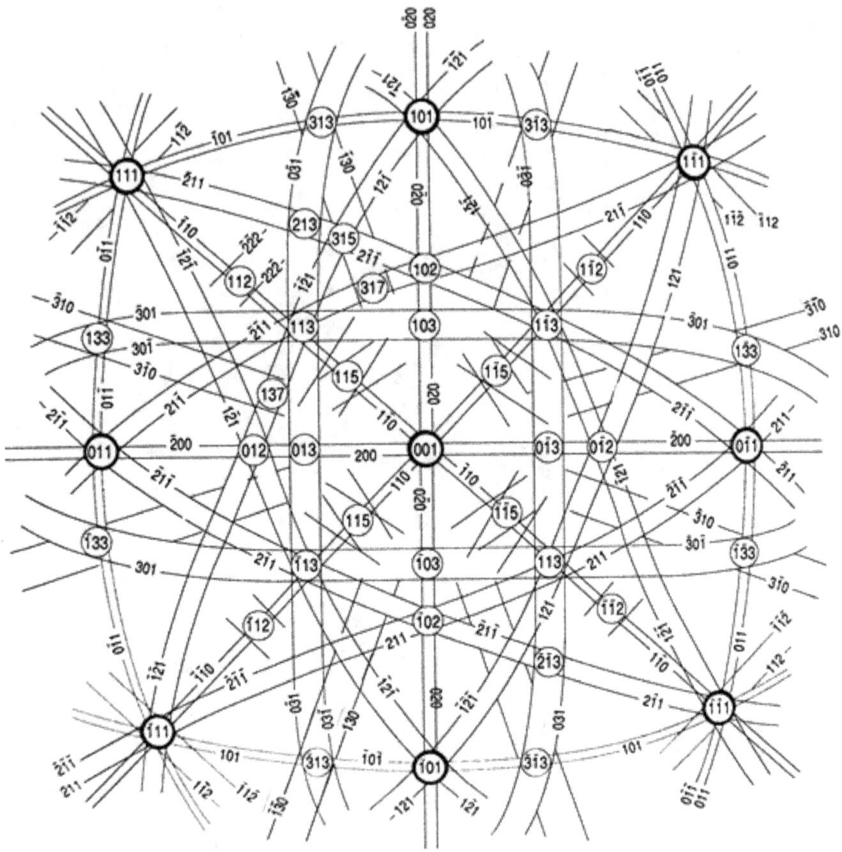

Fig. K.2: Kikuchi map around [001] of BCC crystal.

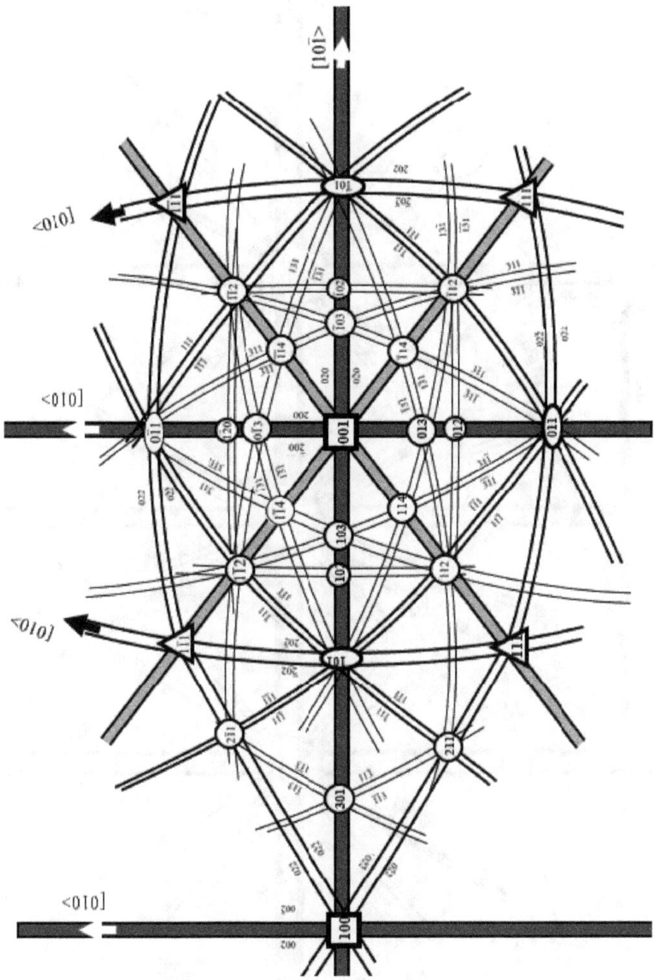

Fig. K.3: Kikuchi map ranging from [001] to [100] in FCC crystal.

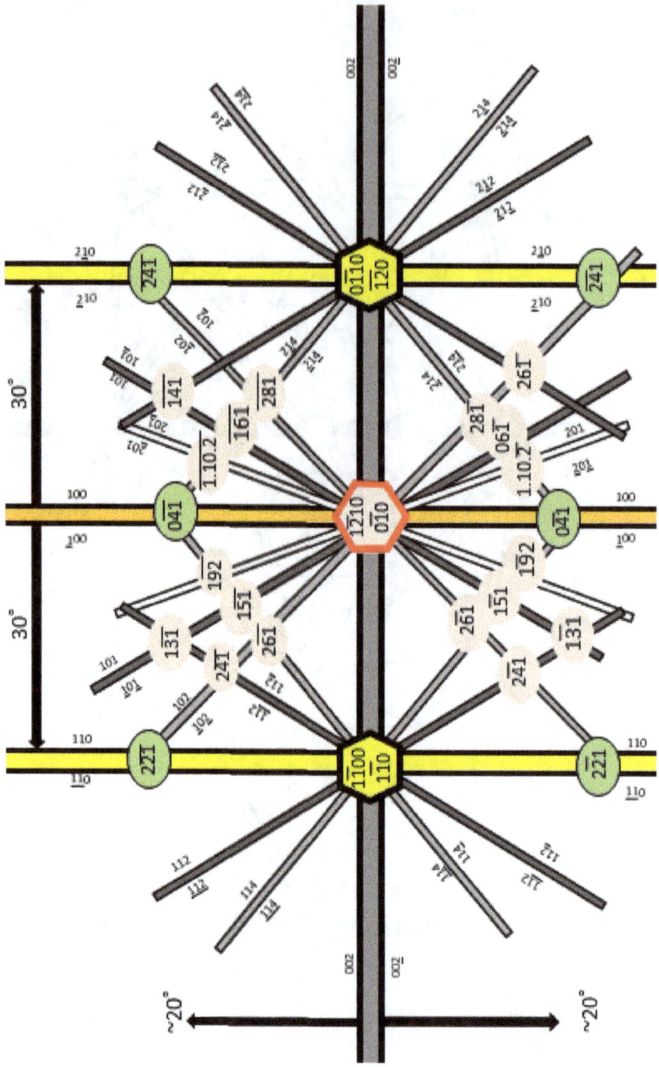

Fig. K.4: Kikuchi map ranging from $[1\bar{1}0]$ to $[\bar{1}20]$ in HCP crystal.

Appendix L: Determining Orientation of Single Crystal by Back-Scattered Laue Method

Figure L.1 shows the set-up of back scattered Laue camera. By analyzing the positions of Laue spot, it is possible to determine the normal of that crystallographic plane (*hkl*) which produces the Laue spot, or pole N. Co-ordinate of N is represented either by (δ, γ) or by (μ, σ). The former corresponds to the Wulff net and the latter to the polar net. Needless to say, they are equivalent, however, (μ, σ) is more convenient when the crystal globe is used, and (δ, γ) is more convenient when the Wulff net is used. Angular relationship between zones and planes in (δ, γ) is read using a chart, Greninger chart: The Greninger chart is shown in Fig. L.2.

In what follows how to determine the orientation by a Wulff net is illustrated using a chart shown in Fig. L.3 as an example.

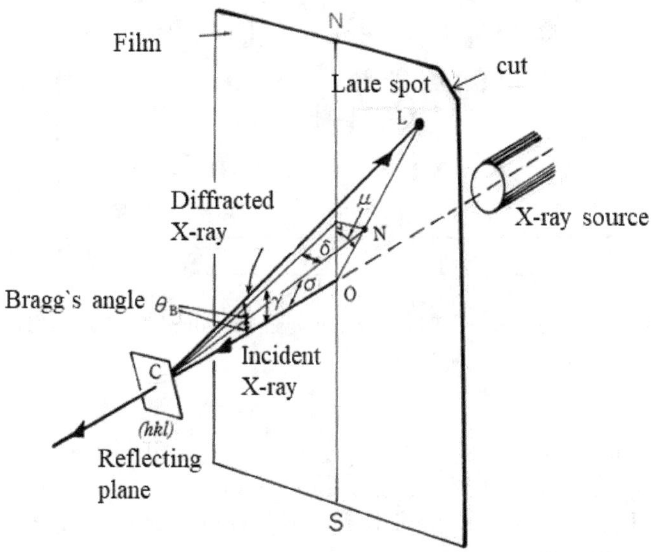

Fig. L.1: Set-up of back-scattered Laue camera. Cut the corner at the upper right when viewed from the crystal. From the position of Laue spot, the normal of (*hkl*) plane, N, which produces the spot is determined. To express N, there are two methods available, that is (δ, γ) and (μ, σ).

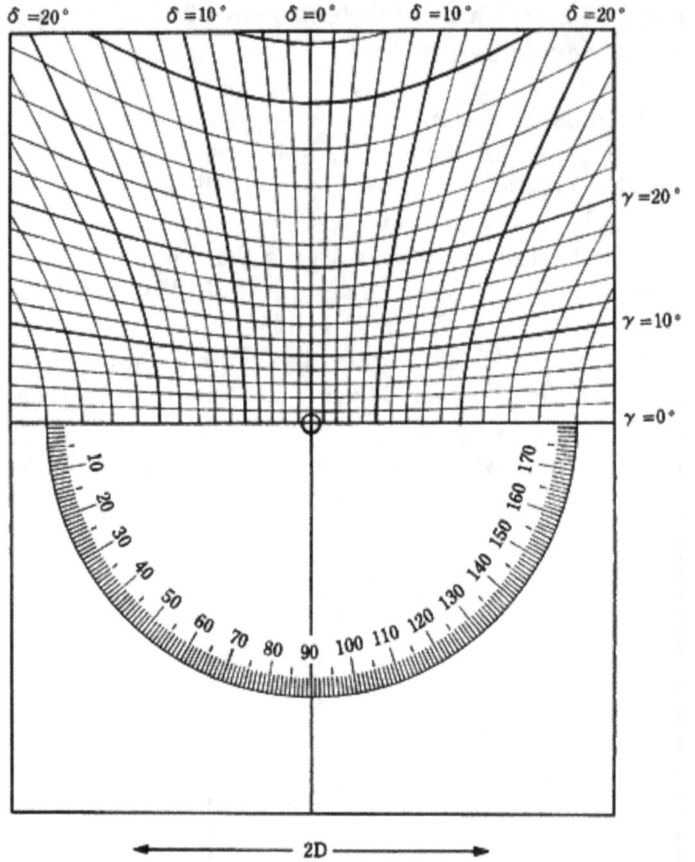

Fig. L.2: Greninger chart. Distance between the film and crystal (D = 3 cm).

(1) In order to ensure the orientation of the film, the upper-right corner is cut and set to the Laue camera. As can be seen from Fig. L.1, this corresponds to the experimenter viewing the crystal from outside.

(2) Put the Greninger chart on the film in such a way that the centers of the Greninger chart and the film coincide.

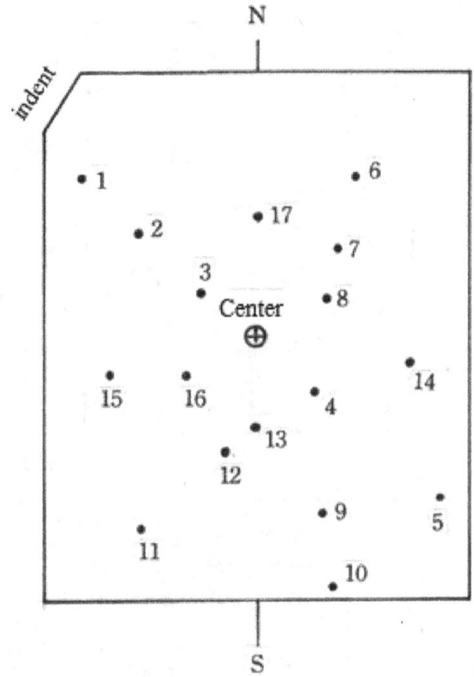

Fig. L.3: An example of Laue spots. The indent is to be at the upper-left.

(3) Read the co-ordinates (δ, γ) for the Laue spots and plot them on a tracing paper which is put on the Greninger chart (Fig. L.4).

(4) Poles projected on the tracing paper lie on several sets of great circles. This is because they belong to one of the zone axes as shown in Fig. L.5. In the case shown in Fig. L.5, four zone axes P_A, P_B, P_C and P_D are determined. Great circles connecting the zone axes P_A, P_B, P_C and P_D give the zone axis P_E, which is coincident with pole 4.

(5) Referring to the standard stereographic projection, assume candidates of concrete orientations for P_A, P_B, P_C and P_D. For

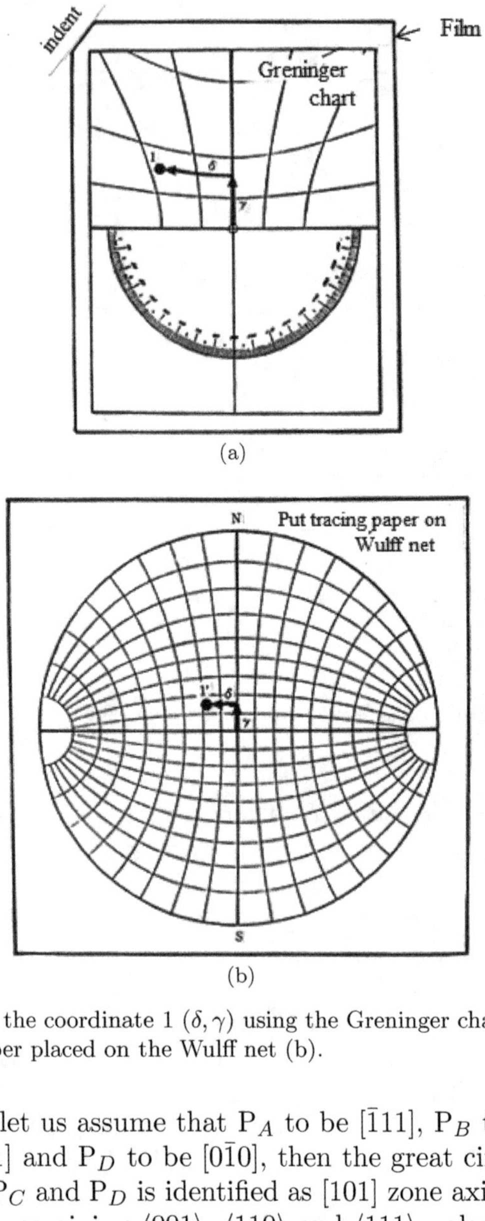

Fig. L.4: Read the coordinate 1 (δ, γ) using the Greninger chart (a) and plot it on a tracing paper placed on the Wulff net (b).

instance, let us assume that P_A to be $[\bar{1}11]$, P_B to be $[\bar{1}01]$, P_C to be $[111]$ and P_D to be $[0\bar{1}0]$, then the great circle connecting P_A, P_B, P_C and P_D is identified as [101] zone axis (Fig. L.6(a)). Draw the remaining $\langle 001 \rangle$, $\langle 110 \rangle$ and $\langle 111 \rangle$ poles and great circles connecting the standard triangle. Here, ■, ⬬ and ▲ are poles identified by the aforementioned operation.

(6) Finally, this is converted to 001 standard projection (Fig. L.6(b)).

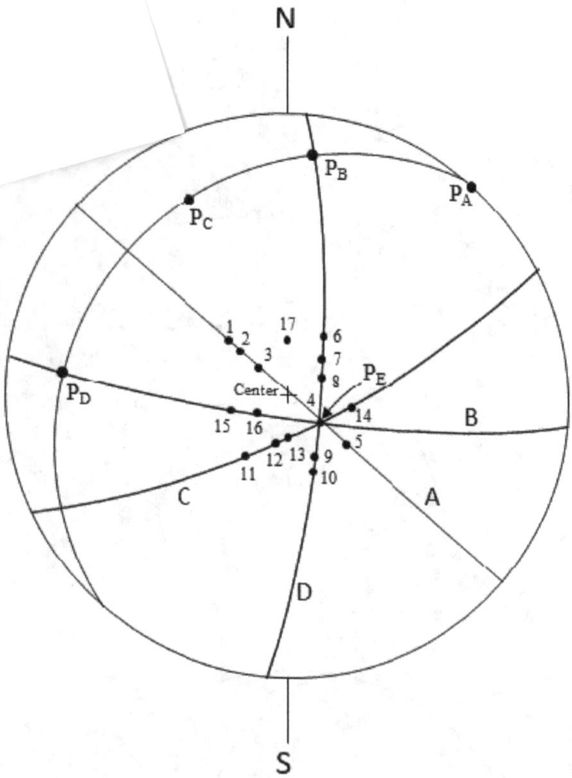

Fig. L.5: Stereographic projection obtained from Laue spots shown in Fig. L.3. The zone axes of great circles A, B, C and D are P_A, P_B, P_C and P_D, respectively.

In the aforementioned operation, it is more important to determine the zone axes P_A, P_B, P_C and P_D than to index individual Laue spots. Therefore, as illustrated in Figs. L.7(a) and L.7(b), finding out a set of Laue spots which lie on a single great circle, the zone axes P_A, P_B, P_C and P_D can be determined directly. Indeed, experienced researcher can determine the orientation by this method swiftly.

Points:
54° is an angle between $\langle 111 \rangle$ and $\langle 001 \rangle$.
35° is an angle between $\langle 011 \rangle$ and $\langle 111 \rangle$.
70° is an angle between $\langle 111 \rangle$ and $\langle 111 \rangle$.

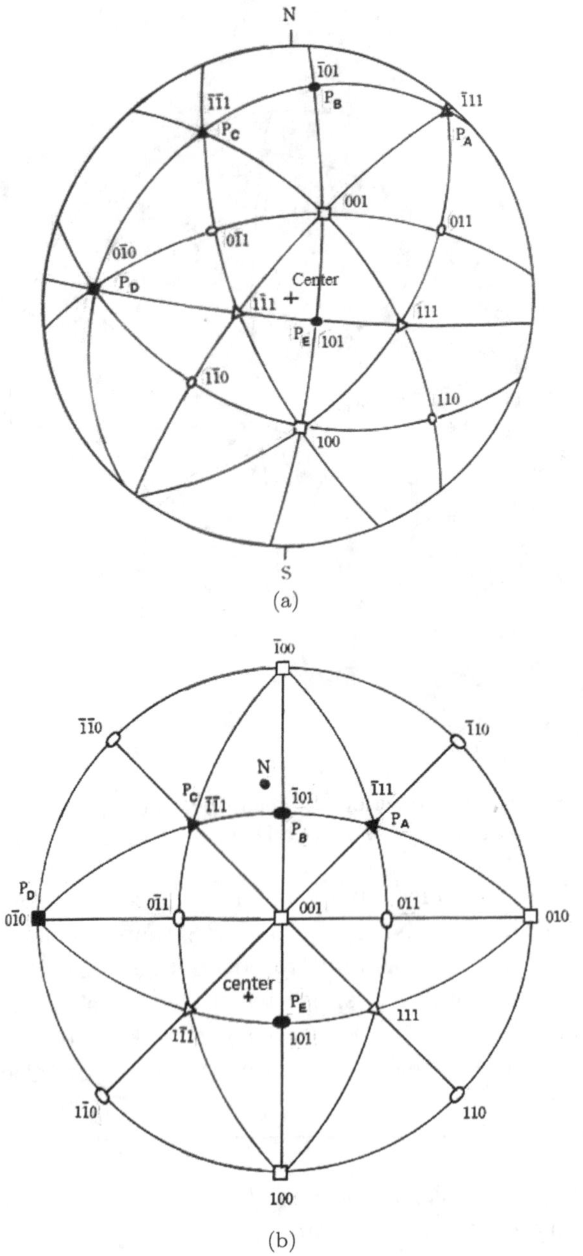

Fig. L.6: (a) Indexing the zone axes P_A, P_B, P_C and P_D, shown in Fig. L.5. (b) Replotting Fig. L.6(a) to the 001 standard projection.

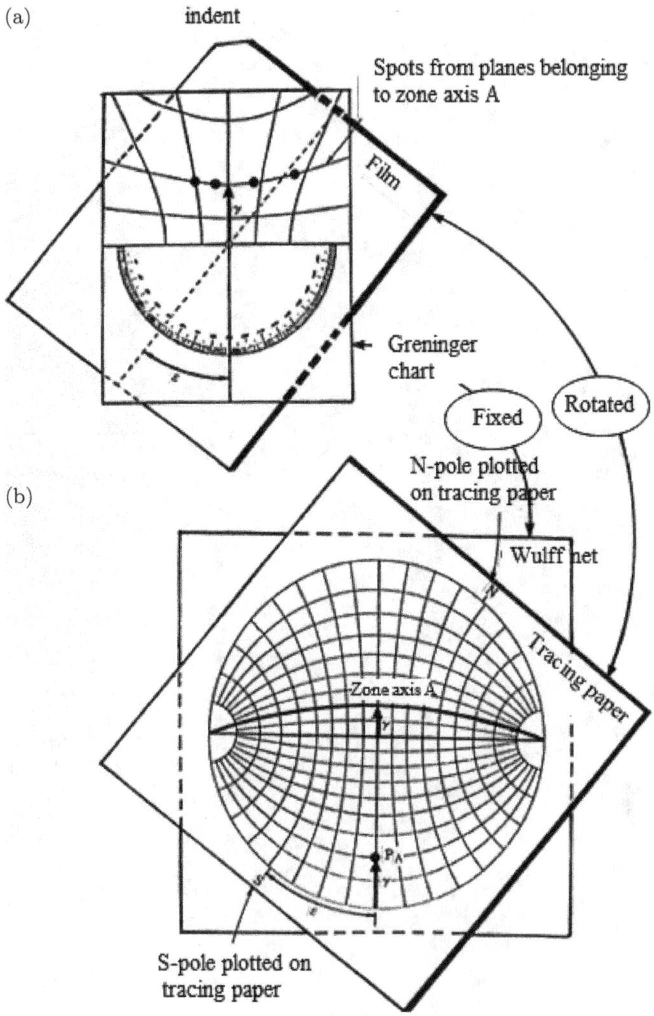

Fig. L.7: Determining zone axes directly.

Bibliography

C. S. Barrett and T. B. Massalxski, *Structure of Metals* (McGraw-Hill, 1971).

D. Cherns and S. R. Preston, *Proc. 11th Intl. Cong. Electron Microscopy*, Vol. 1, Kyoto, 1968 (eds., T. Imura, S. Maruse and T. Suzuki, *Jpn. Soc. Electron Microscopy*, Tokyo, 1986, p. 721).

D. J. H. Cockayne, M. L. Jenkins and I. L. F. Ray, *Philo. Mag.*, **24** (1971), 1385.

J. W. Edington, *Monographs in Practical Electron Microscopy in Materials Science*, Vol. 4, *Typical Electron Microscope Investigations*, (The Macmillan Press Ltd., London, 1976).

J. Friedel, *Dislocations* (Pergamnon, NY, 1967).

R. Gever, A. Art and S. Amelinckx, *Phys. Stat. Soli.*, **3** (1963), 1563.

H. Hashimoto, A. Howie and M. J. Whelan, *Proc. Royal Society (London) A*, **269** (1962), 80.

P. B. Hirsch, A. Howie, R. B. Nicholson, D. W. Pashley and M. J. Whelan, *Electron Microscopy of Thin Crystals*, 1st edn. (Butterworth, 1965) (2nd edn., Robert E. Krieger, 1977).

J. P. Hirth and J. Lothe, *Theory of Dislocations*, 2nd edn. (John-Wiley and Sons, 1982).

C. J. Humphreys, *Understanding Materials* (Maney, London, 2002).

Y. Ishida, H. Ishida, K. Kohra and H. Ichinose, *Philo. Mag. A*, **42** (1980), 453.

H. Iwata, D. Kawaguchi and H. Saka, *Microscopy*, **66** (2017), 328.

H. Iwata and H. Saka, *Philo. Mag. Lett.*, **97** (2017), 74.

H. Iwata and H. Saka, *Philos. Mag.*, **99** (2019), 1849.

H. Kakibayashi, F. Nagata and Y. Ono, *Jpn. J. Appl. Phys.*, **26** (1987), 770.

A. Muroga and H. Saka, *Scripta Material*, **31** (1995), 151.

J. Nishigaki, K. Kuroda and H. Saka, *Phy. Stat. Sol. A*, **128** (1991), 319.

S. Onda, H. Watanabe, T. Okamoto, H. Kondo, H. Uehigashi and H. Saka, *Philo. Mag. Lett.*, **95** (2015), 489.

H. Saka, *Philo. Mag. A*, **49** (1984), 327.

H. Saka, *Classical Theory of Crystal Dislocation — From Iron to Gallium Nitride* (World Scientific Publishing, Singapore, 2017).

H. Saka, Y. Sueki and T. Imura, *Phil. Mag. A*, **37** (1978), 273.

E. Spiecker and W. Jäger, *J. Phys.: Cond. Matter*, **14** (2002), 12767.

Suprijadi, H. Saka and S. Arai, *Phil. Mag. Lett.*, **82** (2002) 133.

M. Tanaka, M. Terauchi and K. Tsuda, *Electron Diffraction and Elementary Crystallography, Indexing of Electron Diffraction Patterns and Use of Convergent Electron Diffraction*, 2nd edn. (Kyoritu Shuppans, Tokyo, 2014).

N. Torazawa, S. Arai, Y. Takase, K. Sasaki and H. Saka, *Material Trans.*, **44** (2000), 1438.

D. A. Williams and C. B. Carter, *Transmission Electron Microscopy* (Plenym, 1966).

Index

A

absolute values of Burgers vectors
  LACBED by, 253–254
  Spiecker–Jäger's method by, 254–256
  thickness contours by, 253
amplitude-phase diagram, 78–80, 82–84, 133–135
anti-phase boundary (APB), 70, 163–166
anti-phase boundary energy (APB energy), 70
atomic scattering factor, 76, 82, 240

B

basis, 2–3
BCC(B2), 4, 14, 47, 83–85, 98, 100, 103, 118–119, 121–122, 163–164, 179, 267–268, 270, 274
B2 (CsCl) structure, 16–17
bend contour, 132–143, 145, 246
  BF image of, 143
  high-order, 201–206
  profile of, 138, 145–146
bend contour method, 255
bending of lattices, 248
biological optical microscope (BOM), 93
Bloch waves, 144, 146
body-centered cubic (BCC) structure, 14
Bragg's condition, 81–83, 86, 111, 113–115, 128–129, 131, 133–134, 137–138, 140, 143–144, 151–152, 169–172, 184, 188–189, 203–204, 206–207, 209, 239, 243, 246–248, 250
Bragg diffraction, 97, 104
Bragg's angle, 96
Bragg's line, 246, 249–253
Bravais lattices, 2–3, 6–7
bright-field (BF) imaging, 127–131, 138–139, 142–143, 145–147, 151, 160–162, 166, 168–170, 172–177, 180, 189–190, 207–209, 211, 215, 217, 219, 221–224, 233–234, 236–237, 240–241
Burgers circuit, 52–53, 55
Burgers vectors, 52–60
  absolute values of, 253–256
  definition of, 52–53
  dissociated dislocation, 66–68
  edge dislocation, 54–55
  FS/RH(Perfect) convention, 53–54
  I-type and V-type dislocation loop, 57–58, 70, 189–196
  Kirchhoff's law of, 55–59
  of mixed dislocation, 54–55

of screw dislocation, 54–55, 183–187
sign of, 187–189

C

camera constant, 97–98, 100–102
camera length, 97–98, 233–234, 239
CAT method, 147–149
Cherns–Preston's rule, 250, 254
coherent interface, 166–169, 178
column approximation, 133
contrast experiment, 185–187, 219–220
contrasts
of dislocations and misleading dislocations, 198–201
of inclined dislocations, 196–198
conventional transmission electron microscope (CTEM), 95
BF imaging, 233
DF imaging in, 233–234
DFWB image in $g^*/3g^*$ condition, 239
dislocation image, 236
features of, 232
image formation, 236
ray paths of, 232
versus STEM, 231
convergent-beam electron diffraction (CBED), 233–245
crystal dislocations
Burgers circuit, 52–53
Burgers vectors, 52–60, 66–68
definition of dislocation, 49–52
dislocation loop, 51
dislocation network, 55–56
displacement around edge dislocation, 60–61
displacement around screw dislocation, 60
edge dislocation, 54–55
elastic deformation, 49–51
extrinsic stacking fault, 68–70
Frank's partial dislocation, 68–70
intrinsic stacking fault, 68–70
Kirchhoff's law, 55–59
mixed dislocation, 54–55
plastic deformation, 49–51
point defects, 57
screw dislocation, 54–55
stacking fault, 65–66
stress–strain curve, 49–50
super-dislocations, 70–72
Thompson tetrahedron, 61–64, 66–68, 113, 161–162, 186, 193–194, 266
crystallographic axes, 23
crystallographic directions, 8, 262
crystallographic planes, 23–24, 262
crystal rotating method, 88
crystal structure, 1
two-dimensional crystal, 3
crystal structures of intermetallic compounds
B2 (CsCl) structure, 16–17
fluorite structure, 20
$L1_2$ (Cu_3Au) structure, 17–18
NaCl structure, 18
wurtzite structure, 19–20
zincblende structure, 19
crystal structures of pure elements
body-centered, 14
diamond structure, 15
face-centered, 13–14
hexagonal closed packed (HCP) structure, 15–16
crystal system
cubic, 263
hexagonal, 263
monoclinic, 264
orthorhombic, 263
rhombohedral, 263
tetragonal, 263
triclinic, 264

D

dark-field (DF) imaging, 129
dark field imaging, 169–170
  of whisker, 177
dark field weak-beam (DFWB) method
  absolute value of Burgers vector, 220–222
  BF and DF images, 207
  dislocation dipoles, 215–218
  dissociated dislocation, 213–217
  elastically anisotropic crystals, 217–220
  $g^*/3g^*$ method, 210–211
  practice of, 209
  principle of, 207–208
  procedures of, 208–212
  systematic reflection, 211
Debye rings, 98–100, 104–105, 180
deformation (*see also* elastic deformation; plastic deformation), 49
deviation parameter, 89–90, 112, 144, 236, 243
  effective, 144
diamond structure, 15
diamond/zinc blende, 271
diffracted beam, 96
diffracting planes, 267–268
diffraction (*see also* electron diffraction; multiple diffraction), 77–79
  dynamical theory of, 143–147
diffraction contrast, 129, 148–152, 233–235
  bright-field imaging, 127–131
  dark-field imaging, 127–131
  general theory, 231–233
  HAADF, 239–242
  objective aperture, 130, 152
  resolution of dislocations, 235–239
  two-beam approximation, 127, 149–150, 201–202
  two-beam condition, 148, 150–151, 199, 237, 243
diffraction disk, 232
diffraction pattern, 127
  of α-Fe, 118–119
  and Kikuchi pattern, 111–112
  from many twins and/or stacking faults, 123
  from single crystal, 99–104
  from specimen containing two crystals, 116–124
  of twin, 119, 120, 122–123
  orientation relationship, 116–124
  Pb particles of, 117–118
diffraction spot, 97
direction and plane, determining, 34–42
dislocation, 51 (*see also* crystal dislocation)
  definition of, 49–52
  helical, 58–59
dislocation loop, 51, 189–196
  I-type and V-type, 57–58, 70, 189–196
dislocation network, 55–56
displacement
  edge dislocation, 60
  screw dislocation, 60
dissociated dislocations
  Burgers vectors of, 66–68
  dark field weak-beam method, 213–217
  extrinsic stacking fault, 68–70
  Frank's partial dislocation, 68–70
  intrinsic stacking fault, 68–70
  Shockley's partial dislocations, 63, 69
  stacking fault, 65–66
  stacking of planes, 61–62
  structure of, 63
  Thompson tetrahedron, 61–64
double diffraction (double reflection), 104
dynamical theory, 143–147, 152–153

Bloch waves, 144–146
transmitted and diffracted beams, 153

E

edge dislocation, 54–58, 60–61, 183–187, 201, 203, 214
displacement around, 60–61
EDX, 231, 240
effects of surface, 226–227
Eshelby twist, 227
slip traces, 227–229
ehp in, 54, 58, 184, 187
elastic deformation, 49–51
elastic scattering, 76, 109
electron beam cone, ray path of, 244
electron diffraction pattern and indexing, 97–107
Debye rings, indexing, 98–100, 104–105, 180
diffraction pattern from single crystal, 99–104
diffraction spot, 97
geometry of reciprocal lattice spot, diffraction spot and Ewald sphere, 97
Laue zone, 105–107, 124–125
multiple diffraction, 104–105
electron microscope (*see also* biological optical microscope (BOM); scanning electron microscope (SEM); transmission electron microscope (TEM))
basic design of, 93–95
electron diffraction and, 95–97
Eshelby twist, 227
Euler's formula, 75
Ewald sphere, 79–82, 86–88, 97, 99, 105–106, 111–112, 130–131, 144, 193, 210, 248
construction, 236, 244
definition of, 81
geometry, 79–80
extended dislocation, 63
extinction law, 84, 98
distance, 136, 144, 147–148, 166, 247
extra half plane, 54, 58, 184, 187
extrinsic stacking fault, 69, 160

F

π-fringes, 166
face centered cubic (FCC), 7, 13–16, 47, 61–62, 65, 69, 72, 84–85, 98–102, 113, 117, 119, 121–122, 125, 159–161, 172, 213, 220–222, 267–269, 273, 275
lattice, 3, 7
structure, 13–14
FCC($L1_2$), 269
finger prints, 98
first order Laue zone (FOLZ), 106
fluorite structure, 20
forbidden reflections, 85
forbidden spot, 104
Frank's partial dislocation, 68–70
Frank's symbol (notation), 69, 119, 163
FS/RH(Perfect) convention, 53–54, 187
fundamental reflections, 85

G

glide dislocation loop, 58
great circle, 23–24
Greninger chart, 277–280

H

HCP/wurtzite, 272
hexagonal closed packed (HCP) structure, 15–16, 268
hexagonal indices, 10–11
high angle annular dark field (HAADF), 239–242
principle of, 239
high-order bend contour and dislocation, 201–206
high-order reflection imaging method, 208
Hooke's law, 49

I

incoherent interface, 168
incoherent precipitate, 168
inelastic scattering, 109
inside/outside method, 195
interstitial (I)-type dislocation loop, 57
intrinsic stacking fault, 69, 159

K

Kikuchi lines, 109–110, 115
Kikuchi map, 113–115, 273–276
Kikuchi pattern, 107–115
  and diffraction pattern, 111–113
  features of, 107–110
  indexing, 110–111
  Kikuchi lines, 109–110, 115
  Kikuchi map and, 113–115
Kirchhoff's law, 55–59
Kurdjumov-Sacks (KS) relationship, 119

L

LACBED (large angle convergent beam electron diffraction)
  absolute values of Burgers vectors, 253–254
  Bragg's line, 246, 249–253
  nature of dislocation, 245–253
  principles of, 247
  two modes of illumination in, 246
lattice constants/parameters, 2
lattice defect, 49, 155
lattice point, 1–2, 18, 88, 97–98, 111
lattice space and unit cell, 5
Laue function, 90, 92, 137
Laue method, 86–87
  co-ordinates, 280
  Greninger chart, 277–278
  Laue spots, 279
  set-up of back scattered, 277
Laue spots, 277, 279, 281
Laue zones, 105–106, 124–125
  indexing, 125
$L1_2$ (Cu_3Au) structure, 17–18

M

measuring angle between two poles, 31–34
Miller indices, 8–10
  crystallographic direction, 8
  definition of (*hkl*) plane, 9
  in hexagonal crystal, 10, 265
  of important planes, 8–9
misfit dislocations, 168
mixed dislocation, 54–55
moiré fringes, 170–171
  parallel moiré, 171
  rotation moiré, 171–172
multiple diffraction, 104–105

N

NaCl structure, 18
Nishiyama's relationship (N relationship), 119

O

optical-axis detector (OAD), 233
orientation relationship, 116, 118–119, 178

P

parallel moiré, 171
path difference, 78
planar defects
  anti-phase boundary (APB), 163–166
  displacement vector of, 156, 159
  general theory, 155–159
  moiré fringes, 170–171
  π-fringes, 166
  stacking faults in FCC crystal, 159–163
  types of, 173–175
plane wave, 73–75
  amplitude of, 73

complex plane in, 75, 78, 134, 137
spherical wave and, 74
wave number vector, 74
plastic deformation, 49–51
point defects, 57–59
polar net, 26, 259
polar sphere, 23–24
pole, 23–24
powder method, 87
precipitates, 166
coherent interface, 166–167
dark field imaging, 169–170
incoherent interface of, 167–168
matrices and, 167, 178
misfit dislocations, 168
of NiAl, 178–179
partial coherent precipitate, 167–168
semi-coherent interface, 167–168
spherical, 166–169
primitive unit cells, 3
prismatic loop, 58
pseudo-coherent interface, 168

R

ray path, 96, 128, 130–131, 134, 223–224, 232, 234, 244, 250
real space, 42–45, 77, 80–81, 101, 110, 113, 130, 245–247
reciprocal lattice, 42, 43, 45–47, 80–81, 85–88, 90, 92, 97–99, 106, 110–111, 125, 130–131, 171
reciprocal lattice points, 88–92
broadening of due to crystal with finite size, 89
deviation parameter, 89–90
relationship between size of crystal, 91
reciprocal space, 42, 44
reciprocal vector
calculating lattice spacing d_{hkl}, 47
definition of vector $\mathbf{g}^*_{\mathrm{hkl}}$, 42–43, 45, 79, 81, 99, 104, 193
real space and, 42–45
reciprocal space, 42, 44
two-dimensional projection, 43
reference sphere, 23–24
crystallographic plane, 24
planes and perpendiculars, 26
polar co-ordinate, 24–25
projection of crystal plane, 24
reflection (Ewald) sphere, 81
residual contrast, 184–185, 193, 201
rhombohedral lattice, 3, 7
rocking curves, 243, 245
beam, 243, 245, 253
crystal, 243, 246, 254
rotation moiré, 171
rotation of crystal, 29–31

S

scanning electron microscope (SEM), 93–95
scanning transmission electron microscope (STEM), 95, 231
advantages and drawbacks of, 231, 240–242
DFWB image in $g^*/3g^*$ condition, 234, 239
diffraction contrast in, 233–235
Ewald sphere construction, 236
features of, 232
general theory, 231–233
image formation, 233, 236
ray paths of, 232, 234
resolution of diffraction contrast in, 235–239
scattering (*see also* elastic scattering; inelastic scattering)
amplitude-phase diagram, 79
atomic scattering factor and scattering angle, 76
by two atoms, 77
by unit cell, 77–79
path difference, 78
single atom by, 76
superimposing two waves, 78

screw dislocation, 54–56, 58–60, 183–184, 189, 227, 247–248, 250, 266
    displacement around, 60
    structure of, 266
second order Laue zone (SOLZ), 106
selected area aperture, 132
selected area diffraction (SAD), 131–132
semi-coherent interface, 168
Shockley's partial dislocations/imperfect dislocations, 63, 69
slip plane, 39
space lattices, 1–2
space projection
    reference sphere, 23–24
    standard projection, 27–29
    stereographic projection, 25–27
spherical precipitates, 166–169
spherical wave, 73–74
Spiecker–Jäger's method, 254–256
stacking fault, 65, 68–70, 122–123, 159, 161, 164–165, 173
    extrinsic, 68–69, 122, 164
    in FCC crystal, 159–163
    intrinsic, 68–69, 122, 164
stacking-fault energy (SFE), 65
stacking sequence, 69
standard projection, 27–29
standard triangle, 29
stereographic observation, 223–225
stereographic projection, 25–27, 261, 279, 281
    application of
        determining direction and plane, 34–42
        measuring angle between two poles, 31–34
        rotation of crystal, 29–31
    Wulff net and polar net, relationship between, 26, 28
stereomicrograph, 224–225
streaks, 119
structure factor, 82–88, 98, 100, 136, 166
    amplitude-phase diagram, 82–83
    crystal rotation method, 88
    of BCC, 83, 85
    of FCC, 85
    of NaCl, 86
super-dislocations, 70–72
superlattice reflections, 85
super partial dislocations (super-partials), 71, 213
systematic reflection, 211

T

TEM images of crystal, 127
thickness contour, 132–143, 253
    bamboo-shoot pattern, 140–142
    effect of s_g^* on, 138–139
    examples of, 139
    profile of, 147
    shift of, 253
Thompson tetrahedron, 66–70, 113, 161–162, 186, 193–194, 266
    model of, 266
three-dimensional model of cubic crystal, 257–258
trace analysis, 37–40, 42, 225, 227–229
    trace analysis
        determining directions, 225
        determining planes, 225–226
transmission electron microscope (TEM), 93–94
    basic design of, 94
    BOM and, 93–94
transmitted beam, 96–97
two-beam approximation, 127

U

unit cell, 2–3
    and Bravais lattices, 2–3, 6–7
    choice of, 5
    face centered cubic lattice, 3, 7
    lattice constants/parameters, 2

lattice space and, 5
primitive, 3
rhombohedral lattice, 3

V

vacancy (V)-type dislocation loop, 57

W

wave number vector, 74
white X-ray, 87
Wulff net, 26, 260
wurtzite structure, 19–20

Z

Z-contrast, 240
0th order Laue zone (ZOLZ), 106
zincblende structure, 19
zone axis, 34–35, 45, 225, 243, 279–280, 282–283

www.ingramcontent.com/pod-product-compliance
Lightning Source LLC
LaVergne TN
LVHW010332220926
848484LV00004BB/886